Optimization Software Class Libraries

OPERATIONS RESEARCH/COMPUTER SCIENCE INTERFACES SERIES

Other published titles in the series:

Greenberg, Harvey J. / *A Computer-Assisted Analysis System for Mathematical Programming Models and Solutions: A User's Guide for ANALYZE*

Greenberg, Harvey J. / *Modeling by Object-Driven Linear Elemental Relations: A Users Guide for MODLER*

Brown, Donald/Scherer, William T. / *Intelligent Scheduling Systems*

Nash, Stephen G./Sofer, Ariela / *The Impact of Emerging Technologies on Computer Science & Operations Research*

Barth, Peter / *Logic-Based 0-1 Constraint Programming*

Jones, Christopher V. / *Visualization and Optimization*

Barr, Richard S./ Helgason, Richard V./ Kennington, Jeffery L. / *Interfaces in Computer Science & Operations Research: Advances in Metaheuristics, Optimization, and Stochastic Modeling Technologies*

Ellacott, Stephen W./ Mason, John C./ Anderson, Iain J. / *Mathematics of Neural Networks: Models, Algorithms & Applications*

Woodruff, David L. / *Advances in Computational & Stochastic Optimization, Logic Programming, and Heuristic Search*

Klein, Robert / *Scheduling of Resource-Constrained Projects*

Bierwirth, Christian / *Adaptive Search and the Management of Logistics Systems*

Laguna, Manuel / González-Velarde, José Luis / *Computing Tools for Modeling, Optimization and Simulation*

Stilman, Boris / *Linguistic Geometry: From Search to Construction*

Sakawa, Masatoshi / *Genetic Algorithms and Fuzzy Multiobjective Optimization*

Ribeiro, Celso C./ Hansen, Pierre / *Essays and Surveys in Metaheuristics*

Holsapple, Clyde/ Jacob, Varghese / Rao, H. R. / *BUSINESS MODELLING: Multidisciplinary Approaches — Economics, Operational and Information Systems Perspectives*

Sleezer, Catherine M./ Wentling, Tim L./ Cude, Roger L. / *HUMAN RESOURCE DEVELOPMENT AND INFORMATION TECHNOLOGY: Making Global Connections*

Optimization Software Class Libraries

Edited by
Stefan Voß
Braunschweig University of Technology, Germany
David L. Woodruff
University of California, Davis, USA

Springer Science+Business Media, LLC

Library of Congress Cataloging-in-Publication Data

Voss, Stefan.
Optimization software class libraries / edited by Stefan Voss, David L. Woodruff.
p. cm. – (Operations research/computer science interfaces series ; 18)
Includes bibliographical references and index.

DOI 10.1007/978-0-306-48126-0
1. Application software--Development. 2. Mathematical optimization--Software. I. Voss, Stefan. II. Woodruff, David L. III. Title. IV. Series.

QA76.76.A65 067 2002
005.3--dc21 2002022215

Printed on acid-free paper.
www.springer.com/mycopy

Contents

Preface

Optimization problems in practice are diverse and evolve over time, giving rise to requirements both for ready-to-use optimization software packages and for optimization software libraries, which provide more or less adaptable building blocks for application-specific software systems. In order to apply optimization methods to a new type of problem, corresponding models and algorithms have to be "coded" so that they are accessible to a computer. One way to achieve this step is the use of a modeling language. Such modeling systems provide an excellent interface between models and solvers, but only for a limited range of model types (in some cases, for example, linear) due, in part, to limitations imposed by the solvers. Furthermore, while modeling systems especially for heuristic search are an active research topic, it is still an open question as to whether such an approach may be generally successful. Modeling languages treat the solvers as a "black box" with numerous controls. Due to variations, for example, with respect to the pursued objective or specific problem properties, addressing real-world problems often requires special purpose methods. Thus, we are faced with the difficulty of efficiently adapting and applying appropriate methods to these problems. Optimization software libraries are intended to make it relatively easy and cost effective to incorporate advanced planning methods in application-specific software systems.

A general classification provides a distinction between callable packages, numerical libraries, and component libraries. Component libraries provide useful abstractions for manipulating algorithm and problem concepts. Object-oriented software technology is generally used to build and apply corresponding components. To enable adaptation, these components are often provided at source code level. Corresponding class libraries support the development of application-specific software systems by providing a collection of adaptable classes intended to be reused. However, the reuse of algorithms may be regarded as "still a challenge to object-oriented programming". Component libraries are the subject of this edited volume. That is, within a careful collection of chapters written by experts in their fields we aim to discuss all relevant aspects of component libraries. To allow for wider applicability, we restrict the exposition to general approaches opposed to problem-specific software.

Acknowledgements

Of course such an ambitious project like publishing a high quality book would not have been possible without the most valuable input of a large number of individuals. First of all, we wish to thank all the authors for their contributions, their patience and fruitful discussion. We are grateful to the whole team at the University of Technology Braunschweig, who helped in putting this book together, and to Gary Folven at Kluwer Academic Publishers for his help and encouragement.

The Editors:

Stefan Voß
David L. Woodruff

1 OPTIMIZATION SOFTWARE CLASS LIBRARIES

Stefan Voß[1] and David L. Woodruff[2]

[1]Technische Universität Braunschweig
Institut für Wirtschaftswissenschaften
Abt-Jerusalem-Straße 7, D-38106 Braunschweig, Germany
stefan.voss@tu-bs.de

[2]Graduate School of Management
University of California at Davis
Davis, California 95616, USA
dlwoodruff@ucdavis.edu

Abstract: Many decision problems in business and engineering may be formulated as optimization problems. Optimization problems in practice are diverse, often complex and evolve over time, so one requires both ready-to-use optimization software packages and optimization software libraries, which provide more or less adaptable building blocks for application-specific software systems.

To provide a context for the other chapters in the book, it is useful to briefly survey optimization software. A general classification provides a distinction between *callable packages*, *numerical libraries*, and *component libraries*. In this introductory chapter, we discuss some general aspects of corresponding libraries and give an overview of available libraries, which provide reusable functionality with respect to different optimization methodologies. To allow for wider applicability we devote little attention to problem-specific software so we can focus the exposition on general approaches.

1.1 INTRODUCTION

New information technologies continuously transform decision processes for managers and engineers. This book is the result of the confluence of recent developments in optimization techniques for complicated problems and developments in software development technologies. The confluence of these technologies is making it possible for optimization methods to be embedded in a host of applications.

Many decision problems in business and engineering may be formulated as optimization problems. Optimization problems in practice are diverse, often complex and evolve over time, so one requires both ready-to-use optimization software packages and optimization software libraries, which provide more or less adaptable building blocks for application-specific software systems. To provide a context for the other chapters in the book, it is useful to briefly survey optimization software.

In order to apply optimization methods to a new type of problem, corresponding models and algorithms have to be "coded" so that they are accessible to a computer program that can search for a solution. Software that can take a problem in canonical form and find optimal or near optimal solutions is referred to as a *solver*. The translation of the problem from its physical or managerial form into a form usable by a solver is a critical step.

One way to achieve this step is the use of a *modeling language*. Such modeling systems provide an excellent interface between models and solvers, but only for a limited range of model types (in some extreme cases, e.g., linear). This is partly due to limitations imposed by the solvers. Furthermore, while modeling systems are an active research topic, it is still an open question whether such an approach may be successful for complex problems. Modeling languages treat the solvers as a "black box" with numerous controls.

Due to variations, for example, with respect to the pursued objective or specific problem properties, addressing real-world problems often requires special purpose methods. Thus, we are faced with the difficulty of efficiently adapting and applying appropriate methods to these problems. Optimization software libraries are intended to make it relatively easy and cost effective to incorporate advanced planning methods in application-specific software systems.

Callable packages allow users to embed optimization functionality in applications, and are designed primarily to allow the user's software to prepare the model and feed it to the package. Such systems typically also include routines that allow manipulation of the model and access to the solver's parameters. As with the modeling language approach, the solver is treated essentially as an opaque object, which provides a classical functional interface, using procedural programming languages such as C. While there are only restricted means to adapt the corresponding coarse-grained functionality, the packages do often offer *callbacks* that facilitate execution of user code during the solution process.

Numerical libraries provide similar functionality, except that the model data is treated using lower levels of abstraction. For example, while modeling languages and callable packages may allow the user to provide names for sets of variables and indexes into the sets, numerical libraries facilitate only the manipulation of vectors and matrices as numerical entities. Well-known solution techniques can be called as subroutines, or can be built from primitive operations on vectors and matrices. These libraries provide support for linear algebra, numerical computation of gradients, and support for other operations of value, particularly for continuous optimization.

Component libraries provide useful abstractions for manipulating algorithm and problem concepts. Object-oriented software technology is generally used to build and deploy components. To enable adaptation these components are often provided at source code level. Class libraries support the development of application-specific software systems by providing a collection of adaptable classes intended to be reused. Nevertheless, the reuse of algorithms may be regarded as "still a challenge to object-oriented programming" (Weihe (1997)). As we point out later, there is no clear dividing line between *class libraries* and *frameworks*. Whereas class libraries may be more flexible, frameworks often impose a broader structure on the whole system. Here we use the term component library or *componentware* that should embrace both class libraries and frameworks, but also other concepts that build on the idea of creating software systems by selecting, possibly adapting, and combining appropriate modules from a huge set of existing modules.

In the following sections we provide a brief survey on callable packages and numerical libraries (Section 1.3) as well as component libraries (Section 1.2). Our survey in this chapter must necessarily be cursory and incomplete; it is not intended to be judgmental and in some cases one has to rely on descriptions provided by software vendors. Therefore, we include several references (literature and WWW) that provide further information; cf. Fink et al. (2001).

As our main interest lies in optimization software class libraries and frameworks for heuristic search, we provide a somewhat more in depth treatment of heuristics and metaheuristics within the section on component libraries to let the reader visualize the preliminaries of this rapidly evolving area; cf. Voß (2001).

1.2 COMPONENT LIBRARIES

Class libraries support the development of application-specific software systems by providing a collection of (possibly semi-finished) classes intended to be reused. The approach to build software by using class libraries corresponds to the basic idea of object-oriented software construction, which may be defined as building software systems as "structured collections of possibly partial abstract data type implementations" (Meyer (1997)). The basic object-oriented paradigm is to encapsulate abstractions of all relevant concepts of the considered domain in classes. To be truly reusable, all these classes have to be applicable in different settings. This requires them to be polymorphic to a certain degree, i.e., to behave in an adaptable way. Accordingly, there have to be mechanisms to adapt these classes to the specific application. Class libraries are mostly based on dynamic polymorphism by factoring out common behavior in general classes and providing the specialized functionality needed by subclassing (inheritance). *Genericity*, which enables one to leave certain types and values unspecified until the code is actually instantiated and used (compiled) is another way - applicable orthogonal to inheritance - to define polymorphic classes.

One approach primarily devoted to the goal to achieve a higher degree of reuse is the framework approach; see, e.g., Bosch et al. (1999), Fayad and Schmidt (1997b) and Johnson and Foote (1988). Taking into account that for the development of application systems for given domains quite similar software is needed, it is reasonable to implement such common aspects by a generic design and embedded reusable software components. Here, one assumes that reuse on a large scale cannot only be based on individual components, but there has to be to a certain extent a reuse of design. Thus, the components have to be embedded in a corresponding architecture, which defines the collaboration between the components. Such a *framework* may be defined as a set of classes that embody an abstract design for solutions to a family of related problems (e.g., heuristics for discrete optimization problems), and thus provides us with abstract applications in a particular domain, which may be tailored for individual applications. A framework defines in some way a definition of a reference application architecture ("skeleton"), providing not only reusable software elements but also some type of reuse of architecture and design *patterns* (Buschmann et al. (1996b), Gamma et al. (1995)), which may simplify software development considerably. (Patterns, such as frameworks and components, may be classified as object-oriented reuse techniques. Simply put a pattern describes a problem to be solved, a solution as well as the context in which the solution applies.) Thus, frameworks represent implementation-oriented generic models for specific domains.

There is no clear dividing line between class libraries and frameworks. Whereas class libraries may be more flexible, frameworks often impose a broader structure on the whole system. Frameworks, sometimes termed as component libraries, may be subtly differentiated from class libraries by the "activeness" of components, i.e., components of the framework define application logic and call application-specific code. This generally results in a bi-directional flow of control.

In the following, we will use the term *component library* or *componentware* that should embrace both class libraries and frameworks, but also other concepts that build on the idea of creating software systems by selecting, possibly adapting, and combining appropriate modules from a large set of existing modules. The flexibility of a component library is dependent on the specific possibilities for adaptation. As certain aspects of the component library application cannot be anticipated, these aspects have to be kept flexible, which implies a deliberate incompleteness of generic software components.

Based on these considerations we chose the title *optimization software class libraries*. In the sequel we distinguish between libraries for heuristic search (Section 1.2.1) and constraint programming (Section 1.2.2).

1.2.1 Libraries for Heuristic Optimization

Most discrete optimization problems are nearly impossible to solve to optimality. Many can be formally classified as $\mathcal{NP}$-hard (Garey and Johnson (1979)). Moreover, the modeling of the problem is often an approximate one, and the data are often imprecise. Consequently, heuristics are a primary way to tackle these problems. The use of appropriate metaheuristics generally meets the needs of decision makers to efficiently generate solutions that are satisfactory, although perhaps not optimal. The common incorporation of advanced metaheuristics in application systems requires a way to reuse much of such software and to redo as little as possible each time. However, in comparison to the exact optimization field, there is less support by corresponding software libraries that meet practical demands with respect to, for example, robustness and ease-of-use. What are the difficulties in developing reusable and adaptable software components for heuristic search? Compared to the field of mathematical programming, which relies on well-defined, problem-independent representation schemes for problems and solutions on which algorithms may operate, metaheuristics are based on abstract definitions of solution spaces and neighborhood structures. Moreover, for example, memory-based tabu search approaches are generally based on abstract problem-specific concepts such as solution and move attributes.

The crucial problem of local search based metaheuristics libraries is a generic implementation of heuristic approaches as reusable software components, which must operate on arbitrary solution spaces and neighborhood structures. The drawback is that the user must, in general, provide some kind of a problem/solution definition and a neighborhood structure, which is usually done using sophisticated computer languages such as C++.

An early C++ class library for heuristic optimization by Woodruff (1997) included both local search based methods and genetic algorithms. This library raised issues that illustrate both the promise and the drawbacks to the adaptable component approach. From a research perspective such libraries can be thought of as providing a concrete taxonomy for heuristic search. So concrete, in fact, that they can be compiled into machine code. This taxonomy sheds some light on the relationships between heuristic search methods for optimization and on ways in which they can be combined. Furthermore, the library facilitates such combinations as the classes in the library can be extended and/or combined to produce new search strategies.

From a practical and empirical perspective, these types of libraries provide a vehicle for using and testing heuristic search optimization. A user of the library must provide the definition of the problem specific abstractions and may systematically vary and exchange heuristic strategies and corresponding components.

In the sequel, we provide a brief survey on the state-of-the-art of heuristic search and metaheuristics before we discuss several heuristic optimization libraries. These libraries differ, e.g., in the design concept, the chosen balance between "ease-of-use" and flexibility and efficiency, and the overall scope. All of these approaches are based on the concepts of object-oriented programming and will be described in much more detail in later chapters of this book.

1.2.1.1 Heuristics: Patient Rules of Thumb and Beyond. Many optimization problems are too difficult to be solved exactly within a reasonable amount of time and heuristics become the methods of choice. In cases where simply obtaining a feasible solution is not satisfactory, but where the quality of solution is critical, it becomes important to investigate efficient procedures to obtain the best possible solutions within time limits deemed practical. Due to the complexity of many of these optimization problems, particularly those of large sizes encountered in most practical settings, exact algorithms often perform very poorly (in some cases taking days or more to find moderately decent, let alone optimal, solutions even to fairly small instances). As a result, heuristic algorithms are conspicuously preferable in practical applications.

The basic concept of heuristic search as an aid to problem solving was first introduced by Polya (1945). A *heuristic* is a technique (consisting of a rule or a set of rules) which seeks (and eventually finds) *good* solutions at a reasonable computational cost. A heuristic is *approximate* in the sense that it provides (hopefully) a good solution for relatively little effort, but it does not guarantee optimality. Moreover, the usual distinction refers to finding initial feasible solutions and improving them.

Heuristics provide simple means of indicating which among several alternatives seems to be the best. And basically they are based on intuition. That is, "heuristics are criteria, methods, or principles for deciding which among several alternative courses of action promises to be the most effective in order to achieve some goal. They represent compromises between two requirements: the need to make such criteria simple and, at the same time, the desire to see them discriminate correctly between good and bad choices. A heuristic may be a *rule of thumb* that is used to guide one's action." (Pearl (1984))

Greedy heuristics are simple heuristics available for any kind of combinatorial optimization problem. They are iterative and a good characterization is their *myopic* behavior. A greedy heuristic starts with a given feasible or infeasible solution. In each iteration there is a number of alternative choices (*moves*) that can be made to transform the solution. From these alternatives which consist in fixing (or changing) one or more variables, a *greedy choice* is made, i.e., the best alternative according to a given evaluation measure is chosen until no such transformations are possible any longer.

Among the most studied heuristics are those based on applying some sort of greediness or applying priority based procedures such as insertion and dispatching rules. As an extension of these, a large number of local search approaches has been developed to improve given feasible solutions. The basic principle of *local search* is that solutions are successively changed by performing moves which alter solutions *locally*. Valid transformations are defined by neighborhoods which give all neighboring solutions that can be reached by one move from a given solution. (Formally, we consider an instance of a combinatorial optimization problem with a solution space S of feasible (or even infeasible) solutions. To maintain information about solutions, there may be one or more solution information functions I on S, which are termed exact, if I is injective, and approximate otherwise. With this information, one may store a search history (trajectory). For each S there are one or more neighborhood structures N that define for each solution $s \in S$ an ordered set of neighbors $N(s) = \{n_1(s), \ldots, n_{|N(s)|}(s)\}$. To each neighbor $n(s) \in N(s)$ corresponds a move that captures the transitional information from s to $n(s)$.) For a general survey on local search see the collection of Aarts and Lenstra (1997) and the references in Aarts and Verhoeven (1997).

Moves must be evaluated by some *heuristic measure* to guide the search. Often one uses the implied change of the objective function value, which may provide reasonable information about the (local) advantage of moves. Following a greedy strategy, *steepest descent* (SD) corresponds to selecting and performing in each iteration the best move until the search stops at a local optimum.

As the solution quality of the local optima thus encountered may be unsatisfactory, we need mechanisms which guide the search to overcome local optimality. A simple strategy called iterated local search is to iterate/restart the local search process after a local optimum has been obtained, which requires some perturbation scheme to generate a new initial solution (e.g., performing some random moves). Of course, more structured ways to overcome local optimality might be advantageous.

Starting with Lin and Kernighan (1973), a variable way of handling neighborhoods is a topic within local search. Consider an arbitrary neighborhood structure N, which defines for any solution s a set of neighbor solutions $N_1(s)$ as a neighborhood of depth $d = 1$. In a straightforward way, a neighborhood $N_{d+1}(s)$ of depth $d+1$ is defined as the set $N_d(s) \cup \{s' | \exists s'' \in N_d(s) : s' \in N_1(s'')\}$. In general, a large d might be unreasonable, as the neighborhood size may grow exponentially. However, depths of two or three may be appropriate. Furthermore, temporarily increasing the neighborhood depth has been found to be a reasonable mechanism to overcome basins of attraction, e.g., when a large number of neighbors with equal quality exist.

The main drawback of local search approaches – their inability to continue the search upon becoming trapped in a local optimum – leads to consideration of techniques for guiding known heuristics to overcome local optimality. Following this theme, one may investigate the application of intelligent search methods like the *tabu search* metaheuristic for solving optimization problems. Moreover, the basic concepts of various strategies like *simulated annealing*, *scatter search* and *genetic algorithms* come to mind. This is based on a simplified view of a possible inheritance tree for heuristic search methods, illustrating the relationships between some of the most important methods discussed below, as shown in Figure 1.1.

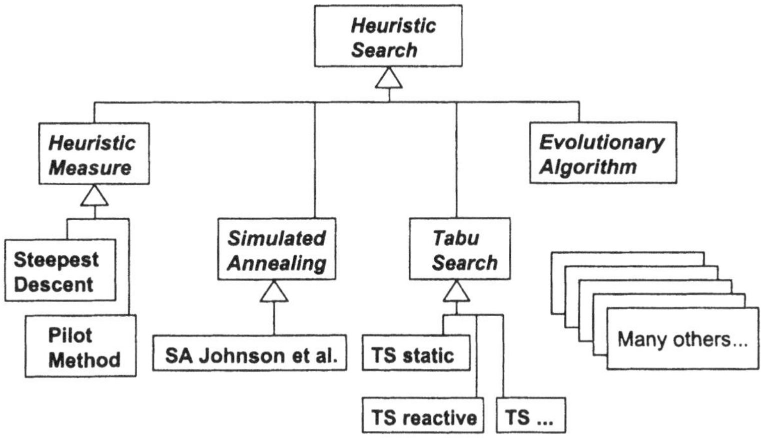

Figure 1.1 Simplified Metaheuristics Inheritance Tree

1.2.1.2 Metaheuristics Concepts. The formal definition of metaheuristics is based on a variety of definitions from different authors going back to Glover (1986). Basically, a metaheuristic is a top-level strategy that guides an underlying heuristic solving a given problem. Following Glover it "refers to a master strategy that guides and modifies other heuristics to produce solutions beyond those that are normally generated in a quest for local optimality" (Glover and Laguna (1997)). In that sense we distinguish between a *guiding process* and an *application process*. The guiding process decides upon possible (local) moves and forwards its decision to the application process which then executes the chosen move. In addition, it provides information for the guiding process (depending on the requirements of the respective metaheuristic) like the recomputed set of possible moves.

To be more specific, "a meta-heuristic is an iterative master process that guides and modifies the operations of subordinate heuristics to efficiently produce high-quality solutions. It may manipulate a complete (or incomplete) single solution or a collection of solutions at each iteration. The subordinate heuristics may be high (or low) level procedures, or a simple local search, or just a construction method. The family of meta-heuristics includes, but is not limited to, adaptive memory procedures, tabu search, ant systems, greedy randomized adaptive search, variable neighborhood search, evolutionary methods, genetic algorithms, scatter search, neural networks, simulated annealing, and their hybrids." (Voß et al. (1999), p. ix)

To understand the philosophy of various metaheuristics, it is interesting to note that adaptive processes originating from different settings such as psychology ("learning"), biology ("evolution"), physics ("annealing"), and neurology ("nerve impulses") have served as a starting point. Applications of metaheuristics are almost uncountable. Helpful sources for successful applications may be Vidal (1993), Pesch and Voß (1995), Rayward-Smith (1995), Laporte and Osman (1996), Osman and Kelly (1996), Rayward-Smith et al. (1996), Glover (1998a), Voß et al. (1999), Voß (2001), just to mention some.

Simple Local Search Based Metaheuristics: To improve the efficiency of greedy heuristics, one may apply some generic strategies that may be used alone or in combination with each other, such as dynamically changing or restricting the neighborhood, altering the selection mechanism, look ahead evaluation, candidate lists, and randomized selection criteria bound up with *repetition*, as well as combinations with other methods that are not based on local search.

If, instead of making strictly greedy choices, we adopt a random strategy, we can run the algorithm several times and obtain a large number of different solutions. However, purely random choices usually perform very poorly. Thus a combination of best and random choice or else biased random choice seems to be appropriate. For example, we may define a *candidate list* consisting of a number of the best alternatives. Out of this list one alternative is chosen randomly. The length of the candidate list is given either as an absolute value, a percentage of all feasible alternatives or implicitly by defining an allowed quality gap (to the best alternative), which also may be an absolute value or a percentage.

Replicating a search procedure to determine a local optimum multiple times with different starting points has been investigated with respect to many different applications; see, e.g., by Feo and Resende (1995). A number of authors have independently noted that this search will find the global optimum in finite time with probability one, which is perhaps the strongest convergence result in the heuristic search literature. The mathematics is not considered interesting because it is based on very old and well-known theory and, like all of the other convergence results in heuristic search, it is not relevant for practical search durations and provides no useful guidance for such searches.

When the different initial solutions or starting points are found by a greedy procedure incorporating a probabilistic component, the method is named *greedy randomized adaptive search procedure* (GRASP). Given a candidate list of solutions to choose from, GRASP randomly chooses one of the best candidates from this list with a bias toward the best possible choices. The underlying principle is to investigate many good starting points through the greedy procedure and thereby to increase the possibility of finding a good local optimum on at least one replication. The method is said to be adaptive as the greedy function takes into account previous decisions when performing the next choice. It should be noted that GRASP is predated by similar approaches such as Hart and Shogan (1987).

Building on simple greedy algorithms such as a construction heuristic the *pilot method* may be taken as an example of a guiding process based on modified uses of heuristic measure. The pilot method builds primarily on the idea to *look ahead* for each possible local choice (by computing a so-called "pilot" solution), memorizing the best result, and performing the according move. One may apply this strategy by successively performing a cheapest insertion heuristic for all possible local steps (i.e., starting with all incomplete solutions resulting from adding some not yet included element at some position to the current incomplete solution). The look ahead mechanism of the pilot method is related to increased neighborhood depths as the pilot method exploits the evaluation of neighbors at larger depths to guide the neighbor selection at depth one. Details on the pilot method can be found in Duin and Voß (1999) and Duin and Voß (1994). Similar ideas have been investigated under the name *rollout method*; see Bertsekas et al. (1997).

Hansen and Mladenović (1999) examine the idea of changing the neighborhood during the search in a systematic way. *Variable neighborhood search* (VNS) explores increasingly distant neighborhoods of the current incumbent solution, and jumps from this solution to a new one iff an improvement has been made. In this way often favorable characteristics of incumbent solutions, e.g., that many variables are already at their optimal value, will be kept and used to obtain promising neighboring solutions.

Moreover, a local search routine is applied repeatedly to get from these neighboring solutions to local optima. This routine may also use several neighborhoods. Therefore, to construct different neighborhood structures and to perform a systematic search, one needs to have a way for finding the distance between any two solutions, i.e., one needs to supply the solution space with some metric (or quasi-metric) and then induce neighborhoods from it.

Simulated Annealing: *Simulated annealing* (SA) extends basic local search by allowing moves to inferior solutions; see, e.g., Kirkpatrick et al. (1983). The basic algorithm of SA may be described as follows: Successively, a candidate move is randomly selected; this move is accepted if it leads to a solution with a better objective function value than the current solution, otherwise the move is accepted with a probability that depends on the deterioration Δ of the objective function value. The probability of acceptance is computed as $e^{-\Delta/T}$, using a temperature T as control parameter. The value of T is successively reduced during the algorithm execution according to a *cooling schedule* that is governed by parameters set by the programmer or user. If T is increased within the search, this may be called reheating.

Threshold accepting (Dueck and Scheuer (1990)) is a modification (or simplification) of SA with the essential difference between the two methods being the acceptance rules. Threshold accepting accepts every move that leads to a new solution which is "not much worse" (i.e., deteriorates not more than a certain threshold which reduces with a temperature) than the older one.

Tabu Search: The basic paradigm of *tabu search* (TS) is to use information about the search history to guide local search approaches to overcome local optimality (see Glover and Laguna (1997) for a survey on TS). In general, this is done by a dynamic transformation of the local neighborhood. Based on data structures that record the properties of recent moves, certain moves may be forbidden. We say that the forbidden moves are *tabu*. As with SA, the search may lead to performing deteriorating moves when no improving moves exist or all improving moves of the current neighborhood are set tabu. At each iteration a best admissible neighbor may be selected. A neighbor and the corresponding move are called admissible, if the move is not tabu or if an aspiration criterion is fulfilled.

In the literature various TS methods may be found that differ especially in the way in which the tabu criteria are defined, taking into consideration the information about the search history (performed moves, traversed solutions). An aspiration criterion may override a possibly unreasonable tabu status of a move. For example, a move that leads to a neighbor with a better objective function value than encountered so far should be considered as admissible.

The most commonly used TS method is based on a *recency-based* memory that stores moves, more exactly move attributes, of the recent past (*static TS*). The basic idea of such approaches is to prohibit an appropriately defined inversion of performed moves for a given period. For example, one may store the solution attributes that have been created by a performed move in a tabu list. To obtain the current tabu status of a move to a neighbor, one may check whether (or how many of) the solution attributes that would be destroyed by this move are contained in the tabu list.

Strict TS embodies the idea of preventing cycling to formerly traversed solutions. The goal is to provide necessity and sufficiency with respect to the idea of not revisiting any solution. Accordingly, a move is classified as tabu iff it leads to a neighbor that has already been visited during the previous part of the search. There are two primary mechanisms to accomplish the tabu criterion: First, we may exploit logical interdependencies between the sequence of moves performed throughout the search process as observed in the reverse elimination method (see, e.g., Glover (1990), Voß (1993b)). Second, we may store information about all solutions visited so far. This may be carried out either exactly or, for reasons of efficiency, approximately (e.g., by using hash-codes).

Reactive TS aims at the automatic adaptation of the tabu list length of static TS (see, e.g., Battiti (1996)). The idea is to increase the tabu list length when the tabu memory indicates that the search is revisiting formerly traversed solutions. Furthermore, there is a great variety of additional ingredients that may make TS work successful. For applications of TS as well as other local search based metaheuristics within class libraries and frameworks the reader will find a large variety of hints and specifications in later chapters, as, e.g., Chapter 4.

Evolutionary Algorithms: *Evolutionary algorithms* comprise a great variety of different concepts and paradigms including genetic algorithms (see, e.g., Holland (1975), Goldberg (1989)), evolutionary strategies (see, e.g., Hoffmeister and Bäck (1991), Schwefel and Bäck (1998)), evolutionary programs (Fogel (1993)), scatter search (see, e.g., Glover (1977), Glover (1995)), and memetic algorithms (Moscato (1993)). For surveys and references on evolutionary algorithms see also Fogel (1995), Bäck et al. (1997), Mühlenbein (1997) and Michalewicz (1999).

Genetic algorithms are a class of adaptive search procedures based on the principles derived from the dynamics of natural population genetics. One of the most crucial ideas for a successful implementation of a genetic algorithm (GA) is the representation of an underlying problem by a suitable scheme. Trial solutions must be represented as *strings* or vectors that will serve as analogs to chromosomes.

A GA starts (for example) with a randomly created initial population of artificial chromosomes (strings), found, for example, by flipping a "fair" coin. These strings in whole and in part are the base set for all subsequent populations. They are copied and information is exchanged between the strings in order to find new solutions of the underlying problem. The mechanisms of a simple GA essentially consist of copying strings and exchanging partial strings. A simple GA requires three operators which are named according to the corresponding biological mechanisms: reproduction, crossover, and mutation. Performing an operator may depend on a *fitness function* or its value (fitness), respectively. This function defines a means of measurement for the profit or the quality of the coded solution for the underlying problem and may depend on the objective function of the given problem. For a more detailed consideration of these topics and related libraries see Chapter 10 by *Pain* and *Reeves*.

GAs are closely related to *evolutionary strategies*. Whereas the mutation operator in a GA serves to protect the search from premature loss of information, evolution strategies may incorporate some sort of local search procedure (such as SD) with self-adapting parameters incorporated within the procedure.

For some interesting insights on evolutionary algorithms the reader is referred to Hertz and Kobler (2000). On a very simple scale many algorithms may be coined evolutionary once they are reduced to the following frame:

1. Generate an initial population of individuals
2. While no stopping condition is met do
   (a) co-operation
   (b) self-adaptation

Self-adaptation refers to the fact that individuals (solutions) evolve independently while co-operation refers to an information exchange among individuals.

Recently it appeared that *scatter search* ideas may establish a link between early ideas from various sides – evolutionary strategies, TS and GAs. Scatter search is designed to operate on a set of points, called reference points, that constitute good solutions obtained from previous solution efforts. The approach systematically generates linear combinations of the reference points to create new points, each of which is mapped into an associated point that yields integer values for discrete variables. The library OptQuest heavily relies on scatter search; see Chapter 7 by *Laguna* and *Martí*.

Path relinking provides a useful means of intensification and diversification. Here new solutions are generated by exploring search trajectories that connect elite solutions, i.e., solutions that have proven to be better than others throughout the search or elite members of a population of solutions. For references on path relinking see, e.g., Glover and Laguna (1997).

Miscellaneous: One of the recently explored concepts within intelligent search is the *ant system*, a dynamic optimization process reflecting the natural interaction between ants searching for food (see, e.g., Dorigo et al. (1996), Taillard (2000)). The ants' ways are influenced by two different kinds of search criteria. The first one is the local visibility of food, i.e., the attractiveness of food in each ant's neighborhood. Additionally, each ant's way through its food space is affected by the other ants' trails as indicators for possibly good directions. The intensity of trails itself is time-dependent: With time going by, parts of the trails "are gone with the wind", meanwhile the intensity may increase by new and fresh trails. With the quantities of these trails changing dynamically, an autocatalytic optimization process is started forcing the ants' search into most promising regions. This process of interactive learning can easily be modeled for most kinds of optimization problems by using simultaneously and interactively processed search trajectories. To achieve enhanced performance of the ant system it is useful to hybridize it with a local search component.

Target analysis may be viewed as a general learning approach. Given a problem we first explore a set of sample instances and an extensive effort is made to obtain a solution which is optimal or close to optimality. The best solutions obtained will provide some *targets* to be sought within the next part of the approach. For instance, a TS algorithm may resolve the problems with the aim of finding what are the right choices to lead the search to the already known solution (or as close to it as possible). This may give some information on how to choose parameters for other problem instances. For more information on target analysis see Glover and Laguna (1997).

Given an initial feasible solution, the *noising method* performs some data perturbation (Storer et al. (1995)) in order to change the values taken by the objective function of a respective problem to be solved. With this perturbed data some local search iterations may be performed (e.g., following a SD approach). The amount of data perturbation (the noise added) is successively reduced until it reaches zero. The noising method is applied, e.g., in Charon and Hudry (1993) for the clique partitioning problem and in Ribeiro et al. (2000) for the Steiner problem in graphs as a hybrid with

GRASP. Another method based on *data perturbation* is called *ghost image processing*; see, e.g. Glover (1994) and Woodruff (1995).

The key issue in designing *parallel algorithms* is to decompose the execution of the various ingredients of a procedure into processes executable by processors operating in parallel. (Multiple processes in this sense may also be called threads.) Ant systems and population based methods are embarrassingly easy to parallelize, at least to some degree. Local search algorithms (and the local search portions of population based methods) require some effort to efficiently unroll the loops that explore the neighborhoods so that the exploration can be done by multiple processors simultaneously. However, some effort has been undertaken to define templates for parallel local search (see, e.g., Voß (1993b), Verhoeven and Aarts (1995), Crainic et al. (1997), Vaessens et al. (1998)). Examples of successful applications are referenced in Aarts and Verhoeven (1997). The discussion of parallel metaheuristics has also led to interesting hybrids such as the combination of a population of individual processes, agents, in a cooperative and competitive nature (see, e.g., the discussion of *memetic algorithms* in Moscato (1993)) with TS.

Of course *neural networks* may be considered as metaheuristics, although we have not considered them in this brief survey; see, e.g., Smith (1999) for a comprehensive survey on these techniques for combinatorial optimization. Furthermore, we have not considered problems with multiple objectives and corresponding approaches (see, e.g., Sakawa (2001) for a great variety of ideas regarding GAs and fuzzy multiobjective optimization).

1.2.1.3 Recent Advances on General Frameworks. Recently, general frameworks have been investigated to explain the behavior and the relationship between various methods. Given general frameworks, commonalities between incorporated metaheuristics may enhance the degree of reuse possible when building corresponding software systems. One of the key aspects regarding metaheuristics in general and, with that, with a considerable relevance for respective libraries, is an intelligent interplay between intensification (concentrating the search into a specific region of the search space) and diversification (elaborating various diverse regions within the solution space). It has very often been appropriate to incorporate a certain means of exploring promising regions of the search space in more detail (intensification) and additional methods of leading the search into new regions of the search space (diversification). An as yet not fully explored related aspect refers to diversity measures which are important when performing diversification. That is, the distance between solutions may be evaluated by means of a certain metric. An important concern when advising an appropriate metric is whether it allows to incorporate the distinction of structural solution properties (cf. Voß (1995), Glover and Laguna (1997), Glover et al. (2000b)). Within intelligent search the exploration of memory plays a most important role in ongoing research.

Many exact methods, such as branch-and-bound and constrained logic programming, keep a complete memory of the search space exploration (although in some cases the memory is implicit rather than explicit). Most metaheuristics do not, so meaningful mechanisms for detecting situations when the search might be trapped in a certain area of the solution space and for when it has left a region of the solution space must be developed. One proposal is referred to as *chunking* (Woodruff (1998)). See Woodruff (1999) for an application of reactive TS embedded in branch-and-bound using chunking within the reactive TS but relying on the branch-and-bound data structures for global memory. Here we may also use the term *cooperative solver*.

The phrase *adaptive memory programming* (AMP) was coined to encompass a general approach (or philosophy) within heuristic search focusing on exploiting a collection of memory components (Glover (1997), Taillard et al. (2001)). That is, iteratively constructing (new) solutions based on the exploitation of some sort of memory may be viewed as AMP process, especially when combined with learning mechanisms, that helps to adapt the collection and use of the memory. Based on the simple idea of initializing the memory and then iteratively generating new solutions (utilizing the given memory) while updating the memory based on the search, most of the metaheuristics described can be subsumed by the AMP approaches. This also includes the idea of exploitation of provisional solutions in population based methods that are improved by a local search approach.

The performance as well as the efficiency of a heuristic scheme strongly depends on its ability to use AMP techniques providing flexible and variable strategies for types of problems (or special instances of a given problem type) where standard methods fail. Such AMP techniques could be, e.g., dynamic handling of operational restrictions, dynamic move selection formulas, and flexible function evaluations. Consider, as an example, adaptive memory within TS concepts. Realizing AMP principles depends on which specific TS application is used. For example, the reverse elimination method observes logical interdependencies between moves and infers corresponding tabu restrictions, and therefore makes fuller use of AMP than simple static approaches do.

To discuss the use of AMP in intelligent agent systems, we may also refer to the simple model of ant systems as an illustrative starting point. Ant systems are based on combining local search criteria with information derived from the trails. This follows the AMP requirement for using flexible (dynamic) move selection rules (formulas). However, the basic ant system exhibits some structural inefficiencies when viewed from the perspective of general intelligent agent systems, as no distinction is made between successful and less successful agents, no time-dependent distinction is made, there is no explicit handling of restrictions providing protection against cycling and duplication. Furthermore, there are possible conflicts between the information held in the adaptive memory (*diverging trails*).

A somewhat different approach within heuristic search starts from the observation that a natural way to solve large combinatorial optimization problems consists of decomposing them into independent sub-problems that are solved with an appropriate procedure. However, such approaches may lead to solutions of moderate quality since the sub-problems might have been created in a somewhat arbitrary fashion. Indeed, it is not easy to find appropriate ways to decompose a problem *a priori*. The basic idea of *POPMUSIC* (Partial OPtimization Metaheuristic Under Special Intensification Conditions; see Taillard and Voß (2002)) is to locally optimize sub-parts of a solution, *a posteriori*, once a solution to the problem is available. These local optimizations are repeated until a local optimum is found. So, POPMUSIC can be seen as a local search working with a special, large neighborhood. Therefore, various metaheuristics may be naturally incorporated into the same framework (see, e.g., Shaw (1998)).

Consider the vehicle routing problem (in short: the vehicle routing problem may be characterized as determining a set of least cost vehicle routes starting and ending at a given depot such that the demand of each member of a given set of customers is satisfied). For this problem a part may be a tour (or even a customer). Suppose that a solution can be represented as a set of parts. Moreover, some parts are more in relation with some other parts so that a relatedness measure can be defined between two parts. The central idea of POPMUSIC is to select a so-called *seed part* and a set P of parts that are mostly related with the seed part to form a sub-problem. Then it is possible to state a local search optimization framework that consists of trying to improve all sub-problems that can be defined, until the solution does not contain a sub-problem that can be improved. In the POPMUSIC framework of Taillard and Voß (2002), P corresponds precisely to seed parts that have been used to define sub-problems that have been unsuccessfully optimized. Once P contains all the parts of the complete solution, then all sub-problems have been examined without success and the process stops. Basically, the technique is a gradient method that starts from a given initial solution and stops in a local optimum relative to a large neighborhood structure.

To summarize, both, POPMUSIC as well as AMP may serve as a general framework encompassing various other approaches.

1.2.1.4 Local Search for Propositional Satisfiability. A considerable amount of work especially within the computer science community refers to local search for *propositional satisfiability* (SAT). For given sets of variables and clauses SAT asks for the existence of an assignment to the variables that satisfies all clauses. While SAT is a decision problem the problems MAX-SAT and weighted MAX-SAT are related problems, which ask for a propositional variable assignment that maximizes the number of satisfied and the weighted sum of satisfied (weighted) clauses, respectively.

One idea regarding the solution of a great variety of problems is to describe them by means of instances of SAT or MAX-SAT (SAT encoded problems) and to apply solution methods available for these problems (see, e.g., Jiang et al. (1995) as well as Selman et al. (1992), Selman et al. (1994), Kautz and Selman (1996), Gu (1999), Walser (1999), Schuurmans and Southey (2000) and Hoos and Stützle (2000)). While many problems formulated under this paradigm have been solved successfully following such approaches, however, this is not necessarily the case for some classical combinatorial optimization problems.

1.2.1.5 Optimization Software Libraries. After having described some of the fundamentals of metaheuristics we are in a position to incorporate them into appropriate libraries. That is, systems may be developed to support the implementation of local search and metaheuristics. Such algorithms can be supported by (high-level) modeling languages that reduce their development time substantially while preserving most of the efficiency of special purpose implementations. For instance, the design of LOCALIZER (see Michel and van Hentenryck (1999), Michel and van Hentenryck (2000) and Chapter 9) may be viewed as such an approach with several extensions being contemplated, including additional support for some metaheuristics and the integration of consistency techniques.

There have been some interesting class libraries developed for sub-domains such as Fleurent and Ferland (1996) and Spinellis and Papadopoulos (2001). While there are some well-known approaches for reusable software in the field of exact optimization, there is, as far as we know, only a very limited number of ready-to-use and well-documented component libraries in the field of local search based heuristics and metaheuristics.

HOTFRAME: HOTFRAME, a Heuristic OpTimization FRAMEwork implemented in C++, provides both adaptable components that incorporate different metaheuristics and an architectural description of the collaboration among these components and problem-specific complements. All typical application-specific concepts are treated as objects or classes: problems, solutions, neighbors, solution and move attributes. On the other side, metaheuristics concepts such as different methods and their building-blocks such as tabu criteria and diversification strategies are also treated as objects. HOTFRAME uses genericity as the primary mechanism to make these objects adaptable. That is, common behavior of metaheuristics is factored out and grouped in generic classes, applying static type variation. Metaheuristics template classes are parameterized by aspects such as solution spaces and neighborhood structures.

HOTFRAME defines an architecture for the interplay between heuristic classes and application-specific classes, and provides several such classes, which implement classic methods that are applicable to arbitrary problem types, solution spaces and neighborhood structures. All heuristics are implemented in a consistent way, which facilitates an easy embedding of arbitrary methods into application systems or as parts of more advanced/hybrid methods. Both new metaheuristics and new applications can be added to the framework. HOTFRAME includes built-in support for solution spaces representable by binary vectors or permutations, in connection with corresponding standard neighborhood structures, solution and move attributes, and recombination operators. Otherwise, the user may derive specialized classes from suitable built-in classes or implement corresponding classes from scratch according to a defined interface. For further information about HOTFRAME see Fink and Voß (1999b), HotFrame (2001) as well as Chapter 4.

Templar: The Templar framework, implemented in C++, provides a method, and software components, for constructing systems to solve optimization problems. The Templar framework is based on problem classes and engine classes. The engine's source code is representation- and problem-independent. The problem class is an abstract base class, which is used to derive specific problem classes. These specific problem classes embody the characteristics of the problem type and are capable of loading problem instances. Problems make operators, such as neighborhood operators or genetic operators, available for use by an engine. The representation used by the problem (for example, permutation) also provides its own operators. Engines can then use these operators to produce good solutions to a given problem instance. The base engine abstract base class is also used for derivation of specific engine classes. Examples include simulated annealing engines, tabu search engines, and genetic algorithm engines. Although a single instance of an engine is connected to a particular problem, each specific engine class is problem- and representation-independent. As well as being easy to control, engines are designed so that a driving process can make a group of engines work cooperatively. For further information, see Templar (2001). A detailed treatment of Templar can be found in Chapter 2 by *Jones, McKeown* and *Rayward-Smith*.

NeighborSearcher: Andreatta et al. (1998) describe NeighborSearcher, an object-oriented framework for local search heuristics. Their main goal is to provide an architectural basis for the implementation and comparison of different local search heuristics. Accordingly, they define a coarse-grained modularization of the domain in abstract classes that encapsulate concepts such as the construction of the initial solution, the local search algorithm, the solution, or the movement model. With this, one may implement a specific search strategy by selecting derived classes that provide the respective functionality. This provides an adaptation mechanism to the higher level client code. A more comprehensive consideration of NeighborSearcher, or Searcher for short, can be found in Chapter 3 by *Andreatta, Carvalho* and *Ribeiro*.

EASYLOCAL++: EASYLOCAL++, which is treated in Chapter 5 by *di Gaspero* and *Schaerf*, is a C++ class library that supports metaheuristics based on local search such as tabu search and simulated annealing. The library provides explicit support for so-called kick moves. A local search is typically based on a particular neighborhood structure induced by a corresponding type of move. A kick move may be made at strategically defined epochs by temporarily adopting a different neighborhood. Another important feature of the library is the inclusion of classes to support algorithmic testing. A batch control language EXPSPEC is supported by classes that collect run-time statistics. The output is generated both in machine- and human-readable format.

iOpt: The Java class library iOpt has support for population based evolutionary algorithms (EA) as well as local search based methods. Such libraries facilitate hybrid algorithms, e.g., the use of tabu search as a mutation operator (this is sometimes referred to as "learning" by EA researchers who yearn for metaphor). Classes for some problem specific domains such as scheduling are also provided.

A key distinguishing feature of iOpt is built-in support for propagation of one-way constraints Zanden et al. (1994). Thus the library provides connections between metaheuristics and constraint logic programming. A one-way constraint is based on a general function C, which is a function of one or more variables and whose value constrains the value of some other variable. For instance,

$$x_i = C(x_j, x_k, \ldots, x_m)$$

fixes the value of x_i as a function of some other variables whose values are not directly affected by the constraint. The iOpt library supports sophisticated methods of propagating changes in variables taking advantage of the fact that if there are many one-way constraints, a change in one variable can result in cascading changes in many variables. The use of one-way constraints requires some modeling sophistication, but the result is that local search neighborhoods involving changes in a single variable become much more powerful. Details on iOpt can be found in Chapter 6 by *Voudouris* and *Dorne*.

Genetic Algorithm Libraries: There exist several libraries for genetic algorithms; surveys, as well as software, can be found at Heitkötter and Beasley (2001). In principle, an advantage of using classic genetic algorithm libraries such as Genitor (2001) or GAlib (2001) is that no neighborhood must be specified. If the built-in genomes of a genetic algorithm library adequately represent one's problem, a user-specified objective function may be the only problem-specific code that must be written. Unfortunately, genetic algorithms without a local search component have not generally proven to be very effective.

As an example we briefly discuss the functionality of the GAlib library, a C++ library, which provides the application programmer with a set of genetic algorithms objects; cf. GAlib (2001). GAlib is flexible with respect to arbitrary data representations and standard or custom selection, crossover, mutation, scaling, replacement, and termination methods. Overlapping (steady-state GA) and non-overlapping (simple GA) populations are supported. Built-in selection methods include rank, roulette wheel, tournament, stochastic remainder sampling, stochastic uniform sampling, and deterministic sampling. One can use chromosome types built-in to the library (bit-string, array, list, tree) or derive a chromosome based on user-defined objects. All chromosome initialization, mutation, crossover, and comparison methods can be customized. Built-in mutation operators include random flip, random swap, Gaussian, destructive, swap subtree, swap node. Built-in crossover operators include arithmetic, blend, partial match, ordered, cycle, single point, two point, even, odd, uniform, node- and subtree-single point.

For a comprehensive overview of genetic algorithm libraries the reader is referred to Chapter 10.

1.2.2 Constraint Programming

Constraint programming (CP) is a paradigm for representing and solving a wide variety of problems expressed by means of variables, their domains, and constraints on the variables (see, e.g., van Hentenryck (1995), Hooker (1998), Heipcke (1999) and Jaffar (1999) for some surveys from slightly different perspectives). Usually CP models are solved using depth-first search and branch and bound. Naturally, these concepts can be complemented by local search concepts and metaheuristics. This idea is followed by several authors; see, e.g., de Backer et al. (2000) for TS and guided local search hybrids. Interestingly, following Rousseau et al. (2000) and Pesant and Gendreau (1999) one may deduce commonalities with the POPMUSIC approach described above.

Of course, the treatment of this topic is by no means complete and various ideas have been developed (see, e.g., those regarding local search within CP; e.g., Nareyek (2001) describes an architecture for constraint-based modeling and local search based reasoning for planning and scheduling). Especially, the use of higher level or global constraints, i.e., those which use domain specific knowledge based on suitable problem representations, may influence the search efficiency. Another idea is to transform a greedy heuristic into a search algorithm by branching only in a few (i.e., limited number) cases when the choice criterion of the heuristic observes some borderline case or where the choice is least compelling. This approach may be called *limited discrepancy search* (see, e.g., Harvey and Ginsberg (1995), Caseau et al. (1999)).

Constraint logic programming may be regarded as a backtracking search that is based on the assumption that it may be advantageous to solve combinatorial optimization problems by investing more effort in a reduction of the solution domain by applying constraint propagation rules; see Marriott and Stuckey (1998). Problems can often be specified more naturally as constraint programs than as integer programs, which can be exploited by modeling languages that have constraint programming capabilities (Fourer (1998), van Hentenryck (1999)). These approaches have been quite successfully applied to problems with a significant number of logical constraints (for example, special scheduling and assignment problems). In the sequel, we discuss several such libraries. For some detailed consideration see also Chapter 8 by *Shaw*, *Furnon* and *de Backer* as well as Chapter 9 by *van Hentenryck* and *Michel*.

ILOG Solver: ILOG Solver is a C++ library, which embodies constraint logic programming concepts such as logical variables, incremental constraint satisfaction and backtracking (Puget and Leconte (1995)). Application-specific add-ons such as ILOG Scheduler or ILOG Dispatcher provide functionality for special types of problems (constraint-based scheduling and vehicle routing / technician dispatching, respectively). For example, ILOG Scheduler provides classes that support modeling of concepts such as resources, activities, and scheduling constraints. Further information can be found at ILOG (2001).

ECLiPSe: ECLiPSe is a development environment for constraint programming applications. It contains several constraint solver libraries and provides a high-level modeling language to facilitate the development of programs to solve combinatorial problems in planning, scheduling, resource allocation, timetabling, transport, etc. ECLiPSe has an open architecture for constraints extensions based on attributed variables, and its constraint solvers allow the users to define and use new constraints at different conceptual levels. Further information can be found at Eclipse (2001).

CLAIRE: CLAIRE is a high-level functional and object-oriented language with rule processing capabilities. It is intended to allow the programmer to express complex algorithms in a concise and natural manner. To achieve this goal, CLAIRE provides means such as a rich type system, parametric classes and methods, an object-oriented logic with set extensions, and dynamic versioning. CLAIRE is a complete programming system with an interpreter, a compiler and a set of tools (tracer, debugger, object inspector) that works both on compiled and interpreted code. CLAIRE can also be used as a C++ pre-processor because it generates human-readable C++ code. On top of CLAIRE, CLAIRE SCHEDULE is a library of constraint propagation algorithms for preemptive, "elastic", and "mixed" scheduling, that is, scheduling with both interruptible and non-interruptible activities. It consists of a set of functions allowing the definition of preemptive and mixed scheduling problems, and the propagation of the basic types of decisions that can be made in the process of solving such problems. As an alternative package above CLAIRE, a simple finite domain constraint solver was developed: ECLAIR. For further information see Caseau and Laburthe (1996), CLAIRE (1999).

1.3 CALLABLE PACKAGES AND NUMERICAL LIBRARIES

Software packages serve different needs with linkages to different systems providing modeling language as well as graphical user interface functionality. Usually, it is assumed that a problem at hand is already modeled by means of an appropriate modeling language (such as AMPL (2001), GAMS (2001) or MPL (2001); see also Fourer et al. (1993) and Bisschop and Meeraus (1982)). The main distinction refers to the type of problem: linear programming, integer or mixed integer programming, non-linear or stochastic optimization. We provide a brief overview in order to provide context for the libraries that are the subject of this book. General information concerning callable packages can be found in More and Wright (1993), Wright (2001) as well as NEOS (2001). The section closes with a brief subsection on numerical libraries.

1.3.1 Linear Programming Packages

Linear programming problems can be efficiently solved to optimality by well-known techniques. Algorithms can be implemented using well-defined low-level schemes such as vectors and matrices and inputs can be provided as declarative specification of problems. This facilitates the use of highly advanced methods that are, in general, independent from specific types of problems. At present, large-scale linear programs having hundreds of thousands of continuous variables are regularly solved by using advanced techniques such as exploiting the presence of sparse matrices. There are several commercial software systems such as, e.g., CPLEX (2001), LINDO (2001), OSL (2001) and XPRESS-MP (2001).

While these products are applicable as ready-to-use packages, they generally include subroutine libraries (such as shared/dynamic link libraries), which can be integrated in application-specific software systems. Corresponding libraries include optimizing functions that may be called with arguments defining problem instances and/or solution vectors. Moreover, such libraries generally include routines to set parameters, set, get and modify problem and/or solution instances, and to store information to or retrieve information from a file system. While these libraries commonly include some type of callback mechanisms, which allows user-specific additions to the solver process, the libraries are generally designed to be utilized as a single entity to solve problems rather than as a set of components to be mixed and matched to solve problems. For additional information and a software survey see, e.g., Fourer (2001).

Additional packages or libraries can be found under MOPS (2001) (Mathematical OPtimization System) and SoPlex (2001) (Sequential Object-oriented simPLEX class library). For instance, the SoPlex class library comprises classes that may be categorized into three different types:

- Elementary classes are provided for general purpose use.
- Linear algebra classes provide basic data types for (sparse) linear algebra computations. Their functionality is restricted to simple operations such as addition and scaling.
- Algorithmic classes are used for complex tasks such as solving linear systems of equations and serve for implementing a variety of algorithms for solving numerical (sub-)problems.

Linear programming algorithms are also available as spreadsheet add-ons which afford users the opportunity to manipulate their data using the spreadsheet interface. Such packages are available from, e.g., LINDO or Frontline; see Savage (1997).

While many linear combinatorial optimization problems, such as minimum-cost network flow problems, can be modeled as linear programs, specialized algorithms generally solve such problems more efficiently. Corresponding methods are often included in linear programming packages. However, there are also specialized packages such as, e.g., NETFLOW (2001).

1.3.2 Integer Linear Programming Packages

Integer and mixed integer programming problems, which are in general $\mathcal{NP}$-hard, can be addressed by well-known combinatorial optimization algorithms. While these problems may in theory be intractable due to high computational requirements, recent advances in both corresponding algorithms and computing technology have led to a broad applicability of a variety of methods. Almost all libraries mentioned in the preceding section also include methods for solving integer programming problems by branch-and-bound algorithms exploiting linear programming relaxations. In the sequel, we discuss some additional specialized libraries, which follow different approaches.

MINTO: MINTO (*Mixed INTeger Optimizer*) is a software system that solves mixed-integer linear programs by a branch-and-bound algorithm with linear programming relaxations (using callable libraries for linear programming). It also provides automatic constraint classification, preprocessing, primal heuristics and constraint generation. Moreover, the user can enrich the basic algorithm by providing a variety of specialized application routines that can customize MINTO to a problem class. Further information can be found at MINTO (1999).

EMOSL: EMOSL (*Entity Modelling and Optimisation Subroutine Library*) of DASH Associates is a combined modeler and optimizer library, which is based on the linear and integer programming capabilities of XPRESS-MP. EMOSL aims at techniques that provide mechanisms for exploiting model structure. The EMOSL toolkit enables developers to access entities expressed in the modeler's notation from a high level programming language. It allows developers to refer to generic model entities such as data tables, parameters, subscripted variables and constraint names, special ordered sets, etc. Further information can be found at XPRESS-MP (2001).

ABACUS: ABACUS (*A Branch-And-CUt System*) is a framework, implemented in C++, that incorporates cutting plane and column generation methods for combinatorial optimization problems. A framework defines a general template, which has to be completed by application-specific functionality. ABACUS provides support for the implementation of branch-and-bound algorithms using linear programming relaxations that can be complemented with the dynamic generation of cutting planes or columns (branch-and-cut, branch-and-price, branch-and-cut-and-price). ABACUS provides a variety of general algorithmic concepts, e.g., enumeration and branching strategies, from which the user of the system can choose the best alternative for her/his application. For further information, see ABACUS (2001) or Thienel (1997). A very detailed description of implementation issues is provided in Jünger and Thienel (2000).

1.3.3 Non-Linear and Global Optimization Packages

Once the assumption of linearity is dropped, the problems can become much more difficult to solve. This is particularly true when convexity is not present. Problems that require a search over a large space for the global minimizer to a non-linear function are often referred to as *global optimization* problems.

BARON: BARON (*Branch And Reduce Optimization Navigator*) is a FORTRAN package/library for solving nonconvex optimization problems to global optimality. It derives its name from its combining interval analysis and duality in its "reduce" arsenal with enhanced branch-and-bound concepts as it searches the solution space. The code consists of a core module for global optimization of arbitrary problems as long as problem-specific lower and upper bounding routines are provided by the user. A variety of specialized modules for which no coding from the user is required are also provided. The list of specialized modules includes: separable concave quadratic minimization, separable concave minimization, programming with economies of scale (Cobb-Douglas functions), fixed-charge programming, fractional programming, univariate polynomial programming, linear multiplicative programming, general linear multiplicative programming, mixed-integer convex quadratic programming, mixed-integer linear programming, indefinite quadratic programming, factorable nonlinear programming. With the exception of the factorable nonlinear programming module, all other modules solve linearly constrained problems. The factorable nonlinear programming module can handle most types of typical optimization problems. All modules can solve problems where some or all of the variables are required to have integral values. Further information can be found at BARON (2001).

LGO: LGO (*Lipschitz Global Optimizer*) is a package for continuous global optimization, which also provides corresponding callable modules. The LGO program system solves (Lipschitz-) continuous global optimization problems on finite n-interval ("box") regions, in the possible presence of additional (Lipschitz-) continuous constraints; cf. Pinter (1996).

1.3.4 Stochastic Programming Package

Stochastic programming is a methodology for bringing uncertainty into the decision making process, especially with respect to models based on linear programming. For example, IBM's *Optimization Library Stochastic Extensions* enhance the OSL package by multistage stochastic optimization algorithms, which may be applied as callable modules; see OSL (2001).

1.3.5 Numerical Libraries

Numerical libraries are typically implemented as classic function libraries, which provide related functionality by the definition of a set of routines that transform data at lower levels of abstraction (for example, vectors and matrices). Function libraries are relatively inflexible, as such routines commonly perform well-defined functionality without mechanisms to be varied or extended. However, function libraries have been proven useful as they provide often needed numerical algorithms as building-blocks of optimization software.

The series of books entitled Numerical Recipes provide a collection of numerical algorithms in the form of a "cookbook", including both mathematical descriptions and actual implementations (available, e.g., in FORTRAN, Pascal and C; see Press et al. (1993)). The functionality includes algorithms of numerical analysis and linear algebra, as well as simple optimization methods.

The NAG and IMSL libraries are collections of routines for the solution of numerical and statistical problems (both available for different languages such as FORTRAN and C); cf. NAG (2001) and IMSL (2001). LAPACK (Linear Algebra Package) is a library of FORTRAN subroutines for solving the most commonly occurring problems in numerical linear algebra (Anderson et al. (1995)).

MATLAB is an integrated technical computing environment that combines numeric computation, visualization, and a high-level programming language. MATLAB thus provides a more abstract and easy-to-use interface to the functionality of the libraries discussed before. Special "toolboxes" extend the functionality of MATLAB. The optimization toolbox provides methods for the optimization of different types of nonlinear problems; cf. MATLAB (2001).

1.4 CONCLUSIONS AND OUTLOOK

Optimization libraries enable the broad application of sophisticated methods by facilitating their inclusion in applications as well as supporting extension and adaptation of the methods. We should note that, at present, the success of applications often critically depends on appropriate adaptations of library components to the specific problem at hand. Clearly, research will continue along various dimensions (such as an increased exploitation of the techniques provided by object-oriented programming to build fine-grained components), which should result in more effective optimization libraries.

Over the last decade metaheuristics have become a substantial part of the optimization tool kit with various applications in science, engineering and management. Ready to use systems such as class libraries and frameworks are under development, although usually still restricted to be used by the *knowledgeable* user. In this respect the development of an adoption path taking into account the needs of different users at different levels is still a challenge (Fink et al. (1999a)).

A final aspect that deserves special consideration is to investigate the use of information within different metaheuristics. While the adaptive memory programming framework provides a very good entry into this area, this still provides an interesting opportunity to link artificial intelligence with operations research concepts.

2 DISTRIBUTION, COOPERATION, AND HYBRIDIZATION FOR COMBINATORIAL OPTIMIZATION

Martin S. Jones, Geoff P. McKeown and Vic J. Rayward-Smith

School of Information Systems
University of East Anglia
Norwich NR4 7TJ, United Kingdom
{msj,gpm,vjrs}@sys.uea.ac.uk

Abstract: In the literature there is a growing evidence that demonstrates the effectiveness of object-oriented frameworks in assisting software development. Furthermore, object-oriented frameworks may not only improve the development process of software, but can also improve the quality and maintainability of software. This chapter looks at the Templar framework from the University of East Anglia, UK, with a specific interest in distribution, hybridization, and cooperation.

2.1 INTRODUCTION

There is a body of growing evidence, both theoretical and practical, that demonstrates the effectiveness of object-oriented frameworks in assisting software development; see, for example, Johnson and Foote (1988), Fayad and Schmidt (1997a), Mamrak and Sinha (1999). In fact, object-oriented frameworks may not only improve the development process of software, but can also improve the quality and maintainability of software.

Although the object-oriented paradigm has been in use for many years, the production of mainstream, quality frameworks is a comparatively new phenomenon. The application of frameworks to combinatorial optimization is even more recent. However, as can be seen from the literature, it can be applied with good results to this domain, see Fink et al. (1999b), Andreatta et al. (1998), Woodruff (1997), Marzetta (1998) and Michel and van Hentenryck (1998) for examples of frameworks, or framework like systems.

This chapter looks at some features of one framework in particular: the Templar framework from the University of East Anglia, UK. There are many features that can be considered desirable when supporting optimization tasks and software. Section 2.2 provides a brief look at these. Much of the remainder of this chapter looks at how the Templar framework supports distribution, hybridization, and cooperation. Most researchers will probably be familiar with the first two of these points, even if they have not had a chance to experiment with them themselves. The third point, cooperation, is strongly linked to both distribution and hybridization. The distinction between the three areas is certainly not clear. However, considering cooperation separately gives a slightly different insight into how to use more than one machine and more than one optimization technique concurrently. By taking this viewpoint, greater advantage may be taken of the benefits offered by object-oriented frameworks. Cooperation is considered in Section 2.4. Distribution and hybridization are considered in Sections 2.3 and 2.5, respectively.

Although object-oriented frameworks offer a number of benefits, they do have some disadvantages. Section 2.6 attempts to assess the impact of using an object-oriented framework on the execution time of an optimization technique. This can be helpful if execution time is a critical factor. One might even choose to do some initial experimentation using a framework, and then develop a customized, problem specific implementation that was designed to run as quickly as possible. The methodology taken to assess any difference in execution speed can be applied to other frameworks.

2.2 OVERVIEW OF THE TEMPLAR FRAMEWORK

This section takes a brief look at the University of East Anglia's object-oriented framework, Templar. A reader wishing to learn more about Templar's design and construction is referred to Jones (2000) and the Templar web site:

`http://www.sys.uea.ac.uk/~templar`

2.2.1 Evolution of Templar

Before delving into some of the design and construction of Templar, it may be beneficial to provide some of the history of Templar. More specifically, the manner in which Templar evolved is of interest.

The origins of Templar can, in part, be traced back to the X-SAmson (Mann (1995b)) and X-GAmeter (Mann (1995a)) applications of Jason Mann. These were C applications that implemented a single technique, such as Genetic Algorithms, and provided a user interface through the X-Window/Motif system (Nye (1995), Nye and O'Reilly (1995), Heller and Ferguson (1991)). In some cases, special parallel versions were available.

The British Defence Evaluation and Research Agency (DERA), who had funded the development of these applications, required a more flexible system. Templar was initially developed to meet their goals. At this time it was a relatively small project. However, it soon become clear that the goals of DERA were only a subset of the features that could be offered by an object-oriented framework. Indeed, this manner of conception may be shared by many other frameworks, e.g., see Codenie et al. (1997). It may not always be easy, or desirable to make this change in scope. For instance, there must be enough time to implement the more general software; in a commercial environment there must be the funds to support the development; and the result must be commercially exploitable. Additionally, experienced, capable developers must be available – the task of developing a framework is considered more complicated than general software development (Johnson (1997)).

The task of designing and implementing Templar was taken on initially as part of a PhD. As Templar evolved, it was clear that the work required was extensive, and an area of research that was just starting to become lively. Therefore, it became the sole subject of the PhD. Because of its research origin, there was far less commercial justification required. However, this does not mean that the commercial exploitability of the framework was ignored, see Section 2.2.4 and Howard and Rayward-Smith (1998).

Like many frameworks, Templar has been through both minor and major changes. The minor stages are due to small changes, and extensions that do not substantially alter the structure of the framework. However, as observed in Schmidt and Fayad (1997), major changes are sometimes required. After experimenting and using the framework, it was clear that there were several areas of improvement, most notably in how the distributed elements could be used together. It was at this point that a far reaching decision was made to fundamentally alter the design of the framework. This alteration was in line with the shift from a predominantly white-box framework to one with a more black-box nature. It also introduced the mechanisms to support cooperation cleanly. As a practical note, this change required a lot of work, but in the end, "throwing away" the first attempt has produced a cleaner, more efficient, and more flexible framework.

2.2.2 High Level Design

This section gives an overview of the components of the Templar framework. Figure 2.1 shows the major base classes, their relationships, and some concrete specializations. The diagram uses OMT notation (Gamma et al. (1995)). TrEngine, TrProblem, and TrRepresentation are the three most fundamental base classes within Templar. (Most of the identifiers within Templar are prefixed by Tr to avoid namespace pollution, particularly on compilers that do not support C++ namespaces.)

Specializations of TrEngine provide optimization technique specific functionality. Figure 2.1 shows both an SAEngine, and a GAEngine concrete class derived from the base TrEngine class. They provide implementations of Simulated Annealing (SA) and Genetic Algorithms (GA), respectively.

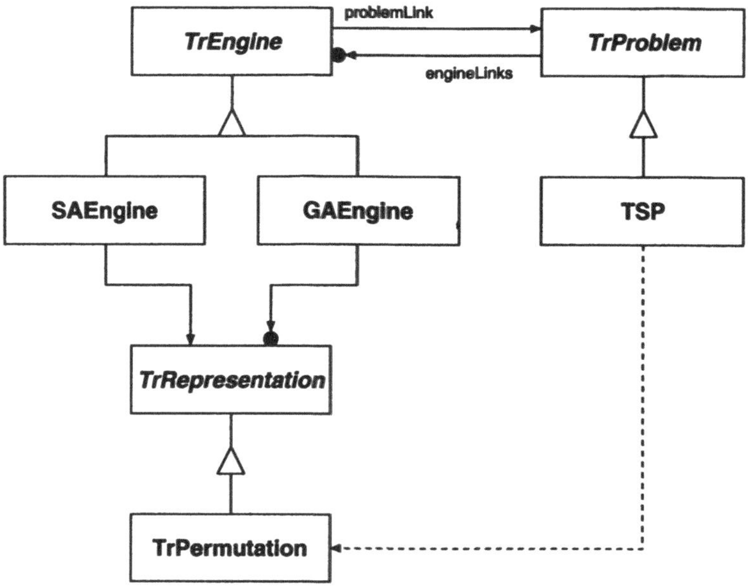

Figure 2.1 Basic Class Relationships

Figure 2.1 also depicts a TSP class derived from the TrProblem base class. This is a concrete implementation that provides functionality to allow a TrEngine to work on an instance of the travelling salesman problem (Lawler et al. (1985), Reinelt (1994)). In order to avoid confusion, we shall write "TrProblem instance" when we refer to an instance of class derived from TrProblem, and "problem instance" when we refer to a problem instance such as att48.tsp from TSPLIB (Reinelt (1991)). Each TrProblem instance is capable of "loading" a problem instance. We say "loading" because the instance is normally loaded from file, but this might not always be the case. For instance, specializations used for data-mining might load the problem instance from a database (Howard and Rayward-Smith (1998)). Once a problem has been loaded, a *linked* TrEngine may perform work against the problem instance. It is the TrProblem instance's job to provide the functionality with which a TrEngine can create initial solutions, evaluate solutions, generate new solutions, genetically combine two solutions, etc.

The third of the fundamental base classes, TrRepresentation, describes representations of solutions in general. As can be seen from Figure 2.1, the TSP class creates instances of TrPermutations. The TSP implementation has chosen to represent tours using permutations. Templar provides a number of basic representation classes: TrPermutation, TrIntArray, TrDoubleArray, and TrBitString. New specializations of TrRepresentation, and indeed TrEngine and TrProblem, can be constructed in a straightforward manner. Therefore, if one of the basic representations does not match the desired representation very well, then a new one can be created.

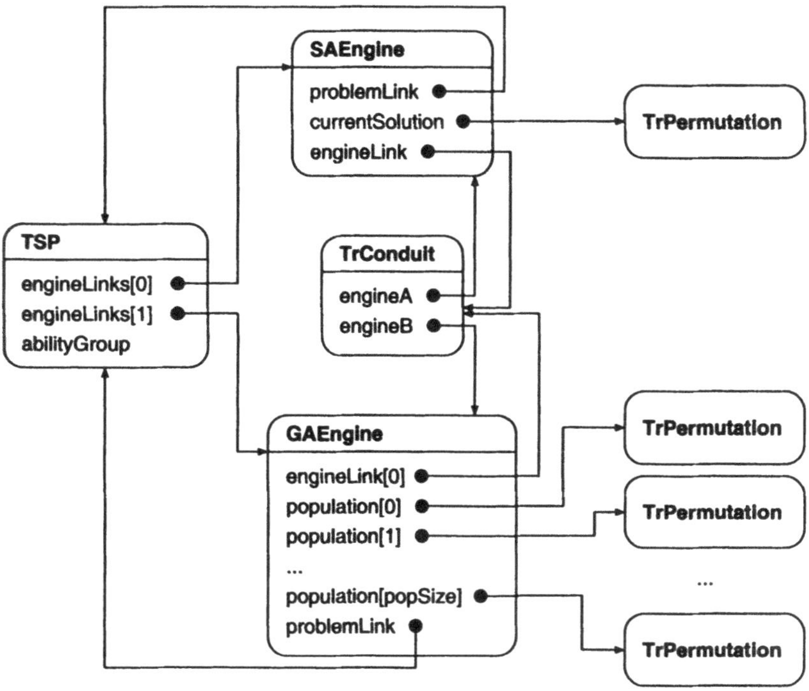

Figure 2.2 Basic Interaction

Figure 2.2 shows the relationships between these specializations as they might appear in a running application. Two TrEngines, an SAEngine and a GAEngine, are both linked to a single TSP object. Both engines are able to work concurrently on a the problem instance loaded into the TSP object. Concurrency is dealt with in more detail in Section 2.3. The SAEngine has a single, current solution with which to work. The GAEngine has a potentially large number of solutions – its solution pool. The new class, TrConduit, is explained in more detail in Sections 2.3 and 2.4. For now, it should be considered as a channel through which TrEngines may communicate with each other.

2.2.3 Maintaining Abstraction

Classes similar to Templar's three fundamental classes may be found in other frameworks, see, for example, Woodruff (1997). However, the way in which they work, and preserve the abstraction differs.

In Templar, TrEngines may only address a TrProblem instance through the TrProblem interface. It is not aware of the full class of the TrProblem on which it is working, and it has no easy means of finding this information. The same is true for a TrRepresentation: the TrEngine is unaware that is dealing with anything other than some form of representation. Where the TrProblem is aware of a TrEngine, it too is unaware of the true nature of the TrEngine. A TrProblem is, however, fully aware of the full type of representation that it uses. The TSP must know that it stores solutions as TrPermutations. TrRepresentations are, in general, aware of neither the TrEngine or TrProblem in any context. However, specialized TrRepresentations may utilize knowledge about the TrProblem for which they were developed.

This simple situation leads to TrEngine specializations that can, theoretically at least, be used to solve any TrProblem. It also means that any TrProblem may be solved using any TrEngine, and that it is free in its choice of representation.

Clearly, the information which has been hidden must be known at some level. This can be achieved in different ways. Templar takes a black-box, dynamic approach. This contrasts to the compile-time polymorphic approach as used in HOTFRAME (Fink et al. (1999b), Fink and Voß (1998)). The latter approach is considered, intuitively at least, more efficient in terms of run-time execution. While the approach taken by Templar may not be as efficient, it is more flexible at run time (Johnson and Foote (1988)). Section 2.6 investigates the "price" one might expect to pay when using a black-box framework rather than hand-coding an optimization technique.

A TrEngine is, in general, completely reliant on the TrProblem to provide it with the functionality to manipulate the solutions provided by the TrProblem. The implementations of these functions are called the *abilities* of the TrProblem. There are different types of abilities, such as neighbourhood abilities, evaluation abilities, genetic recombination operators, etc. A TrProblem can support as many, or as few, of the available ability types as its developers wish. However, without some of the fundamental abilities, such as evaluation abilities, no TrEngine will be able to perform useful work. When a TrEngine is connected to a TrProblem, it will query the TrProblem about which abilities it supports. For example, the SAEngine will query a connected TrProblem about neighbourhood move abilities. Of course, if a given TrProblem does not provide these abilities then the SAEngine will not be able to perform useful work on this particular problem. The TrProblem to which a TrEngine is connected may be changed at run-time. In this situation, the TrEngine must gather ability information once more.

In many situations the abilities supplied by a TrProblem have not been implemented specifically for that particular TrProblem, but are generally applicable to the TrRepresentation used by the TrProblem. For instance, in the case of the TSP using the TrPermutation representation, neighbourhood move operators such as swapping two elements of the permutation or reversing the ordering between two points of the permutation, are not specific to the TSP. The implementation of these abilities have been supplied by TrPermutation. The TrProblem is free to pass on whichever abilities it sees fit. It can also replace and augment these abilities.

A TrProblem may replace the implementation of an ability with one that is considered more efficient for that type of TrProblem (e.g., TSP), and even the currently loaded problem instance (e.g., att48.tsp). As an example, the abilities offered by the TSP TrProblem differ depending on whether the problem loaded is symmetric or asymmetric. The black-box nature of the framework makes this sort of reconfiguration possible at run-time.

An augmentation might occur when a TrProblem supplies abilities that are not necessary for a particular optimization technique, but can improve the speed. For instance, the SAEngine only requires an evaluation, initialization, and a neighbourhood ability. However, if a TrProblem augments this set of abilities with a *delta evaluation* ability – an ability that allows changes in cost due to a neighbourhood move to be calculated – then speed can be improved dramatically. A TrEngine will typically query whether these additional abilities are available. If not, then it will still be able to perform work on solving the problem, just not as efficiently as if it had the additional abilities. As the implementation of a TrProblem evolves, so more of these augmenting abilities are created.

In some cases, augmentation may be taken further, and an ability is provided that is more oriented to a given optimization technique. The basic neighbourhood abilities mentioned before basically involve providing an iterator that iterates over the neighbourhood of moves for a given representation object. This iterator may jump to random locations, and can even move through the neighbourhood moves in a random order. However, the basic neighbourhood move ability does not provide any functionality for directly finding the neighbourhood move that results in the greatest, best change in cost. Such an ability would not generally be of use to a technique such as Simulated Annealing. However, for hill climbing using steepest descent (or ascent) it would be ideal. The implementation of the TSP object does indeed provide such an ability for hill climbers, and other TrEngines that wish to use this functionality. It is important to note that TrProblems should initially concentrate on providing only the most basic, generally applicable abilities. The more specialized abilities should be provided as the TrProblem implementation evolves.

2.2.4 Goals of Templar

Templar was designed to meet a number of goals. These goals were modified as more was understood about the nature of the framework, and the sort of areas in which it could assist. These goals include:

Rapid Experimentation This is of particular importance when applying combinatorial optimization to a new problem type. One may wish to get quickly some idea of which optimization techniques perform the best, and the sort of results that can be achieved. In order to support this, the framework should not only provide a re-usable design and components, but it should also provide a methodology for extending the framework and adapting it to the needs of the problem being investigated. This methodology should promote and facilitate the development of prototypes. From a commercial perspective, early experimentation allows design decisions to be moved towards the front of the development period, potentially saving both time and money. When creating a new TrProblem, the only functionality that is typically required allows a problem instance to be loaded, solutions to be evaluated, and new solutions (TrRepresentation subclasses) to be created.

Incremental Development Rapid Experimentation should not be considered in isolation. It should be possible to take the prototypes and apply modifications to refine and improve the functionality. It is here that the benefits of ability augmentation may be seen readily. As the implementation evolves, so more abilities are provided. These abilities are specialized for the problem under consideration. In some circumstances, and once suitable optimization techniques are known, abilities may be provided that are more efficient for certain optimization techniques.

Uniformity The uniformity of applications produced using a framework is well known, and considered to be an advantage (Fayad and Schmidt (1997a)). However, it is not only the applications that can benefit from this uniformity. Many of the optimization techniques used in combinatorial optimization share features or ideas. These shared ideas or features naturally transfer to general abilities, such as those for neighbourhood movements. Such abilities might be used in less obvious circumstances. For instance, neighbourhood operators can be used by GA implementations to provide mutation operators. Even the less general abilities may be used. For example, consider a GA that makes use of a hill climbing stages (Radcliffe and Surry (1994)), a TrEngine that implemented this technique directly could make use of the abilities that have been specialized for hill climbing. This uniformity increases the potential for reuse.

Education Because of the framework's highly structured nature, and support for rapid development, it became a good tool for education. In a post-graduate course, it was found that students were able to rapidly code their initial TrProblem implementations and experiment easily. The Templar framework also supports GUIs, which is of great benefit when trying to visualize the progress of a technique, and the effects of changing a technique's parameters.

Adaptable Frameworks need to be adaptable – it is unrealistic to believe that a framework is ever complete (Codenie et al. (1997)). As new optimization techniques are developed and implemented, Templar will need to adapt to support the requirements of these optimization techniques. The framework's point of adaption are the abilities – by creating new types of abilities, new optimization techniques may be implemented and treated with the same level of abstraction.

Support for Distribution When solving large combinatorial problems, it may be advantageous to use parallel computers or a distributed arrangement. Clearly, this should be supported by a framework for combinatorial optimization. This point is considered in more detail in Section 2.3.

Support for Hybridization It has been found that where two optimization techniques perform poorly in isolation, they may perform better if they are are used together, possibly as a hybrid algorithm (Ryan et al. (1999), French et al. (1997)). Hybridization is supported by three mechanisms in Templar: TrEngine composition, TrEngine-TrEngine cooperation, and the fine-grain decomposition offered by abilities. Hybridization is addressed in more detail in Section 2.5.

Supports Cooperation Cooperation allows TrEngines to work as part of a loosely coordinated team. Cooperation is both facilitated by, and facilitates, distribution. It can also facilitate one form of hybridization. This simple, but effective concept is investigated in Section 2.4.

Useful for Comparison Assuming that each TrEngine has been programmed by a competent programmer, with a good understanding of the optimization technique, it should be possible to compare techniques fairly. Obviously this is of great benefit when deciding which technique is the "best" for your problem. If necessary, once comparisons have been made, then the "best" optimization technique may be hand-coded if speed is the ultimate requirement. However, the results presented in Section 2.6 suggest that a hand-coded solution may not be worth the development effort for the limited gains in performance.

Useful for Experimentation In order to be useful for research purposes, the framework should make it easy to experiment with different optimization techniques. This means that it should be easy for a researcher to change optimization technique, and to easily change each technique's parameters. Due to the abstraction over TrEngines, it is easy to replace one TrEngine with another, and because of the black-box nature of Templar, this may be done at run-time. An optimization technique's parameters would, for example, include values such as initial temperature and cooling schedule for Simulated Annealing. A base class of both TrEngine and TrProblem, TrComponent (not shown in the diagrams) provides a means by which parameters may be set and retrieved, even on a TrEngine that is currently performing work. To facilitate running a large number of experiments, a scripting language has been devised to automatically control TrProblems and TrEngines. This means that hours, days, or even weeks worth of experimentation can be controlled automatically. An implementation of this scripting language exists for the initial implementation of Templar but has yet to be ported to the more recent implementation.

Embeddability By providing components that can be embedded in other applications, the power of the optimization techniques can be used by non-experts in other domains.

2.2.5 Implementation of Templar

The Templar framework has been implemented in C++. The original request for the provision of Templar stipulated that C++ was used. However, even without that stipulation, C++ would have been the language of choice. Although some authors, such as Meyer (1997) do not see C++ as a fully fledged object-oriented language, it is clearly sufficient to implement an object-oriented framework. Features such as assertions and garbage collection almost certainly have the potential to improve a framework in terms of usability, reliability, and enforcement of the framework's constraints. However, for a system such as Templar which must be both as efficient as possible, and portable, the choice of object-oriented languages is very limited. C++ is typically seen as one of the most efficient object-oriented languages Wu and Wang (1996). It is also a superset of C, and this made initial implementation relatively straightforward because it would have been possible to base the implementation of Templar on the code offered by X-GAmeter and X-SAmson. C++ compilers can be found on most platforms. They are relatively mature and can generate efficient code.

Although C++ is a suitable language, it is by no means perfect. For instance, the lack of support of assertions means that it is more difficult to enforce the framework's constraints on the framework's clients. (The use of the word assertion here refers to assertions as defined in Meyer (1997), not the ASSERT macro.)

Although the C++ language is portable, the implementation of some compilers leaves a little to be desired, particularly at the time that the framework was initially created (1995). One particular point that must be taken into consideration is that different compilers have different template instantiation strategies. What works with one compiler may lead to missing or multiply defined symbols on another. In order to circumvent this problem, a special build script, written in Perl, controls the compilation process and sets the flags needed for a given compiler.

Graphical extensions, designed for use with the Templar framework are available for X/Motif (Nye (1995), Nye and O'Reilly (1995), Heller and Ferguson (1991)). An equivalent set of classes has yet to be implemented for Microsoft platforms.

All of the core framework is documented through the strict use of comments in the header files. A program similar to the Javadoc (Flanagan (1999)) program reads the header files and automatically generates HTML documentation pages.

Although Templar is primarily designed to be a framework for the implementation of optimization techniques and problems, there is an implementation of a simple application that allows users to create, manipulate, and experiment with TrEngines and TrProblems. Although this is an extremely small application (most of the implementation is based in the framework itself), it is powerful enough to meet most experi-

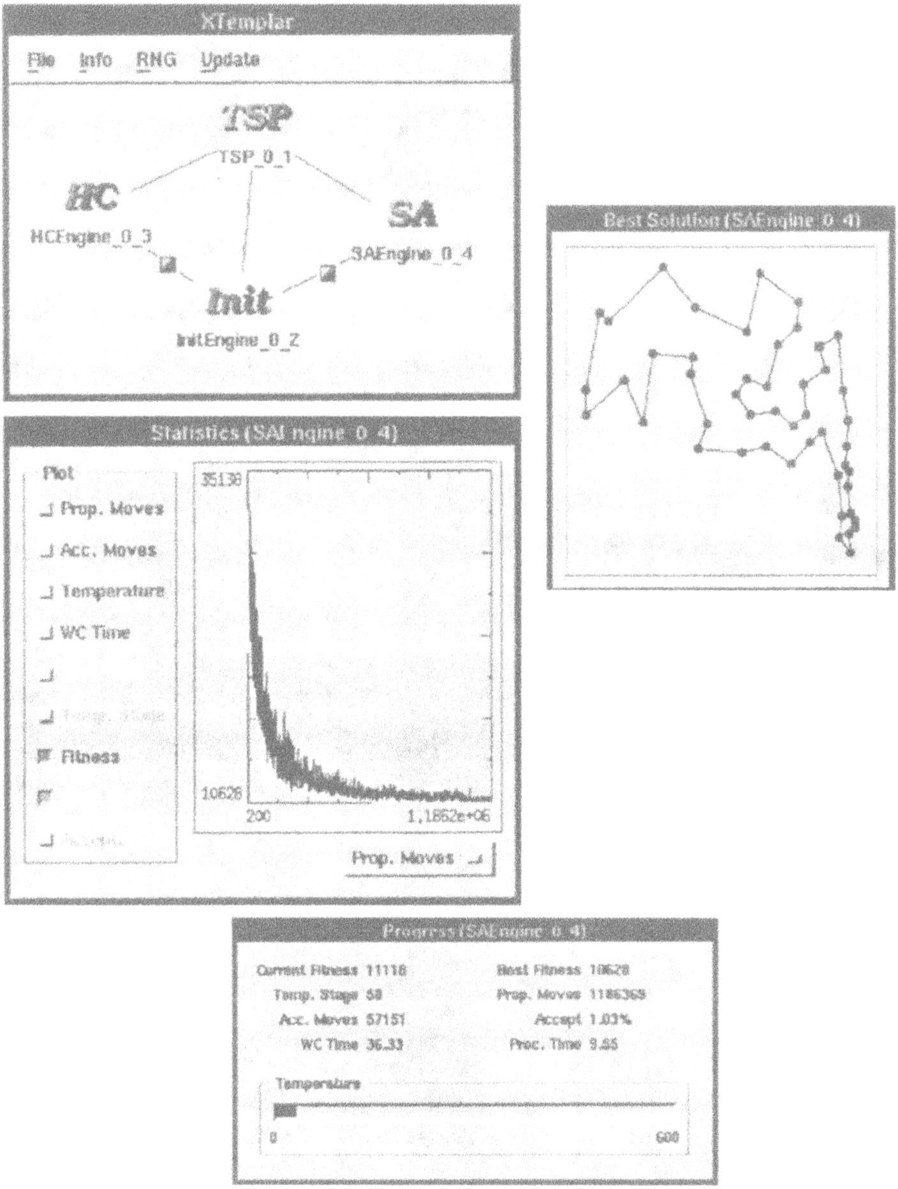

Figure 2.3 Application Screen Shot

mentation and education needs. Figure 2.3 shows a screen shot of this application in use.

2.3 DISTRIBUTION

Due to the increasing availability of powerful desktop machines and fast network connections, distributed computing is currently a highly active area of computer science (see Panda and Ni (1997a), Panda and Ni (1997b)). A networked arrangement may be both cheaper, and more flexible, than larger parallel machines and super-computers. A distributed environment consists of a number of networked processing nodes, each with its own local memory. Communication through the network between nodes is taken to be considerably more expensive in terms of latency than accessing a node's local memory (Bacon (1997)). A distributed system of computers may be *heterogeneous*, i.e., it may contain machines with different architectures, or using different operating systems.

Distribution is seen as such a fundamental requirement that there are a number of well-developed frameworks to support this alone. Most readers will have heard of systems such as CORBA (Seetharaman (1998)) and DCOM (Redmond (1997)). However, these frameworks are not utilized by Templar. There are three driving factors in this decision. The first is purely pragmatic: initial funding for the work required an implementation that made use of MPICH over a Myrinet (Lauria and Chien (1997)). The second is due to practical constraints of the two frameworks. CORBA requires the use of an Interface Description Language (IDL), this would have added to the complexity of the system. Microsoft's DCOM is proprietary and is not as freely available as MPICH – an implementation of the Message Passing Interface (MPI) standard. The third reason, due fortunately to the initial use of message passing, is that simple, asynchronous message passing was found to be both powerful, and led naturally to cooperation, a subject discussed later.

Before describing the distribution mechanisms, this section will show how Templar supports concurrency on a single uni-processor, or multi-processor machine. Then, the mechanisms for supporting message passing are described in Section 2.3.2. Section 2.3.3 describes how Templar handles the distributed objects using the message passing layer. A case study into the use of distribution is provided for simulated annealing in Section 2.3.4.

2.3.1 *Single Machine Concurrency*

The Templar framework makes use of threads (Bacon (1997)), but uses abstraction to hide machine or operating systems specifics. It is, however, assumed that the target operating system offers threads. The layer of abstraction provides thread classes based on the POSIX standard (Nichols et al. (1996)). This practice is by no means new to Templar, for example, see Schmidt (1995). On Unix based systems, the concrete implementation of these classes contain very little code because these systems tend to provide POSIX compliant threads. However, with a Microsoft platform, the native thread mechanisms must be re-mapped to exhibit POSIX like behaviour.

Mapping from one thread implementation to another forces the abstraction provided by Templar to be very simple. For instance, there is no way to "kill" a thread – it must willingly die. However, this is not seen as a problem because, due to the constraints of the framework, a user of the framework does not have general access to the threads used for TrEngines, TrProblems, etc. The simple thread abstraction also makes the implementation easier to test and debug – a problem that is seen as challenging for frameworks in general (Fayad and Schmidt (1997a)), let alone frameworks utilizing threads.

Due to the potential additional complexity of systems using threads over those that do not use threads, the benefits of threads should be thoroughly investigated. It should be clear that an alternative solution, such as an event queue, would not be suitable.

Threads were initially considered for Templar because they can allow an application to block on more than one resource. For Templar, this can be seen when the application needs to wait for GUI events or communications from other processing nodes. Although it is possible, under some operating systems, to perform this sort of blocking, the mechanisms are not portable. The simplest solution was to use multiple threads.

It soon became apparent that, if each TrEngine was given its own "work" thread, then more than one TrEngine could be at work concurrently on a single machine. These TrEngines could possibly be working together, cooperatively, on the same TrProblem. Furthermore, if the machine running the application was a multi-processor machine, then the use of threads would potentially allow the maximal use of processors. This would not be the case with an event queue.

The use of threads also assists in the provision of GUIs. One of the features that is clearly desirable within a GUI is the ability to inspect the current state of a TrEngine, including things such as the TrEngine's current parameters, the solution(s) on which it is currently working, etc. This can be provided relatively easily although care must be taken to ensure that the inspecting thread, and the TrEngine's work thread have exclusive access to the state information where necessary. Templar provides for support for ensuring exclusivity on this data. The use of threads makes the source code for a TrEngine and the GUI easier to write and understand.

A more subtle, but equally compelling reason for the adoption of threads is to support the goal of incremental development. The use of a single machine in a concurrent manner is a natural stepping-stone from a non-distributed to a distributed system. Templar makes little distinction between TrEngines running on the same machine and TrEngines running on different machines.

2.3.2 Supporting Message Passing

Message passing is a common paradigm for writing parallel applications (McBryan (1994)). The paradigm allows processes to communicate with each other by sending and receiving messages. It is applicable to Multiple Instruction, Multiple Data (MIMD) architectures, and hence can also be used in distributed environments.

Standard message passing libraries such as PVM (Geist et al. (1994)), or those conforming to the MPI standard (MPI (1995)) provide message passing constructs for inter-process communication. However, Templar requires a finer destination resolution than processes if messages are to be sent between TrEngines.

To facilitate this form of communication, the framework provides classes that allow messages to be passed between objects. These objects may be in the same process and the same thread, in the same process but in different threads, or in different processes (possibly on different machines in a distributed system). Message passing in this way is by no means unique to Templar; see, for example, Tödter et al. (1995).

The messages are created by a process often called *marshalling* (*unmarshalling* is the complementary operation). This process is sometimes known by other names, e.g., many Java programmers would probably call the process serialization. In essence, the job of marshalling is to break a complicated piece of data into smaller chunks that can be sent across a network, or stored in a serial manner. Unmarshalling takes this information and uses it to construct objects that are identical copies of the original. The closest real-world analogy is probably sending a document through a fax machine. The fax machine marshalls the data into squeaks and squawks, and a complementary machine creates the copy of the document.

Underneath Templar's message passing abstraction lies a concrete implementation that makes use of an existing message passing library. Currently, MPICH is used as the message passing library, i.e., the MPICH implementation of MPI provides the inter-process communication functionality. However, it is possible to use other message passing systems. For example, DCOM (Distributed COM) is another system that allows objects to communicate across a network. DCOM is far more than just a message passing system, but it can be used as such on Microsoft platforms. Indeed, it has been used at the University of East Anglia for an SA in a data-mining application.

Nearly all message passing within Templar is non-blocking and asynchronous. Messages are guaranteed to arrive in the order that they were sent. Templar does not make use of group communication or synchronization primitives. This arrangement keeps the communication layer simple enough to be ported easily, yet provides enough functionality to facilitate the implementation of distributed, cooperative and hybrid techniques. Issues such as network failures are not considered by the Templar framework.

Message types are identified by simple *tags*. New tags can be created easily at run-time for new types of messages. Not all message tags are usable by clients of the framework. Some of the tags have been reserved for control purposes for Templar.

In order to send messages between TrEngines and TrProblems, a communication link must be created. This link is seen as conceptually travelling through a TrConduit, as seen in Figure 2.2. The TrConduit is more than just a conduit, it can also act as a filter to prevent certain messages between the two objects. This filtering of messages is not unique to Templar, see, for example, Aksit et al. (1994).

2.3.3 Creating Objects Remotely

An application that uses Templar in a distributed environment consists of a number of processes. Each process, termed a *host*, may be on any one of the nodes within a distributed environment, although in a distributed environment consisting of uniprocessor nodes there will probably only be a one-to-one mapping. One of the hosts is identified as the *controlling* host. It is to this host that all I/O is directed, and it is this host that contains any GUI elements. All of the other hosts, termed *remote* hosts, are treated as vessels for remote TrEngines and TrProblems.

Figure 2.4 shows an example configuration of three TrEngines, and a TrProblem in a distributed environment. There is one controlling host and two remote hosts.

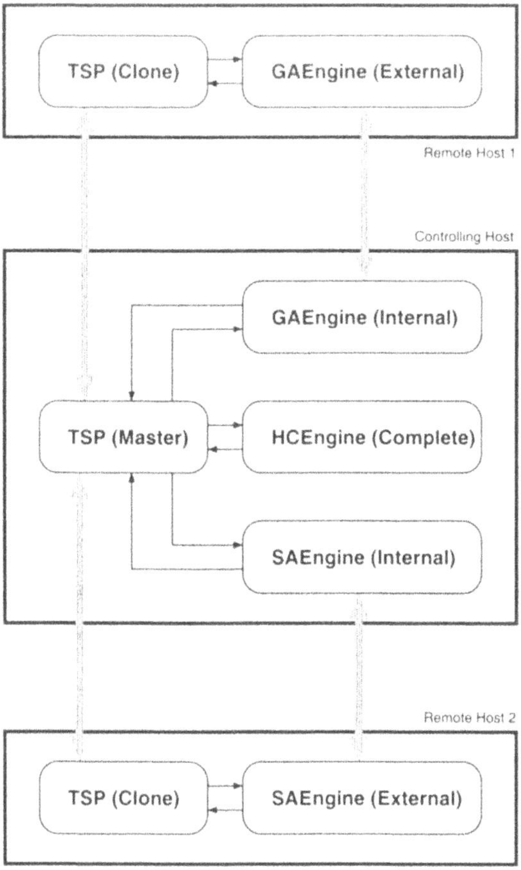

Figure 2.4 Distributed TrEngines and TrProblems

In a distributed environment, TrEngines may be of one of three forms: complete, internal, and external. *Complete* TrEngines can also be found when working in a non-distributed environment. This sort of TrEngine is located wholly within a single host. An *internal* TrEngine provides all of the functions of a TrEngine, but it serves as a proxy for a counterpart external TrEngine. An *external* TrEngine performs the work requested of an internal TrEngine. Hence, the controls for a TrEngine are on the controlling host, yet the work is performed within another host. It is important to note that all of these TrEngines are of the same class; the internal/external/complete status of a TrEngine is reflected by a data member. Internal and external TrEngines can communicate through the message passing mechanisms. Special channels are set up to enable control messages to be sent separately.

To an application using the Templar framework there is only one difference when creating a TrEngine in a distributed environment: the host for the external TrEngine must be specified. The framework ensures that both the internal and external TrEngines are created, or if the host requested for the external TrEngine is the controlling host, a complete TrEngine is created. The application on the controlling host uses internal, and complete TrEngines in the same way. There are no special considerations when using an internal TrEngine. The framework provides most of the additional functionality required behind the scenes to implement the paired nature of distributed TrEngines. The only change required to make a non-distributed TrEngine into one suitable for a distributed environment is to implement methods that allow the state of a TrEngine to be sent and received as a message. This step has been kept small deliberately. It is yet another area in which the goal of incremental development is fulfilled.

In order to allow TrEngines to work in the fashion outlined above, each external TrEngine must be able to work on a local copy of the TrProblem to which the internal TrEngine is connected. It is normal for this sort of replication to be used to create local copies of a remote object in order to preserve efficiency (Briot et al. (1998)). When a TrProblem is created, it is always created on the controlling host. A remote host is never specified because the remote, or *clone*, versions of the TrProblem are only ever created in response to an internal TrEngine being connected to the TrProblem master. The framework ensures that the clone TrProblems are copied and that their state is updated whenever the state of the TrProblem master is altered, i.e., when a new problem instance is loaded. Like TrEngine, the only additional work required by a TrProblem developer to allow a TrProblem to be used in a distributed environment is to provide methods that allow the state to be sent through the message passing system. Every time a new problem instance is loaded, the framework will send the TrProblem master object's state as a message to be received by all of the master's clones.

2.3.4 Implementing a Parallel Technique

This section will look at how parallel simulated annealing may be implemented using the message passing mechanisms of Templar. There have been a number of attempts to parallelize Simulated Annealing (PSA), for example, see Greening (1990) and Ram et al. (1996) for overviews. We will be looking at implementing the coarse-grain parallel method of multiple Markov chains (Graffigne (1992), Lee and Lee (1996),

Gallego et al. (1997)) – a technique already shown to yield good results. A fine-grain parallelism implementation, such as Speculative Simulated Annealing (Witte et al. (1991)), would not be well suited for implementation within the Templar framework.

The basic method used in Graffigne (1992), Lee and Lee (1996), Gallego et al. (1997) involves a number of SA searches that communicate occasionally. There is some theoretical evidence to show that the best solutions are obtained when the SA searches do not communicate (Azencott and Graffigne (1992), Lee (1995)) and one merely chooses the best solution after they have all terminated. It should be possible to see that this "poor-man's parallelism" can be achieved easily within Templar. One need only create the required number of SAEngines, possibly using more than one host. If all of the SAEngines are started, at the same time, to work on a single problem instance, one need only wait for them to terminate and decide on the best solution. Although this example does indeed show that it is trivial to achieve poor-man's parallelism, it does little to demonstrate how the message passing mechanism can be used. In order to do this, the asynchronous multiple Markov chain (MMC PSA) method presented in Lee and Lee (1996) is considered. With MMC PSA, there are a number of processing elements (PEs) each performing simulated annealing. A globally visible best solution is also used. Each PE is allowed to run for a "segment", this could be a period of time, a temperature stage, etc. At the end of the segment, the PE inspects the globally visible best solution in order to compare it with its best solution. If the globally visible best solution is better than the PE's best solution, then it is taken as the PE's working solution. However, if the PE's best solution is better than the globally visible best solution then the globally visible best solution is updated. Each PE starts from a different initial solution and the only synchronization mechanism involved ensures that race conditions do not exist over the globally visible best solution. Speed-up is achieved by modifying the length of a segment. If a segment had length l when a single PE was used then, when n PEs are used, a segment has length l/n. The justification for this is based on the observation that the probability of finding the optimal solution is dependent on the number of proposed moves (Azencott and Graffigne (1992), Lee (1995), Dodd (1990)).

Figure 2.5 shows one possible arrangement of TrEngines that provide a parallel implementation. Four SAEngines are connected to a SAMaster TrEngine. All of the engines are connected to a single TrProblem. There is a communication link between the SAMaster and all of the subordinate SAEngines. Note that the diagram does not specify hosts for each of the TrEngines.

The SAMaster is a new TrEngine that has been created specifically for the parallel implementation. Its purpose is to control all of the SAEngines, and to hold the globally best solution. In order to achieve this, it is necessary to set up some messages that the TrEngines understand. These include: "Start", "Stop", "Here's my best solution", "Here's my current temperature", etc. When the SAMaster is started, it sends out a message to all of the SAEngines indicating that they should start. The SAEngines are requested to stop at the end of each segment and send their best solution to the SAMaster. The SAMaster will then compare the solution with the globally best solution and either request the SAEngine to continue (after resetting its own solution)

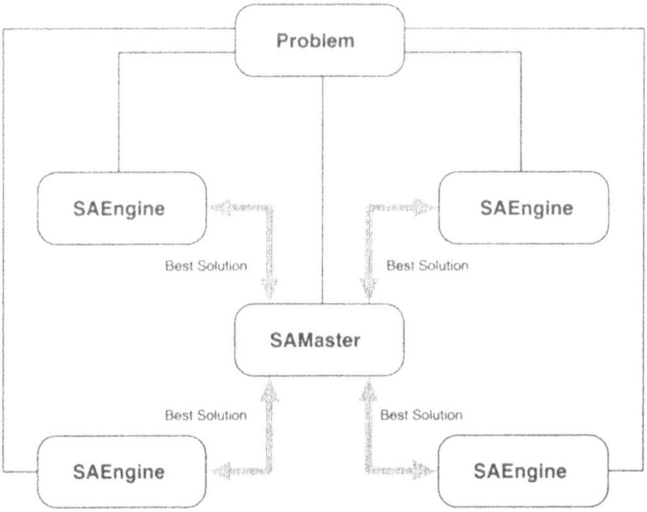

Figure 2.5 Distributed Arrangement of SAEngines

with its current solution, or request the SAEngine to copy the globally best solution and then continue from that point.

Note that this solution is extremely simple, the Simulated Annealing algorithm has not had to change significantly, and the existing functionality present in the SAEngines need only be enhanced slightly. It would, of course, be possible to implement a distributed optimization technique that did not use existing sequential TrEngines. This situation might still utilize a controlling TrEngine, and a number of subservient TrEngines, but the subservient TrEngines would not be able to function correctly without the controlling TrEngine; they would only be a part of the technique.

Although this solution appears to be clean, simple, and is a good example of code reuse – the existing sequential implementation becomes part of the parallel implementation – it is possible to do better. Section 2.4 describes how.

2.3.5 Potential Pitfalls

There are several areas that, if not addressed correctly, will limit the usefulness of the framework with respect to distributed implementations.

Firstly, as far as possible, the implementer of a TrEngine or TrProblem should be shielded from the control messages used by the framework – they should be well hidden. If this is not the case, then the work required to implement a TrProblem or TrEngine may be unduly complicated.

There are a number of different methods of parallelization and of taking advantage of distributed environments. However, a decision should be made as to which of the available methods should be supported (Lee (2000)). Templar supports asynchronous message passing because it is simple, promotes cooperation and maps well to distributed architectures. It does not even attempt to provide any support for publish/subscribe (Lee (2000)), RMI (Flanagan (1999)), etc.

Care must be taken when distributing a randomized algorithm. If the (Pseudo) Random Number Generators (RNGs) used by each of the TrEngines are not independent, then there may be some unwanted correlation (Matteis and Pagnutti (1995)). Using an RNG such as the one supplied with the standard C library (Kelley and Pohl (1990)) is not suitable. Templar creates independent RNGs by using RNGs based on the ZRan generator of Marsaglia and Zaman (1994). This RNG is parameterized and so different RNGs can be initialized with different parameters, not just different seeds. The random number streams for each of these RNGs remains independent. The initialization of each of the RNGs is handled by the framework and requires no user intervention.

2.4 COOPERATION

The previous section described the main constructs used to support distribution. This section introduces cooperation, in the context of Templar, by taking the concept of message passing for distributed techniques and generalizing it to form cooperative message passing. The parallel simulated annealing implementation mentioned before is revisited and made into a cooperative technique. And, an assessment is made of the effectiveness of the resulting cooperative parallel simulated annealing implementation.

2.4.1 Cooperative Message Passing

The message passing primitives introduced in Section 2.3 show that it is easy to implement a coarse-grain parallel optimization technique. Whether the implementation is run on a single machine or several machines, each with any number of processors, makes very little difference. The core concept is that a number of TrEngines are working in some controlled manner on the same TrProblem.

The PSA implementation described previously utilized an SAMaster TrEngine to control a number of SAEngines. Here the SAEngines were, in a sense, subservient to the SAMaster: they did as the SAMaster instructed. However, a cooperative arrangement of TrEngines dispenses with the idea of a centralized controlling TrEngine, and allows each TrEngine to work autonomously. The concept of cooperation is realized by the periodical transfer of information between linked TrEngines. If, when, and what, information is sent from a TrEngine is at its own discretion. What it does with any information it receives is also at its own discretion. It could, for example, choose to act on the information immediately, save the information for later investigation, or just ignore the information.

The types of messages that are sent between TrEngines can be split into two categories: those of a general, TrEngine independent nature, and messages that are specific to a set of TrEngines. Messages of the first type include messages in the following areas:

- initial solutions,
- current solutions,
- best solutions, and

- control messages (i.e., start, stop, etc).

Messages that are SAEngine specific include:

- current temperature,
- number of proposed moves,
- number of accepted moves,
- temperature stage, etc.

Obviously, each type of TrEngine will have its own specific messages.

Each TrEngine has both a *send vocabulary* and a *receive vocabulary*. These are the lists of messages that the TrEngine is able to send, and willing to receive, respectively. When two TrEngines are connected to each other through a communication link, they declare their vocabularies to one another. The messages that can be sent from one TrEngine, a, to the other TrEngine, b, is the intersection of a's send vocabulary and b's receive vocabulary. Two TrEngines connected by a communication link do not have to be of the same class. This is one of the fundamental concepts of cooperation; different classes of TrEngines may cooperate with each other without breaking abstraction. Obviously, they will probably only be able to communicate through the TrEngine independent messages. Two TrEngines of the same class, however, may also communicate using their TrEngine specific messages.

Although cooperation is a very simple idea, it allows communities of different TrEngines to work to a common goal. It also facilitates one method of hybridization, as shown in Section 2.5. Templar defines a number of standard cooperation messages and suggests suitable reactions for them.

It is possible to view TrEngines as being similar to *agents* (Rich and Knight (1991)). Good use has been made of cooperative techniques and agents in the AI community. See for example, Liddle and Hansen (1997), Yang and Ho (1997), Durfee et al. (1989), Decker (1987).

2.4.2 Revisiting Parallel Simulated Annealing

A technique such as MMC PSA is essentially a cooperative technique that makes use of a shared best solution. A system of SAEngines that operates in a similar manner can be created with a minimum of effort. The work needed amounts to creating the correct number of SAEngines and fully connecting them with cooperation links. Each SAEngine has a role similar to that of the PEs in MMC PSA. The arrangement can be seen in Fig 2.6.

In order to understand why this configuration works, one must understand the types of cooperation messages that SAEngines can, by default, send and receive. An SAEngine can send and receive most of the basic cooperation messages. The three that are relevant to this discussion are: a best fitness message, a best solution message, and a best solution request message. Best fitness messages are broadcast by SAEngines with a frequency governed by the SAEngine's parameters. This frequency can be based on time, number of proposed moves, temperature stage, etc. Any TrEngine to

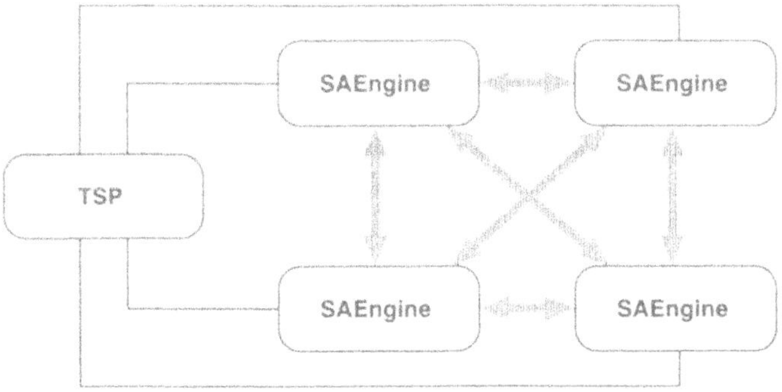

Figure 2.6 Cooperative Arrangement of SAEngines

which the SAEngine is connected through a cooperation link can receive these messages if it wishes. Different TrEngines will react differently to these messages. However, upon receiving a best fitness message, an SAEngine will examine the message and determine whether the fitness value it contains is better than the fitness of the best solution that it has seen. If it is better, it replies with a best solution request message, otherwise it replies with the best solution that it has seen. An SAEngine will react to a request for a best solution by replying with the best solution that it has seen. If an SAEngine receives a best solution message, then the solution that it contains is examined to determine whether it is better than the best solution seen by that SAEngine. If it is, then the SAEngine's best solution, and the solution on which it is working is updated. Otherwise, the message is ignored. With these simple responses, good solutions are propagated amongst the SAEngines. The frequency with which an SAEngine broadcasts its best fitness is set in accordance with the length of a segment.

One might consider that the way the SAEngine handles these messages has, in some way, been designed for this particular form of parallelism. There is some truth in this, but only in respect of the fact that this manner of handling the messages promotes cooperation. It has not been implemented in this manner especially to support MMC PSA. In fact, other TrEngines, such as the HCEngine (Hill Climbing), deal with these messages in a similar manner. Hence, as long as a TrEngine has been implemented to handle the basic cooperation messages used within the framework, it may be used to form part of a cooperative technique.

Other parallel techniques can also be implemented quite naturally in this manner. Take, for example, the island model for parallel genetic algorithms (Lobo and Goldberg (1996)). This model uses a number of GA "islands" that allow their best solutions to "swim" periodically between islands. An island maps quite naturally to a GAEngine.

It is clear that implementing a MMC PSA like technique in this manner is trivial and does not require the implementation of a new TrEngine, or alteration of an existing TrEngine. Also, because of the support for distribution, the same arrangement is suited for single processor machines, multiprocessor machines, and distributed systems.

2.4.3 Evaluating a Cooperative Technique

As well as the ease with which the cooperative technique can be implemented, the performance of such an arrangement is also of interest. Specifically, what can be expected in terms of speed-up and solution quality? In order to assess this, eight problems from TSPLIB were chosen as test problems. For each problem, ten runs were performed with a single SAEngine, two cooperating SAEngines, four cooperating SAEngines, and five cooperating SAEngines. A run consisted of allowing the TrEngine or TrEngines to run until termination. In the case of more than one TrEngine, the best solution obtained for that run was the best solution of all of the TrEngines, and the time for that run was the greatest wall clock time that any of the TrEngines had been running.

It can be difficult to determine both the parameters and the stopping condition associated with SA (Dowsland (1993)). The parameters were chosen empirically, as was the stopping condition. However, great care was needed to determine a suitable stopping condition. If a stopping condition was chosen that allowed the single TrEngine experiments to continue unnecessarily long then this could bias the results in favour of the cooperative techniques. To see why this would be so, consider a set of parameters that were known to give good results for a given problem. This set of parameters will determine how many moves, say k, are proposed at each temperature stage. A poor set of parameter values may cause an SAEngine to spend twice as long at each temperature stage ($2k$ proposed moves) yet not to generate results significantly better than when the good set of parameters were used. If an experiment were to be performed using the poor parameters with two cooperating SAEngines, then, due to the reduction in segment length, each SAEngine would propose k moves at each temperature stage – the same as would be used by a single SAEngine using the good parameters. Therefore, if the single SAEngine had also used the poor parameters, any measure of speed-up would be unfair.

The termination condition was determined by running a single SAEngine until it obtained a solution within 2% of the optimal solution. This test was performed a number of times for each problem instance, and the mean total number of proposed moves was calculated. This value was doubled to provide a limit on the number of total proposed moves. Hence, in the tests, if p proposed moves were allowed before a single TrEngine terminated, then when n SAEngines were cooperating, each was allowed to propose p/n moves. The length of each temperature stage was also reduced accordingly.

Tables 2.1, 2.2, and 2.3 present the results pertaining to solution quality of two cooperating SAEngines, four cooperating SAEngines, and five cooperating SAEngines, respectively. For each problem instance, eight values are provided. The first four are for a single SAEngine, and the last four are for the cooperating arrangement. The values have been calculated over the ten runs and give the best fitness found, the average fitness, the worst fitness and the standard deviation (S.D.) of the fitnesses over the ten runs. The results from the single SAEngine are included so that comparisons may be made easily. The purpose of these tests is not to demonstrate that the cooperative arrangement is better, but to demonstrate that the implementation using Templar behaves

TSPLIB	1 SAEngine				2 SAEngines			
Instance	Best	Mean	Worst	S.D.	Best	Mean	Worst	S.D.
att48	10628	**10661.6**	10738	**36.38**	10628	10666.6	**10736**	37.20
eil101	**629**	638.2	**647**	5.76	630	**637.1**	650	**5.70**
bier127	**118659**	**119243**	**120292**	**629.08**	118741	119822.9	122656	1116.27
u159	42435	42719.7	43669	451.54	**42082**	**42308.1**	**42441**	**148.72**
rat195	2333	**2347**	**2360**	**8.15**	**2327**	2347.3	2378	15.36
tsp225	3953	3979.4	4014	20.46	**3936**	**3967.2**	**3980**	**11.44**
lin318	42496	42618.4	**42888**	**134.33**	**42281**	**42584.6**	43090	222.49
pcb442	**51149**	51511.7	52090	294.40	51167	**51387.2**	**51597**	**148.29**

Table 2.1 Cooperative Arrangement with 2 SAEngines

TSPLIB	1 SAEngine				4 SAEngines			
Instance	Best	Mean	Worst	S.D.	Best	Mean	Worst	S.D.
att48	10628	**10661.6**	**10738**	**36.38**	10628	10681	10835	57.07
eil101	**629**	638.2	647	5.76	630	**637.3**	**642**	**4.15**
bier127	**118659**	119243	**120292**	629.08	118713	**119221.6**	120738	**567.26**
u159	42435	42719.7	43669	451.54	**42080**	**42460.1**	**42824**	**193.22**
rat195	**2333**	**2347**	2360	8.15	2336	2348.7	**2356**	**5.75**
tsp225	**3953**	**3979.4**	**4014**	**20.46**	3957	3982.3	4025	21.39
lin318	42496	**42618.4**	**42888**	**134.33**	42496	42785.4	43095	184.50
pcb442	**51149**	**51511.7**	**52090**	**294.40**	51173	51601	52264	350.05

Table 2.2 Cooperative Arrangement with 4 SAEngines

TSPLIB	1 SAEngine				5 SAEngines			
Instance	Best	Mean	Worst	S.D.	Best	Mean	Worst	S.D.
att48	10628	**10661.6**	**10738**	**36.38**	10628	10684	10772	60.31
eil101	**629**	638.2	647	**5.76**	630	**634.9**	**646**	6.02
bier127	118659	119243	**120292**	629.08	**118461**	**119044.7**	120462	**589.07**
u159	42435	42719.7	**43669**	**451.54**	**42080**	**42641.2**	44013	525.2
rat195	**2333**	**2347**	2360	8.15	2338	2348.3	2360	**7.47**
tsp225	**3953**	**3979.4**	**4014**	20.46	3959	3980.6	4017	**18.91**
lin318	**42496**	**42618.4**	**42888**	**134.33**	42618	42881.5	43626	305.64
pcb442	**51149**	**51511.7**	52090	294.40	51225	51589.6	**51912**	**206.74**

Table 2.3 Cooperative Arrangement with 5 SAEngines

TSPLIB	1 SAEngine	2 SAEngines		4 SAEngines		5 SAEngines	
Instance	Time	Time	PSU	Time	PSU	Time	PSU
att48	1.97	1.78	2.21	1.74	4.53	1.76	5.45
eil101	12.28	12.04	2.04	11.93	4.18	12.12	5.07
bier127	14.63	14.68	1.99	14.72	3.96	14.99	4.88
u159	29.21	29.57	1.98	29.67	3.94	29.86	4.89
rat195	67.76	68.1	1.99	68.46	4.04	68.62	4.93
tsp225	168.32	170.60	1.97	172.14	3.91	171.83	4.90
lin318	338.01	339.15	1.98	339.51	3.98	341.95	4.94
pcb442	614.12	596.63	2.06	620.44	3.96	623.11	4.93

Table 2.4 Cooperative Arrangement Execution Times

as expected. A figure in bold indicates whether the single SAEngine or cooperative arrangement produced a better result on that value.

Tables 2.1, 2.2, and 2.3 do not show that an SAEngine allowed to run for p proposed moves is better than n cooperating engines each allowed to run for p/n proposed moves. Nor do they demonstrate the converse. The results are in line with the findings in Graffigne (1992), Lee and Lee (1996), Gallego et al. (1997).

All of the tests were performed on Linux system with a 166MHz Pentium processor and 32Mb of memory. All of the SAEngines were created on this single processor system. It was not possible at the time to obtain a multi-processor machine, or distributed environment that could be controlled sufficiently to perform these experiments. The experiments require that no additional jobs be run while the experiments are taking place. However, a single processor system is able to provide an indication of the speed up that one could potentially achieve on a multi-processor machine or distributed environment. It is important to note that this is *potential* speed-up because it will be affected by communication delays that may not be present in a single processor system. Table 2.4 shows the execution times for all of the problem instances, and all of the cooperative arrangements. The values for time presented in Table 2.4 are calculated as the mean time over the ten runs. One would expect the execution time of a cooperative arrangement to be approximately that of a single SAEngine because one is running n engines but only for $1/n$ of the number of proposed moves. The values in the PSU columns in Table 2.4 are the potential speed-ups for the given arrangement. Of particular note is that the potential speed-up is occasionally greater than the number of cooperating TrEngines. This could possibly be due to the fact that, because the temperature drops faster in the cooperating SAEngines; they will jointly accept fewer moves than a single SAEngine. To accept a move takes more CPU time than to reject a move. Hence, this speed-up can be explained in terms of the annealing process.

2.5 HYBRIDIZATION

Hybrid optimization techniques can, for some problems, generate better solutions, generate good solutions faster, or behave more robustly, than the constituent algorithms (Rayward-Smith et al. (1996)). This section looks at three ways in which hybrid techniques can be constructed using the Templar framework. Firstly, the use of abilities in hybridization is discussed. Then, TrEngine composition – building new engines from existing engines – is discussed. Finally, the use of message passing and cooperation to generate hybrid techniques is investigated.

2.5.1 Hybridization Through Abilities

Abilities help Templar to achieve several of its design goals, one of which is to support the generation of hybrid techniques. Abilities represent core elements of functionality for a given technique, or class of techniques. Looking at abilities from the perspective of TrEngines, some perform atomic actions such as "generate the next neighbourhood move in the neighbourhood', while others perform composite actions such as "find the neighbourhood move with the best change in fitness'. Although an ability may perform an atomic action in the eyes of a TrEngine, its actual implementation may be arbitrarily complex. As mentioned previously, the composite type of ability can be simulated by the TrEngine through a composition of atomic abilities – the atomic abilities are the building blocks for the technique.

Just as composite actions may be built from these building blocks, so too can new optimization techniques. Where possible, the applicability of abilities should not be limited to a single engine. Hence, neighbourhood abilities are for neighbourhoods, not specifically for hill climbing, simulated annealing, or any other specific optimization technique. This means that any engine might use these, combined with genetic recombination operators and make use of feature information of the solutions. This hypothetical TrEngine might implement a hybrid of GAs and Guided Local Search (GLS) (Voudouris and Tsang (1995b), Voudouris and Tsang (1995a)).

It is not just the atomic abilities that may be used in this way. For instance, if implementing a GA with some form of hill climbing, then the ability that chooses the best change in fitness may also be used. This would then form what is sometimes termed a *memetic* algorithm (Radcliffe and Surry (1994)).

The construction of hybrid techniques provides an example of how code re-use is promoted through Templar. However, this is code re-use at a very low level. The next section demonstrates that it is possible to hybridize techniques at a higher level.

2.5.2 Hybridization Through Composition

Another option when creating hybrid TrEngines is to build a new TrEngine out of existing TrEngines. Because TrEngines are implemented as objects, they may contain, or control directly (i.e., through method invocation), other TrEngines. Hence the memetic algorithm may be implemented by a MAEngine that contains both an HCEngine and a GAEngine. The implementation of the MAEngine delegates most of the work to the two controlled TrEngines. It operates in a management role and

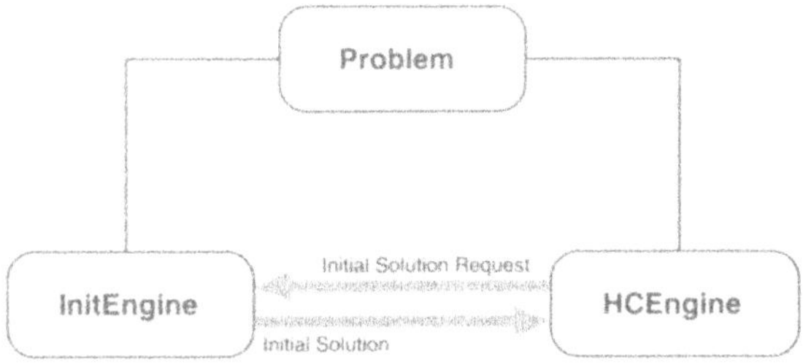

Figure 2.7 Creating GRASP from Existing Components

presents a unified interface to these two TrEngines – a user of the MAEngine does not even need to be aware of the fact that it is built of more than one TrEngine.

This method of constructing hybrid engines makes greater use of existing code but does not have the same flexibility of building a hybrid TrEngine using abilities. A slightly less obvious advantage possessed by composition, is that it allows the MAEngine to take advantage of the concurrent execution of TrEngines. It might be possible for it to run both the TrEngines at the same time, allowing their calculations to overlap. Hence, execution time may be reduced when using a multi-processor machine, or a distributed system.

2.5.3 Hybridization Through Cooperation

Templar can also support hybridization through the use of cooperation messages. The implementation of a GRASP style technique will be used to demonstrate hybridization in this manner.

GRASP is a technique that combines a randomized, greedy initialization stage followed by a local search stage (Feo et al. (1994)). Although it is broken into four stages in Feo et al. (1994), three of the stages merely form a certain style of initialization. An implementation of a TrEngine for GRASP could be created. However, this would not take advantage of the potential for reuse offered by a framework. It would be better to arrange a solution initializing TrEngine (InitEngine) with the desired local search engine, e.g., the SAEngine or the HCEngine.

In order to work correctly, the TrProblem must provide an initialization ability that works in a GRASP manner. The results from the initialization stage are fed into the local search engine. This is depicted in Figure 2.7 using a hill climbing engine for the local search engine. Both the InitEngine, and the HCEngine must be linked to the same problem.

As mentioned before, there is a set of standard cooperation messages each with suggested responses. A subset of these messages deals explicitly with initial solutions. Both the InitEngine and HCEngine have been implemented to react to these messages appropriately.

When the HCEngine is started, it broadcasts a request for an initial solution. The InitEngine duly responds with an initial solution by running one of the TrProblem's initializer abilities. The HCEngine is then able to perform local search on this initial solution. This simple arrangement performs just a single GRASP iteration. Normally, with GRASP one will perform a number of iterations. Some of the parameters of the HCEngine govern how many times it will perform a search. Each search starts from a new solution, and the HCEngine keeps track of the best solution seen since it was started. This gives the basic GRASP algorithm: the HCEngine performs hill climbing a given number of times, each time starting from a solution initialized by the InitEngine, using a GRASP style initializer. One could justifiably claim that any latency present in the communications between the two engines may affect the speed of the technique (Section 2.6.2 investigates this). The HCEngine avoids this problem, to some extent at least, by requesting a new initial solution before it is actually required. This means that the initialization and hill climbing stage may overlap. Execution may be performed concurrently on both stages. This has the added benefit that, if this arrangement was being used on a machine with multiple processors sharing memory, then the work thread for the two engines could be scheduled on different processors; parallelism is achieved without any additional effort.

GRASP is quite a small, but easy to comprehend, example of how techniques may be built from existing TrEngines that can cooperate. It is possible to conceive and construct much more complicated systems. For example, techniques that oscillate between optimization techniques can be constructed in a similar manner to that of the GRASP example presented above. A hybrid technique using GAs, and branch-and-bound is presented in French et al. (1997). This technique oscillates between the two optimization techniques until a solution is obtained or the search is abandoned. In this case, a controlling TrEngine may need to be constructed to ensure that the hybrid technique performs correctly. This controlling TrEngine can communicate with the GAEngine and the branch-and-bound engine through the same mechanisms for communication used in the GRASP example. If necessary, the controller may use any messages that are specific to a given type of TrEngine. This hybrid may also be able to benefit from the concurrent nature of TrEngines.

The InitEngine has been introduced through the concept of hybridization. However, InitEngines, and engines that perform initialization in general, are more important than they initially seem because most TrEngines are not capable of generating initial solutions themselves. This is certainly true for both the SAEngine and HCEngine. Instead, these engines rely on another TrEngine to supply its starting solution. Although it would be possible to make use of the initialization ability in any TrEngine, this would be missing a good opportunity for code reuse. The InitEngine already exists, so it should supply the initial solution or solutions. Furthermore, it is possible for a TrEngine to take the results of another TrEngine as its initial solution. Because the exchange of solution happens using the standard cooperation messages, the receiving TrEngine is unaware of the actual source of the initial solution. As well as implementing oscillating techniques as in French et al. (1997), this technique allows the work of TrEngines to be "cascaded" through a number of stages.

2.6 COST OF SUPPORTING A FRAMEWORK

It would be unrealistic to expect an optimization technique, implemented within the constraints of a framework, to be as efficient as one implemented only with regards to the technique, i.e., where every optimization can be taken in order to make the technique run quickly. This section attempts to evaluate the costs of using Templar in two regards. Firstly, the cost of abstraction and the use of abilities is determined. Then, the cost of using cooperation messages for initialization is also determined to give some idea of the effect of communication latency.

2.6.1 Evaluation of Efficiency

Efficiency is often an important requirement of an optimization technique. Faster optimization techniques may result in an optimal solution being obtained sooner, or in a better solution being obtained within a certain duration. Some optimization techniques will require more memory than others. The memory consideration may also affect the speed of a technique. However, memory usage is not considered here because it is assumed that an optimization technique requires approximately the same amount of memory regardless of whether it is written in an abstract manner, or hand-coded. The relative efficiencies of different optimization techniques are also not of interest here. What is of interest, is the relative execution speed of two implementations of a single optimization technique, where one has been hand-coded for efficiency and the other has been developed within the constraints of the framework.

Templar is intended to be as efficient as possible, while meeting its other design goals. This does, of course, require a tradeoff. This means that, if the developer of a TrEngine within the framework, and the developer of an equivalent hand-coded implementation for a given problem both have experience in the optimization technique, then the implementation within the framework will probably be slower than the hand-coded implementation. This is due to the fact that with a hand-coded implementation, one can, to a certain extent, sacrifice good programming practice, such as modularity and abstraction in order to improve performance. However, if an engine is implemented to behave correctly within the context of the framework, then some of the abstractions used are likely to affect performance.

In order to test the efficiency of the framework, a test problem is required. The TSP was chosen for this purpose because it has a very simple evaluation function ($O(n)$, where n is the size of the TSP), and very effective delta evaluation functions. This makes it harder for an implementation within the framework to be as efficient as a hand-coded implementation. Another problem, such as the QAP (Burkard (1984)), could have been used, and this would have yielded more impressive results. However, by using the TSP, the tests give what can be considered to be a near worst-case assessment. As an added benefit, the TSP is so well-known that most researchers have some knowledge of the problem. A variety of TSP instances from TSPLIB (Reinelt (1991)) were used, varying in size from 29 to 318 towns.

Steepest descent was chosen as a suitable method for testing the efficiency of the framework. It is a very simple technique, and it is quite easy to develop very efficient hand-coded implementations. Unlike SA and GAs, there are no parameters to set. This means that the techniques cannot be "tweaked" to favour the framework. The hand-coded versions were written to run as quickly as possible, therefore, they do not display any results until they complete the work. However, for the framework runs, a progress window was kept open throughout. This updates about once a second to

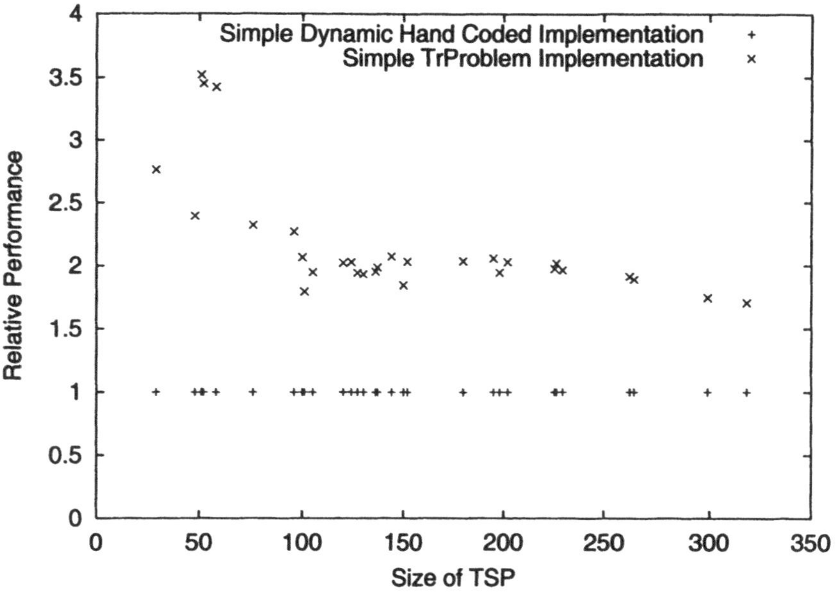

Figure 2.8 Basic Implementation Performance

reflect the state of the search. Due to multi-threading, the time taken to draw this window should not affect the performance of the hill climber because it is working in a separate thread.

All of the tests were performed on a machine running Linux, with a 120MHz Pentium processor, and 64Mb of memory. The only heavy loading on the machine during the test phase, was the test programs themselves. All of the source files were compiled using the gnu compiler, g++, at the highest level of optimization.

The graph in Figure 2.8 shows the relative performance of an implementation of the TSP object that does not provide any additional abilities designed to reduce the search time. The relative performance is calculated using the length of time for the technique to terminate. If an engine in the framework has a relative performance of 2, then it took twice as long to terminate. The framework's hill climber is compared against a simple dynamic hand-coded implementation. This implementation is simple in the sense that it too does not take advantage of any problem specific features to reduce the search time, and is dynamic in the sense that it can load any problem from TSPLIB. The graph shows that when using the framework for TSPs of these sizes, one can expect it to take 1.70 – 3.52 times longer than a hand-coded solution. One can see from the graph the downward trend of these values, and it is quite probable that for larger TSP instances, the reduction in performance is less.

The graph in Figure 2.9 assesses the performance of the framework when the TrProblem uses additional, problem specific abilities to increase the speed of the optimization technique. Some of these problem specific abilities may normally be used by

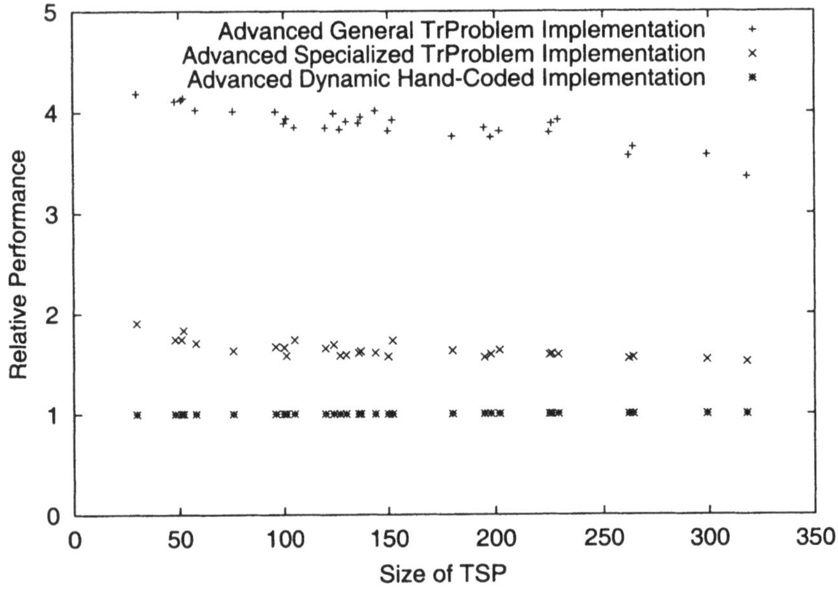

Figure 2.9 Advanced Implementation Performance I

more than one optimization technique. This is termed the "Advanced General TrProblem Implementation" for the purpose of these experiments. The "Advanced Specialized TrProblem Implementation" includes abilities that are less likely to be reusable with different optimization techniques. It is important to point out that the specialized implementation still contains the more general abilities, and these can still be used by other techniques. The hand-coded solution now takes advantage of any problem specific information that can be used to increase the speed of the search. It is guaranteed that the two programs perform the same steps in reaching a local optimum. It is not surprising that the general advanced implementation does not perform as well as the basic implementation. Now, instead of regularly performing a complete evaluation ($O(n)$), the partial evaluation functions are being used which are very fast and require constant time. This means that the overhead paid for abstraction is much more noticeable. However, the implementation of the TSP TrProblem class can be made more efficient by the addition of a solution technique specific ability. One can see from the graph that it is not more than twice as slow as the hand-coded technique, and, as with the basic implementation, exhibits a downward trend as the size of the problem instance increases.

If speed is an absolute necessity, then the hand-coded optimization technique can be compiled to work with a specific problem. Figure 2.10 shows the performance of the framework when competing against a static hand-coded implementation. It is static in terms of the fact that it is only capable of solving TSPs of a certain size. In order to solve TSPs of a different size, it must be recompiled. As would be expected,

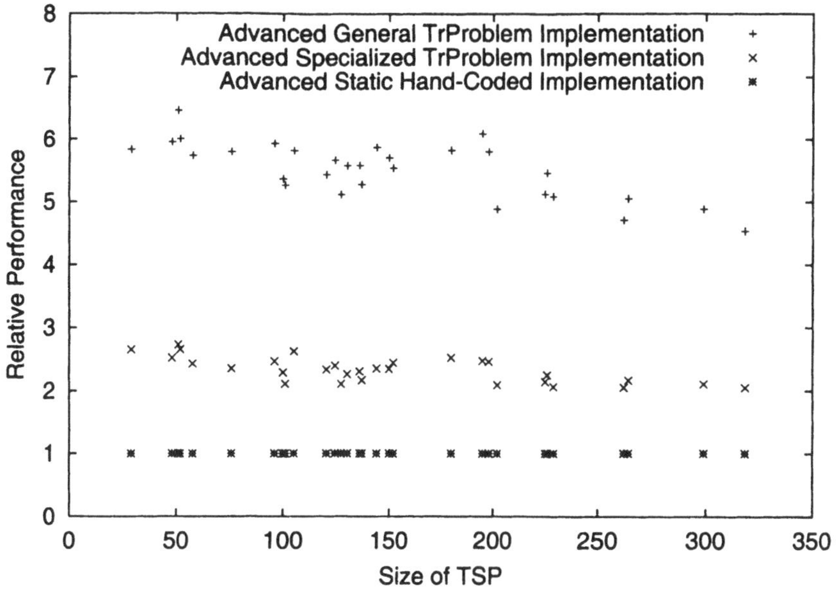

Figure 2.10 Advanced Implementation Performance II

the relative performance of the framework is not as good as when it was compared to the dynamic implementation. However, as a near worst-case result, the framework achieves an execution time that is always within three times of a hand-coded solution. The results can be considered as near worst-case results for the framework due to the fact that the optimization technique considered involves small simple steps, as opposed to larger, more complex steps. This means that the penalty in terms of execution speed due to abstraction is likely to be high. In fact it is possibly as high as one would expect because hill climbing is possibly the simplest of heuristic optimization techniques. If one investigated a technique such as branch-and-bound, then the results would be more impressive for the framework, but not as honest.

2.6.2 Cooperation Overhead

Although TrEngines are capable of initializing their own solutions, it is recommended that most TrEngines rely on an initial solution provided by another TrEngine. This allows distributed and hybrid optimization techniques to be created relatively easily. In Section 2.5.3 it was suggested that GRASP style techniques can be created by using an InitEngine, with a suitable initializer, cooperating with a local search optimization technique. This arrangement will serve as the basis for assessing to what extent the performance of the technique is affected. It is worth noting that the results from Section 2.4.2 do not make any adjustment for cooperation, i.e., part of the reason for the Templar implementation being slower could be due to the cooperation mechanism.

QAPLIB Instance	No of Runs	Best Fitness Found	TOMS 754 Best Fitness	No Coop. Seconds	With Coop. Seconds	Degra-dation
chr12a	62000	9552	9552	581.5	652.51	1.12
nug15	26000	1150	1150	598.4	643.06	1.07
esc16a	35000	68	68	588.62	710	1.21
chr18a	12000	11098	11098	575.25	580.56	1.01
chr20a	8050	2196	2224	556.8	555.34	0.98
chr22a	5000	6282	6320	567.16	578.11	1.02
chr25a	3000	4040	3884	589.66	604.34	1.02
kra30a	1400	88900	88900	603.21	630.45	1.05
ste36a	500	9624	9596	550.45	578.93	1.05
lipa40a	500	31897	31884	563.38	609.69	1.08
lipa50a	220	62748	62704	632.07	679.6	1.08
lipa60a	100	108221	108077	650.31	703.9	1.08

Table 2.5 Degradation Due to Cooperation

The QAP is used as the basic problem for these tests. This is because a well-known GRASP implementation has been applied to this problem (Resende et al. (1996)). The implementation of the QAP TrProblem class provides the initialization stage presented in Resende et al. (1996) as a problem specific initializer. As in Resende et al. (1996), hill climbing was used as the local search TrEngine cooperating with the InitEngine. A special hill climbing engine was created that did not obtain its initial solution from another TrEngine. Instead, it creates its initial solution internally and does not use cooperation in any way. This special hill climbing engine was used to determine how quickly the technique would run without cooperation.

Tests were performed on a total of twelve problem instances from QAPLIB (Karisch et al. (1997)). For each test, a number of runs were chosen so that the test would last approximately ten minutes. The best fitness found for each test was recorded, as well as the time, measured by the wall clock, that the TrEngine, or TrEngines took to complete the number of runs. Each run was a GRASP style initialization followed by hill climbing until no better solution could be found. Each test was performed for both the special HCEngine, and the InitEngine, standard HCEngine pair. Both of these times for all of the problem instances are presented in Table 2.5. This table also presents the best fitness obtained from 2048 runs of TOMS algorithm 754 (Resende et al. (1996)), and the degradation in performance when using cooperation.

The first thing to note about the results is that the best fitness obtained is indeed comparable to the results presented in Resende et al. (1996). In that research, a set number of runs (2048) was used. This may account for the fact that in the experiment, the Templar implementation did better on a problem instance such as chr20a where more runs were performed, and did not find as good a solution on the larger problem instances because there were fewer runs. Unfortunately, the execution time of the algorithm in Resende et al. (1996) is presented as a graph on a logarithmic scale.

This makes it quite hard to determine execution times accurately, and so perform an accurate comparison.

In order to determine the execution time of both the special HCEngine, and the InitEngine, standard HCEngine pairing, wall clock time was used. In order for this to be reliable, it was performed on a machine on which the only load was the experiment. The machine in question was a 166Mhz Pentium processor in a Linux based system with 32Mb of RAM. It was not connected to a network. It is necessary to measure wall clock time for this experiment because it is intended to show the degradation that is due to cooperation. One of the mechanisms that facilitates cooperation is the use of threads. This may incur a higher cost in terms of OS scheduling and hence affect the overall time. As described in Section 2.5.3 the work of the InitEngine and HCEngine are overlapped to obtain speed up in a multiprocessor or multiple machine system.

As can be seen from Table 2.5, there is quite a degree of variation in the amount of degradation one can expect. Of particular interest are the results for instance chr20a, in which using cooperation has resulted in a shorter execution time. This reduction in execution time is unlikely to be attributable to the use of cooperation. There are two, more likely, explanations for this behaviour. Firstly, due to the fact that using cooperation involves using two TrEngines, this means that two random number generators are in use. Without cooperation, only a single RNG is in use, and hence identical work is not being performed. With this in mind, it is conceivable that the cooperation based test for chr20a "got lucky" in the initialization stage, more so than the non-cooperation based test. However, after 8050 runs it might be reasonable to expect that chance is less likely to sway the results. The other possible explanation is due to the scheduler within the OS. It is the scheduler's job to determine which thread, or process, may use the CPU. It may be that the cooperation tests for chr20a worked particularly well with the scheduler and hence less time was spent inside the OS Kernel.

Even when one looks at the other levels of degradation, one can see that the effects of cooperation are far from drastic, the worst recorded being, for instance, chr12a. With a small problem instance, one would expect the degradation due to cooperation to be worse because cooperation is more frequent.

Although using cooperation can degrade performance, the degradation is not substantial, and is potentially outweighed by the benefits offered. The results in Table 2.5 are for a single problem type on a single machine. The degradation experienced is likely to vary between problems and architectures. However, the tests are based on a known GRASP method for a well studied problem and so should provide a realistic guide for the effects of cooperation.

2.7 SUMMARY

Object-oriented frameworks are applicable to many domains. In recent years they have started to be applied to the area of combinatorial optimization. This chapter has investigated one of these frameworks and described some of the benefits that frameworks can offer to researchers, developers, and instructors in this field. A framework may also make it easier for a developer in a different field to embed elements of combinatorial optimization within their applications without needing to become an expert.

Distribution and parallelization are important areas of current computer science research. It has been shown that the message passing paradigm can support distributed or parallel optimization techniques, and that much of the complexity associated with this can be hidden by the framework. Not only is the act of writing a distributed or parallel technique made easier by the framework, but the mechanisms introduced can encourage code reuse and so become a fundamental part of the framework, not just an extension.

The idea of cooperation between optimization techniques has been discussed. Future work is likely to concentrate on experimenting with the construction of cooperative systems of optimization techniques.

This chapter has also attempted to assess the impact of using this framework with respect to execution time. It has shown that, in a near worst-case situation, one might be able to expect an optimization technique to take at most three times as long to run. The values may be different for other optimization frameworks. The assessment was near-worst case because the optimization technique chosen requires many function calls to go through a layer of abstraction. This is not the case for all optimization techniques.

3 A FRAMEWORK FOR LOCAL SEARCH HEURISTICS FOR COMBINATORIAL OPTIMIZATION PROBLEMS

Alexandre A. Andreatta[1], Sérgio E.R. Carvalho[2] and Celso C. Ribeiro[2]

[1]Department of Applied Computer Science
University of Rio de Janeiro
Rua Voluntários da Pátria 107
Rio de Janeiro 22270, Brazil
andreatt@uniriotec.br

[2]Department of Computer Science
Catholic University of Rio de Janeiro
R. Marquês de São Vicente 225
Rio de Janeiro 22453-900, Brazil
{sergio, celso}@inf.puc-rio.br

Abstract: In the study of heuristics for combinatorial problems, it is often important to develop and compare, in a systematic way, different algorithms, strategies, and parameters for the same problem. This comparison is often biased not only by different implementation languages, but also by different architectures. This paper proposes a framework, described using design patterns, modeling different aspects involved in local search heuristics, such as algorithms for the construction of initial solutions, methods for neighborhood generation, and movement selection criteria. Using this framework, we fix a basic architecture and thus increase our ability to construct and compare heuristics. We also describe implementation issues and a successful application of this framework to the development and comparison of heuristics for the phylogeny problem.

3.1 INTRODUCTION

Hard combinatorial optimization problems most often have to be solved by approximate methods. Constructive methods build up feasible solutions from scratch. Basic local search methods are based on the evaluation of the neighborhood of successive improving solutions, until no further improvements are possible. Metaheuristics are higher level procedures that guide subordinate local search heuristics specifically designed for each particular problem. As an attempt to escape from local optima, some of them allow controlled deterioration of the solutions in order to diversify the search. Many ideas are common to most local search based methods and metaheuristics. Among them, we cite algorithms for construction of initial solutions, neighborhood definitions, movement selection criteria, local search procedures, and termination criteria.

In the study of heuristics for combinatorial problems, it is often important to develop and compare, in a systematic way, different algorithms, strategies, and parameters for the same problem. The central issue in this paper is that, by modeling the different concerns involved in local search in separate classes, and by relating these classes in a framework, we can increase our ability to construct and compare heuristics, independently of their implementations. Its main goal is to provide an architectural basis for the implementation and comparison of different local search heuristics. The proposed **Searcher** framework, based on the use of design patterns, encapsulates, in abstract classes, different aspects involved in local search heuristics, such as algorithms for the construction of initial solutions, methods for neighborhood generation, and movement selection criteria. Encapsulation and abstraction promote unbiased comparisons between different heuristics, code reuse, and easy extensions. The classification of different aspects of heuristics simplifies their implementations and invites new extensions in the form of subclasses. The reuse of code we can accomplish, implementing common aspects of a family of heuristics, offers us a better platform for comparison, since a large part of the code is common to all members of the family.

This paper is organized as follows. In the next section, we present the basic concepts concerning the object-oriented programming paradigm and design patterns, which are later identified as framework components. In Section 3.3, we begin the presentation of the **Searcher** framework for the implementation and development of local search heuristics for combinatorial optimization problems. To describe the **Searcher** framework, we use an adaptation of a pattern language that provides intent, methodology, motivation, applicability, structure, participants, and collaborations. In Section 3.4, we describe how we use design patterns to assemble our framework. In Section 3.5, we describe the application of this framework to the phylogeny problem. Related work is reviewed in Section 3.6. Finally, we present some concluding remarks concerning the main issues involved in this work, as well as the description of some extensions and related work.

3.2 DESIGN PATTERNS

3.2.1 Fundamental Concepts in Object Orientation

An *object* is a run-time entity that packages both data and the procedures which operate on those data. These procedures are called *operations*, *methods*, or *member functions*. The set of all public operations is the object's *interface*, which describes the set of requests to which it can respond.

An object may be implemented in several ways without changing its interface. Handling an object solely through its interface assures a high degree of *encapsulation*, which is the result of hiding both the representation and the implementation of an object. The representation is not visible and cannot be accessed directly from outside the object, since its operations are the only way to modify its representation. The object's interface and implementation are defined in its class. Objects and classes can be seen as variables and abstract data types, respectively. Actually, the class concept is an extension of the abstract data type concept, and is distinct from it for the use of *inheritance* and *polymorphism*.

Inheritance is a relationship among classes, in which the *derived class* is defined in terms of another one, its *super-class* or *base class*, in such a way that an object of the derived class is also an object of the base class. We distinguish between *abstract* and *concrete classes*. The main objective of an abstract class is the definition of an interface (all operation signatures are declared). Part of (or even all) the implementation of the interface is deferred to subclasses. If a class has an interface where all the operations have already been defined (i.e., there are no abstract operations), it is said to be a concrete class. Inheritance allows the construction of class organizations shaped like trees (single inheritance) or acyclic graphs (multiple inheritance).

Classes can also be *generic*, depending on one or more parameters to actually be able to model objects. Generic classes define class families and are conveniently used in the modeling of data structures, such as lists (of "integers", of "solutions", and so on).

The basic relationships among classes are inheritance and *composition*, also known as "*is a*" and "*has a*" relationships. A class is related to another by composition if it contains at least one object of this other class or a reference thereof. A frequent technique used in object-oriented design is the mechanism of *delegation*, where an object delegates a requisition to another object to whom it is related by composition. The receiving object carries out the request on behalf of the original object.

A *request* (sometimes called a *message*) results in the activation of an object's operation. It is said that the object receives the request, or the request is applied to the object. *Dynamic binding* is the run-time association of a request to an object and one of its operations. *Polymorphism* is the ability of substituting in run-time objects whose interfaces conform. For example, if an operation was declared as having an object of class A as parameter, then an object of a class B (subclass of A) can be an argument for this operation. This is possible because this object *is* also a class A object (due to inheritance).

Summarizing, objects provide a clear frontier between their interface (what users must know) and their implementation (what must be hidden). System extensions and modifications are easier to develop, often changing (through inheritance) only object interfaces. The ability to define abstract object models adds flexibility; entire systems may be developed by instantiating abstract, high-level object models and their relationships.

3.2.2 Design Patterns

The framework for the implementation and development of local search heuristics is a combination of design patterns, which we introduce now and later identify as framework components in Section 3.3. Frameworks are essentially prototypical architectures for related systems or systems in the same application domain (Rumbaugh et al. (1991)). Their strength lies in the fact that they offer abstract solutions to problems within the domain, solutions that can be instantiated, for example, with the refinement of the abstract classes in the framework, and with the particularization of generic classes. Frameworks fundamentally differ from class libraries in that they offer more than resources to use in a given situation; framework classes are bound to one another via relationships also defined as part of the framework.

Design patterns are being extensively studied and reported in the literature (see Buschmann et al. (1996a), Gamma et al. (1995), Pree (1994), Schmidt and Coplien (1995), Schmidt et al. (1996), Vlissides et al. (1996)). They are also architectural solutions to problems, but in a smaller scale: a design pattern usually involves just a few classes. They reflect good solutions to problems repeatedly encountered, perhaps in different application domains, and are documented to allow reuse. Describing a framework (or even a design pattern) with the use of design patterns is not a development method by itself. Techniques such as delegation, inheritance, and polymorphism are used in patterns to solve specific design problems. Design patterns can be seen as a kind of design vocabulary. The use of design patterns to describe a framework helps the designer recognize what problem was addressed and what solution was proposed.

In our framework we use three behavioral patterns, introduced by Gamma et al. (1995): *strategy, template method,* and *state,* each briefly described in what follows. In all three, there is a pattern inside: the behavior of an object depends on a delegation of responsibilities to one of its components. We use the OMT notation (see Rumbaugh (1995a) and Rumbaugh (1995b)) to describe classes and their relationships. Classes are represented by rectangles divided in two parts. The upper part indicates the name of the class (in italics if abstract and in bold if concrete) and the lower part contains class operations. Details of operation implementations can be shown in separate rectangles linked to the operation's name by a dotted line. The inheritance relationship is denoted by a triangle whose base is linked to the subclasses. Composition is represented by line links. The presence of a diamond in a link indicates ownership, while its absence indicates that there is just a reference to an object of the other class. The presence and color of a disk in the link indicate the cardinality of the composition. The absence denotes *one*, a black disk denotes *at least one*, and a white disk means *zero or more*.

3.2.3 The Strategy Pattern

The strategy pattern is an object-oriented version of a collection of related algorithms, or procedures, that can otherwise be found scattered in a conventional program. A strategy allows the encapsulation of each algorithm in a concrete class, all descending from an abstract class representing the algorithm collection. An object of this abstract class is declared as a component of some client. At execution time this component object can change classes, and thus execute any encapsulated algorithm. New "algorithms" can be included in the hierarchy, and their interchangeability is transparent to their clients. The diagram in Figure 3.1 illustrates the strategy pattern. A Client has a *strategy* object, which receives the method M when it is responding to the request R. Since this object is modeled by a concrete subclass of *strategy*, the M applied is actually the method defined in this subclass.

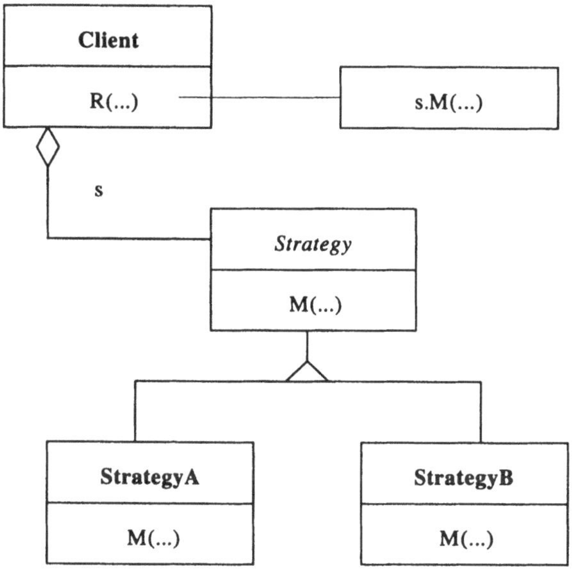

Figure 3.1 Strategy Pattern

3.2.4 The Template Method Pattern

The template method allows the definition, in an abstract class, of common steps in an algorithm, and the deferral of specialized steps to subclasses. In this way code can be saved and better implementations added without client modification. The diagram in Figure 3.2 illustrates this pattern. A Client has an object modeled by the *TemplateMethod* class; when responding to R, this object suffers the application of TM, a template method. At this time this object is modeled by some concrete subclass of *TemplateMethod*. As in each of these subclasses TM steps (Operation1, Operation2, ...) may be redefined, the object t is handled accordingly. It should be noted that operations defined in *TemplateMethod* may be concrete, needing no redefinition in subclasses, and that different subclasses may redefine different concrete operations of *TemplateMethod*.

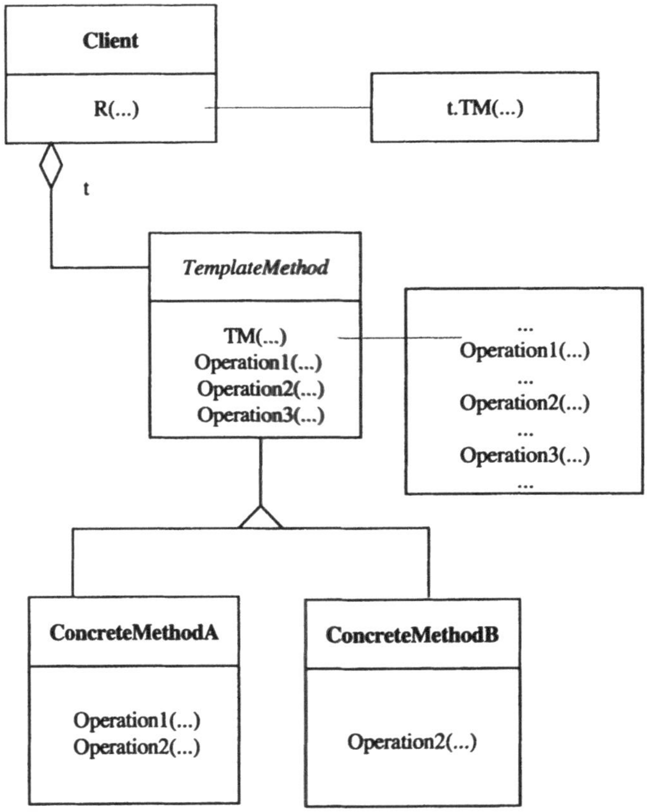

Figure 3.2 Template Method Pattern

3.2.5 The State Pattern

The state pattern allows an object to change its behavior when its internal state (the values of its components) changes. To accomplish this, the object contains a component modeled by an abstract class, which roots a hierarchy of subclasses where behavior redefinition takes place. Delegating to this state component the responsibility to respond to external stimuli, the object appears to change classes; in fact this happens only to its state component. Figure 3.3 illustrates this pattern. When responding to R1, a client object delegates to its state component the execution of the correct method M1. This operation is redefined in state subclasses, which also implement state change, if necessary.

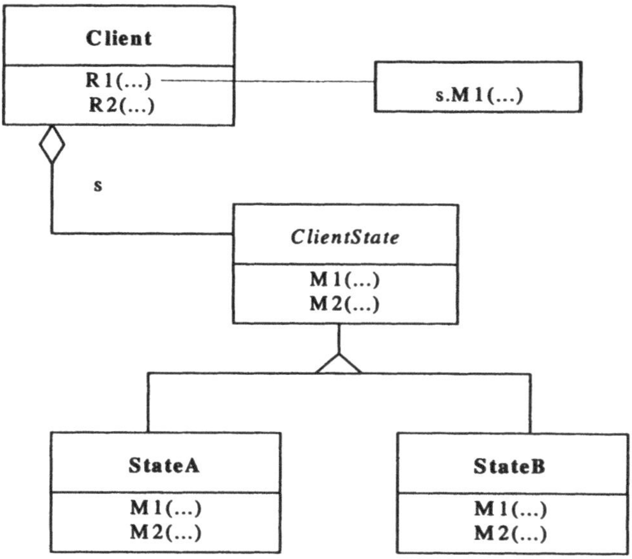

Figure 3.3 State Pattern

Actually, the class diagrams for the state and strategy patterns are the same. The common technique used is delegation. The difference is due to the kind of delegation being used. In the state pattern, the issue consists in providing a variable behavior to an object, which suffers the application of different methods. This is reflected in the diagram by the presence of two methods. In the strategy pattern the main issue is to encapsulate the different variations or implementations of a unique task in different classes.

Patterns create vocabulary, in such a way that, instead of talking about the use of a specific technique to solve a particular problem, designers refer to it by the name of a design pattern. We use the patterns *strategy, template method,* and *state* in the definition of our local search framework, presented in the next section. The role of each pattern in the framework is described later.

3.3 THE SEARCHER FRAMEWORK

To describe the framework, we use an adaptation of the pattern language employed in Gamma et al. (1995), providing intent, motivation, applicability, structure, participants, and collaborations. We discuss implementation issues in the next section.

Intent: To provide an architectural basis for the implementation and comparison of different local search strategies.

Motivation: In the study of heuristics for combinatorial problems, it is important to develop and compare, in a systematic way, different heuristics for the same problem.

It is frequently the case that the best strategy for a specific problem is not a good strategy for another. It follows that, for a given problem, it is often necessary to experiment with different heuristics, using different strategies and parameters. Modeling the different concerns involved in local search in separate classes, and relating these classes in a framework, we increase our ability to construct and compare heuristics, independently of their implementations. Implementations can easily affect the performance of a new heuristic, for example, due to programming language, compiler, and other platform aspects. The classification of different aspects of heuristics simplifies their implementations and invites new extensions in the form of subclasses. The reuse of code that we can accomplish, implementing common aspects of a family of heuristics, allows us a better platform for comparison, since a large part of the code is common to all members of the family.

Applicability: The **Searcher** framework can be used and has particularly appropriate features for situations involving:

- Local search strategies that can use different methods for the construction of the initial solution, different neighborhood relations, or different movement selection criteria.

- Construction algorithms that utilize subordinate local search heuristics for the improvement of partially constructed solutions.

- Local search heuristics with dynamic neighborhood models.

Structure: Figure 3.4 shows the classes and relationships involved in the **Searcher** framework. We use the OMT notation (see Rumbaugh (1995a), Rumbaugh (1995b)) for static views.

Participants:

- *Client:* Contains the combinatorial problem instance to be solved, its initial data and the pre-processing methods to be applied. Contains also the data for creating the *SearchStrategy* that will be used to solve the problem. Generally, it can have methods for processing the solutions obtained by the *SearchStrategy*.

- *Solution:* Encapsulates the representation of a solution for the combinatorial problem. Defines the interface the algorithms have to use to construct and to modify a solution. Delegates to *IncrementModel* or to *MovementModel* requests to modify the current solution.

- *SearchStrategy:* Constructs and starts the *BuildStrategy* and the *LocalSearch* algorithms, and also handles their intercommunication, in case it exists.

- *BuildStrategy:* Encapsulates constructive algorithms in concrete subclasses. Investigates and eventually requests to *Solution* modifications in the current solution, based on an *IncrementModel*.

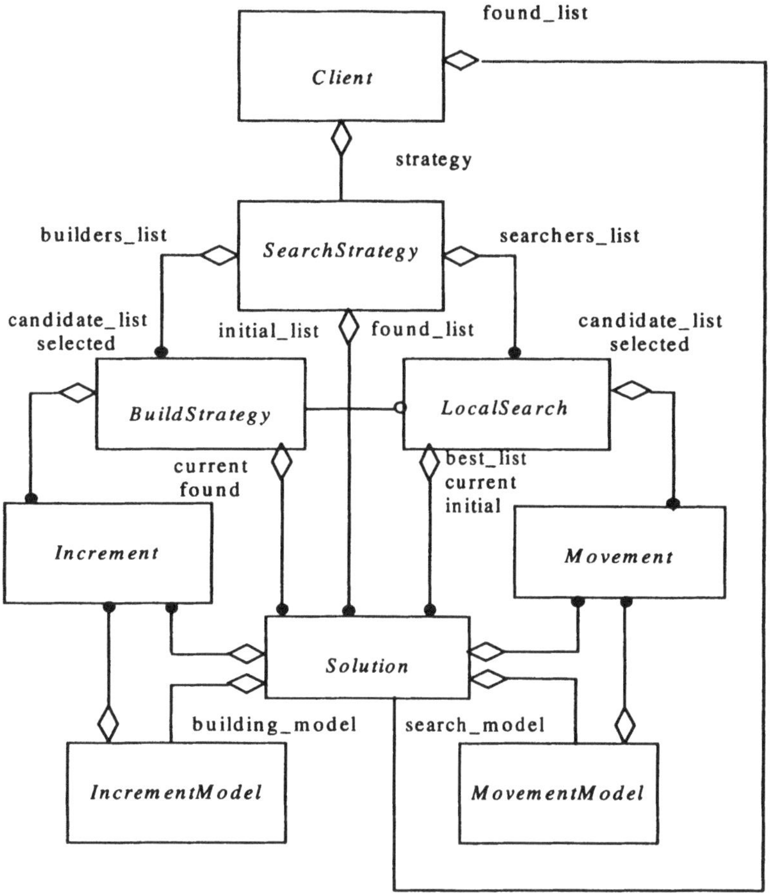

Figure 3.4 The Searcher Framework

- *LocalSearch:* Encapsulates local search algorithms in concrete subclasses. Investigates and eventually requests to *Solution* modifications in the current solution, based on a *MovementModel.*
- *Increment:* Groups the data necessary to modify the internal representation of a solution for constructive algorithms.
- *Movement:* Groups the data necessary to modify the internal representation of a solution for local search algorithms.
- *IncrementModel:* Modifies a solution according to a *BuildStrategy* request.
- *MovementModel:* Modifies a solution according to a *LocalSearch* request.

Collaborations: The *Client* wants a *Solution* for a problem instance. It delegates this task to its *SearchStrategy*, which is composed by at least one *BuildStrategy* and one *LocalSearch*. The *BuildStrategy* produces an initial *Solution* and the *LocalSearch* improves the initial *Solution* through successive movements. The *BuildStrategy* and the *LocalSearch* perform their tasks based on neighborhood relations provided by the *Client*. The implementation of these neighborhoods is delegated by the *Solution* to its *IncrementModel* (related to the *BuildStrategy*) and to its *MovementModel* (related to the *LocalSearch*). The *IncrementModel* and the *MovementModel* are the objects that will obtain the *Increments* or the *Movements* necessary to modify the *Solution* (under construction or not). The *IncrementModel* and the *MovementModel* may change at run-time, reflecting the use of a dynamic neighborhood in the *LocalSearch*, or of having a *BuildStrategy* that uses several kinds of *Increments* to construct a *Solution*. The variation of the *IncrementModel* is controlled inside the *BuildStrategy*, and the *LocalSearch* controls the variation of the *MovementModel*. This control is performed using information made available by the *Client* and accessible to these objects. Figure 3.5 illustrates this scenario.

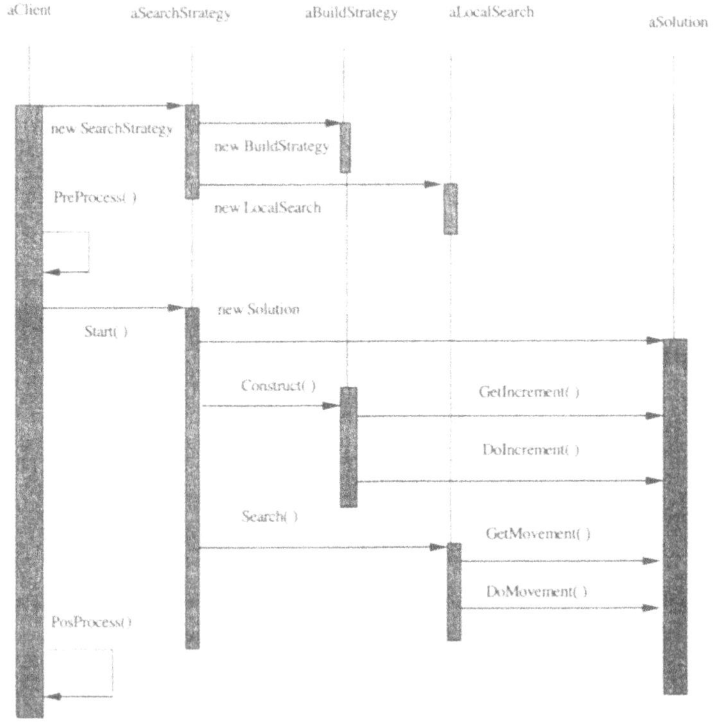

Figure 3.5 Collaborations in the Framework

3.4 USING THE DESIGN PATTERNS

To construct the framework, we use the design patterns presented in Section 3.2. We identify these patterns and present some class hierarchies that could be used in the framework instantiations.

3.4.1 Strategy

We examine the application of this pattern to the framework classes *Client*, *SearchStrategy*, *BuildStrategy*, and *LocalSearch*. Figure 3.4 shows the relationships among them. The *Client* class contains an instance of a combinatorial problem to be solved by a local search heuristic. Instance data is usually pre-processed, to verify whether simplifying hypotheses are satisfied and/or to perform problem reductions (variable fixations and constraint elimination). In addition, after the search is performed, specific post-optimization techniques are usually applied. This suggests the encapsulation of search strategies in other classes, not in *Client*. Even the pre- and post-processing techniques should be isolated in distinct class hierarchies, once more using the strategy pattern. However, to simplify our analysis, we will consider only the separation of local search heuristics (from the context of the combinatorial problem).

In our model, the *SearchStrategy* is responsible for finding solutions to the problem indicated by the *Client*. The operations and data structures used by the *SearchStrategy* are of no concern to the *Client*. The *Client* needs information to instantiate the *SearchStrategy*. The *SearchStrategy* class represents a meta-procedure that uses constructive and local search heuristics to generate solutions. For example, GRASP could be represented as a concrete subclass of *SearchStrategy*.

In summary, we isolate in distinct classes the problem and the search strategy, thus encapsulating search information and allowing greater facility in the construction and extension of a family of search strategies. Using the strategy pattern, the *Client* defines the desired local search via a data structure visible to *SearchStrategy*, which then creates a corresponding *LocalSearch* object, which encapsulates the chosen local search algorithm. Code 1 illustrates a possible implementation for this class diagram. This piece of code should be taken just as an example, since one could implement the Start procedure without constructing the complete set of initial solutions before beginning the exploration by the searchers. This would be the case when several initial solutions should be constructed for being improved by a few searchers, or by a unique searcher, as is the case with GRASP. Also, there is usually only a single *BuildStrategy* and a single *LocalSearch* and not several objects of each class.

The same motivation can be applied to the constructive and local search heuristics related to the *SearchStrategy* class: they can also be modeled in distinct classes, again using the strategy pattern. In this way, families of algorithms for building initial solutions and families of algorithms for local search can be constructed with ease. Figure 3.6 shows an example of a class hierarchy modeling local search algorithms considered from the point of view of the movement selection heuristic. The algorithms in the concrete classes of the above hierarchy, if implemented via conditionals within the same procedure, would create confusing and hard to extend code. This would prevent the inclusion of new heuristics and the extension of those already included, better modeled as their subclasses.

Code 1: *Applying the Strategy Pattern*

```
class Client{
   public:
      List<Solution> Solve( );...
   protected:
      SearchStrategy strategy;
      List<Solution> found_list;
      virtual void PreProcess( );
      void LookFor( );
      virtual void PosProcess( );...
};
List<Solution> Client::Solve( ){
   PreProcess( );
   LookFor( );
   PosProcess( );
   return found_list;
}
void Client::LookFor( ){ found_list = strategy.Start( this ); }

class SearchStrategy{
   public:
      virtual List<Solution> Start( Client * owner );
   protected:
      List<BuildStrategy> builders_list;
      List<LocalSearch> searchers_list; ...
};
List<Solution> SearchStrategy::Start( Client * owner ){
   BuildStrategy builder;
   LocalSearch searcher;
   List<Solution> initial_list; ...
   for( ; ; ){
     builder = builders_list.Next( );
     initial_list.Append( builder.Construct( owner ) );
     if( builders_list.IsEmpty( ) ) break;
   }
   for( ; ; ){
     searcher = searchers_list.Next( );
     for( ; ; ){
       found_list.Append(searcher.Search(initial_list.Next()));
     if( initial_list.IsEmpty(  )) break;
     }
    if( searchers_list.IsEmpty( )) break;
   }   ...
```

3.4.2 Template Method

In the last section we used the strategy pattern to implement *LocalSearch*. The template method pattern could also be used in this derivation, provided we could factor out common steps from the local search algorithm. This can be done, for example, in the specialization of *IterativeImprovement*. To specialize *LocalSearch*, however, the different control mechanisms used in different algorithms do not allow such common-

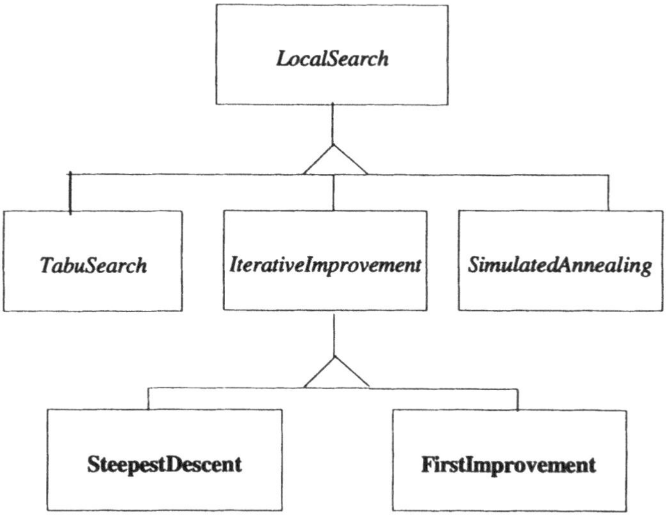

Figure 3.6 A *LocalSearch* Hierarchy

alities to be detected (unless we make the algorithm so abstract that we lose the code reuse facilities that the template method pattern offers).

We illustrate an attempt to use the template method in Code 2. In this example, the implementation of the Selection method declared in *LocalSearch* procedures is deferred to *IterativeImprovement*, *TabuSearch*, and any other concrete subclass of the hierarchy.

Code 2: *Applying the Template Method Pattern*

```
class LocalSearch{
   public:     ...
      List<Solution> Search( Solution initial );
   protected:
      Solution current;      ...
      virtual Boolean StopCriteria( );
      virtual Movement Selection(List<Movement> movement_list);
}
List<Solution> LocalSearch::Search( Solution initial){
      Movement selected, obtained ;
      List<Movement> candidate_list;    ...
      for ( ; ; ){
         if ( StopCriteria( )) break;    ...
         for ( ; ; ){
            if ( current.NbhAlreadyScanned( )) break;
            obtained = current.GetMovement( );
            candidate_list.TryInsertion(obtained);
         }
         selected = Selection(candidate_list);
         current.DoMovement(selected);    ...
      }
```

```
}
class IterativeImprovement: public LocalSearch{
   protected:
   Boolean StopCriteria( );
   Movement Selection(...);   ...
};
class TabuSearch: public LocalSearch{
   protected:   ...
   List<Attribute> tabu_list; //Attribute known by TabuSearch
   Boolean StopCriteria( );
   Movement Selection(...);
};
```

3.4.3 *State*

The state pattern is used in the implementation of dynamic neighborhoods. A neighborhood is an attribute of a solution, hence the construction of different neighborhood types could be specified in the class *Solution*. However, due to the diversity of neighborhood relations that can be defined, it is convenient that *Solution* delegates this task, thus avoiding the inclusion, in its definition, of specialized data structures and generating algorithms.

This delegation of algorithms could again be considered as an application of the strategy pattern. However, because the object responsible for the construction of the neighborhood changes classes during the construction process (in other words, the neighborhood relation changes), this delegation is better identified as an application of the state pattern. We show the classes and relationships involved in this pattern in Figure 3.7. The corresponding Code 3 follows.

A *Movement* (or an *Increment*) is the information that is used by a *Solution*'s operation to modify its internal representation values or to provide another *Solution* with modified representation values. A *Movement* is just the data structure that contains the necessary information to implement the kind of movement associated to a particular neighborhood relation. Each concrete *MovementModel* is related to exactly one *Movement* concrete class. If the *MovementModel* changes during execution, we have a dynamic neighborhood. Delegating the construction of a neighborhood to a *MovementModel* results in the localization of algorithmic peculiarities of the neighborhood relation. This is an application of the strategy pattern. But we do not have a unique method to localize (we must also localize the modification of the solution itself). Moreover, we can use polymorphism on the *MovementModel*. So, the delegations from *Solution* to *MovementModel* are better implemented as applications of the state pattern. Actually, the implementation of dynamic neighborhoods consists in the use of polymorphism on a *MovementModel* inside its class hierarchy.

Code 3: *Applying the State Pattern*

```
class Solution{
   public:
      Movement GetMovement( );
      void DoMovement( );...
   protected:
```

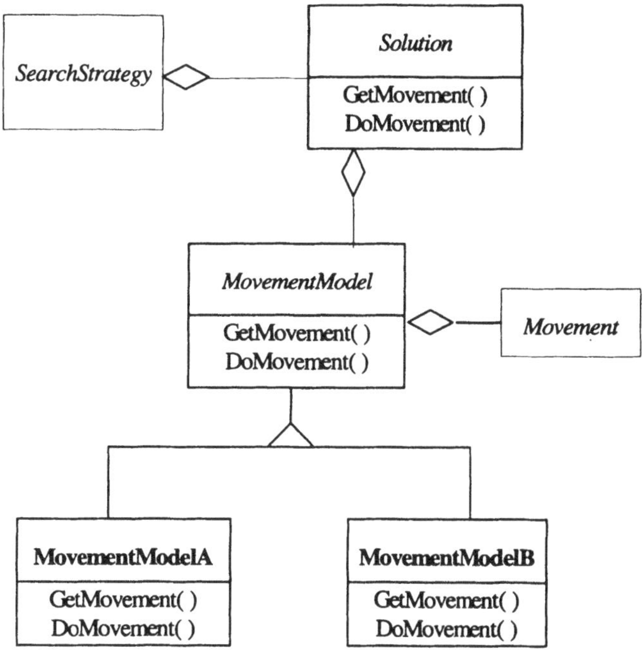

Figure 3.7 Application of the State Pattern

```
        MovementModel search_model;
};
Movement Solution::GetMovement( ){
   return search_model.GetMovement( );
}
void Solution::DoMovement(Movement move){
   search_model.DoMovement(move);
}
class MovementModel{
   public:
   Movement GetMovement( );
   void DoMovement(Movement move);
};
```

Families of construction algorithms can be defined according to different "growth" models. As in local search algorithms, the neighborhood of a solution in a construction algorithm is an attribute of a solution, which delegates the construction of the neighborhood to a subclass of *IncrementModel*, as shown in Figure 3.8.

In the implementation of local search heuristics (via *LocalSearch* objects) that use dynamic neighborhoods, it is convenient that changes in neighborhood relationships be requested to *Solution* objects by *LocalSearch* objects, following a pre-defined tran-

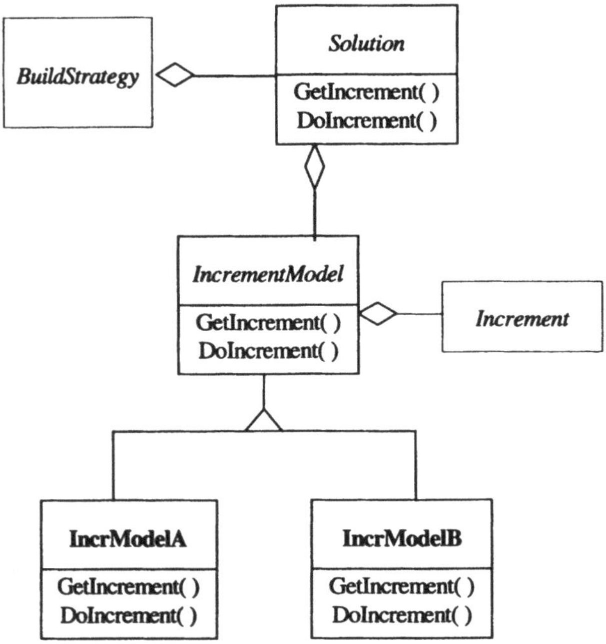

Figure 3.8 Other Application of the State Pattern

sition model. *LocalSearch* controls transitions among neighborhood relationships. This control can use parameters made available by the *Client*.

3.5 IMPLEMENTATION ISSUES

We used the **Searcher** framework to develop, implement, test, and compare a number of heuristics for the phylogeny problem under the parsimony criterion (Swofford and Olsen (1990)). The phylogeny problem is one of the main problems in comparative biology. A phylogeny (or an evolutionary tree) is a tree that relates groups of species, populations of distinct species, populations of the same species, or homologous genes in populations of distinct species, indistinctly denoted by *taxons* (Ayala (1995), Wiley et al. (1991)). This relation is based on the similarity over a set of characteristics. The parsimony criterion states that the best phylogeny is the one that can be explained by the minimum number of evolutionary steps. It is frequently said that this criterion is the best if the mutation probability on the characteristics considered is very small (see Sober (1987)). This problem is $\mathcal{NP}$-hard in general and in the common restricted cases (Bodlaender et al. (1992), Day et al. (1986), Foulds and Graham (1982a), Foulds and Graham (1982b)).

Table 3.1 illustrates an instance of the phylogeny problem, defined by taxons $\{O, A, B, C\}$ and binary characters $\{a, b, c, d, e, f, g\}$. Figure 3.9 illustrates an optimal solution for this instance, with nine evolutionary steps. We indicate on each branch of this tree the characters which change from one taxon to another, together with the associated number of evolutionary steps.

	a	b	c	d	e	f	g
O	0	0	0	0	0	0	0
A	1	1	0	0	1	1	1
B	1	1	1	1	1	1	1
C	1	1	1	1	0	0	0

Table 3.1 Instance of the Phylogeny Problem

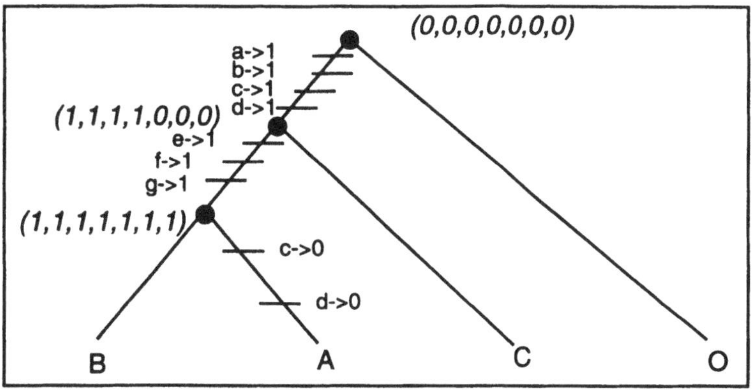

Figure 3.9 Optimal Phylogeny

The phylogeny problem can be modeled as a minimum Steiner tree problem in a metric space (Sankoff and Rousseau (1975)). Each characteristic used to describe and compare the taxons is a metric space by itself. The characteristics are independent from one another. For each metric space, a different algorithm exists to evaluate the number of evolutionary steps for a given evolutionary tree. The object-oriented programming paradigm allows the dynamic binding of the adequate algorithm for each specific characteristic. We developed a set of constructive algorithms (purely greedy and greedy randomized, among others) and local search heuristics (iterative improvement, tabu search, GRASP, and variable neighborhood search, among others) for the phylogeny problem. The instantiation of the proposed framework to the phylogeny problem allowed a fast reproduction of the algorithms already published in the literature. More important, it allowed the development of a family of local search methods and their effective comparison in terms of solution quality and computation times.

SearchStrategy is a purely virtual class where the methods to instantiate concrete *BuildStrategy*s and *LocalSearch*s are defined. We implemented a concrete subclass, LoneSchStrtgy, which uses in its Start(...) method just one *BuildStrategy* and one *LocalSearch*, each time the method is requested by the *Client*. The basic procedure to construct a phylogeny relating n taxons is shown below:

Step 1. Let φ be the phylogeny to be constructed and W the set of n taxons to be related by a phylogeny. Set $\varphi \leftarrow \emptyset$, $U \leftarrow \emptyset$ and $k \leftarrow 0$.

Step 2. While $k < n$ do:
(a) Select one taxon $t \in W \setminus U$
(b) Set $U \leftarrow U \cup \{t\}$ and $k \leftarrow k + 1$
(c) Select a branch of phylogeny φ and insert t on it.

We have implemented and tested four variants of this procedure, using different criteria in the two selection steps: (2a) choice of the taxon to be inserted, and (2c) choice of the branch in which the selected taxon is inserted. In the first algorithm, both steps are purely random. In the second one, the taxon t is selected randomly at step (2a), while a greedy criterion is used to select the most adequate branch where this taxon should be inserted (cheapest insertion). In the case of the third algorithm, steps (2a) and (2c) are jointly performed in a greedy manner: the pair (taxon, branch) is randomly selected among all those pairs that induce a minimal cost insertion. The last variant is a greedy randomized procedure similar to the previous one, in which the universe of random choice also allows pairs (taxon, branch) that induce insertion costs with some deviation from the minimal value. The reader is referred to Andreatta (1998) and Andreatta and Ribeiro (2001) for a more detailed description of these algorithms, including implementation and complexity issues.

To implement these algorithms within the **Searcher** framework, we created and used two concrete *BuildStrategys*. Both of them allow more than one *IncrementModel* in the construction of feasible solutions, and allow using subordinate *LocalSearchs* to improve partial solutions along the construction. The first concrete *BuildStrategy* uses the first *Increment* given by the *Solution* (through its current *IncrementModel*), whereas the second one randomly selects an *Increment* from a restricted list of candidate increments (uniform distribution). Approximately 1000 lines of C++ code were generated in this hierarchy. We have implemented three concrete *IncrementModels* related to the phylogeny problem. The first one generates the *Increments* in a pre-defined coded order. The second *IncrementModel* generates the increments by randomly selecting the taxon, while the branch of the current partial phylogeny is chosen according to a pre-defined coded order. The third one generates them randomly per taxon and, subsequently, in random order per branch of the current partial phylogeny. These *IncrementModels* were implemented with 2000 lines of C++ code. The *IncrementModel* services are not limited to a GetIncrement-like method, but include the methods Get1stBest, GetAllBest and GetBestInRange.

Approximately 30% of the code generated to implement all these constructive algorithms for the phylogeny problem is totally reusable for other problems. However, it must be stressed that the major benefit of using the **Searcher** framework is its flexibility to change the increment models required in the construction of solutions.

With respect to iterative improvement procedures, we implemented a concrete *LocalSearch* with two types of movements: first improvement and best improvement. We implemented four *MovementModels* supporting different neighborhood relationships: "Nearest Neighborhood Interchanges" and three variants of "Sub-tree Pruning and Regrafting", see Andreatta (1998), Andreatta and Ribeiro (2001) for details. Coding such *MovementModels* in accordance with the relationships defined by the **Searcher** framework, allowed the implementation and test of several constructive algorithms using subordinate iterative improvement procedures to improve partial phylogenies. The use of construction algorithms that utilize subordinate local search heuristics (for the improvement of partially constructed solutions) is one of the major features of the **Searcher** framework.

We implemented algorithms based on three metaheuristics: GRASP (Feo and Resende (1995), Resende and Ribeiro (1997)), variable neighborhood search (VNS; Mladenović and Hansen (1996)), and tabu search (Glover and Laguna (1997)). The VNS heuristic was implemented as a concrete *SearchStrategy*. To further illustrate the flexibility of the framework, we also tested a parameterization of a VNS heuristic where its subordinate iterative improvement procedure also has a dynamic neighborhood model. To implement tabu search, we created a class hierarchy called *Memory*, independent of the *MovementModel* hierarchy. A *LocalSearch* has a *Memory*. *LocalSearch* requests a *Movement* to a *Solution*, allowing the latter to access this *Memory*. The *MovementModel* delegates to the *Memory* the verification of the tabu status of a *Movement*. This modularized design of a tabu search memory allows handling several *Memorys* for performing search diversification or intensification. This is done by mapping regions of the search space to *Memorys*, which are associated to *Solutions*. The sequence of transformations of a *Solution* during the search will depend on its attributes and on the associated *Memory*. The implementation of a GRASP heuristic was immediate for the phylogeny problem. Since the implemented constructive algorithms were greedy randomized ones, it was just a matter of creating a concrete *SearchStrategy* that could start up a constructive algorithm several times and apply an iterative improvement procedure to each solution obtained.

Besides allowing code reuse, the use of the framework gave flexibility in programming and enabled the fast development of different algorithms, variants and combinations. Detailed computational results concerning the application of the **Searcher** framework to eight benchmark instances of the phylogeny problem, as well as a thorough comparative study of algorithms and search strategies, are reported in Andreatta (1998), Andreatta and Ribeiro (2001). We identified two particularly effective versions of the construction procedure, each of them with a good trade-off in terms of solution quality and computation time. Three neighborhood structures have been compared, showing that the one with greater size performed better than the others in terms of solution quality. The three metaheuristics lead to equivalent results in terms of the average solution quality and of the number of instances for which the best known value was attained.

3.6 RELATED WORK

ABACUS (Thienel (1997)) is possibly the first successful example of an object-oriented tool that encapsulates exact algorithms for combinatorial problems. The reuse potential of frameworks, however, outweighs that of class libraries (see Fayad and Schmidt (1997a)), being semi-complete implementations of particular domains.

The development of object-oriented framework-like tools in the domain of neighborhood search metaheuristics is a current trend. Ferland et al. (1996) proposed a generic system in the domain of local search heuristics. They describe the use of object orientation in a system which implements four local search heuristics essentially related to resource allocation problems. They propose two class hierarchies. In relation to our work, the first hierarchy can be said to define *Solution* descendants, which can be incorporated in the **Searcher** framework committing to a concrete *Solution* class dynamically (for example, based on the *Client* pre-processing operation). The second hierarchy is actually an application of the strategy pattern, or the definition of a hierarchy of search heuristics, just as we propose for the class *LocalSearch*. Design patterns are not used to describe their architecture. Our framework presents some improvements with respect to their solution:

- It contains a hierarchy of local search strategies and allows the inclusion of a hierarchy of solutions, thus containing the architecture proposed there.
- It allows the use of dynamic neighborhood models and uses a class hierarchy in the definition of several neighborhood relations, while in the previous work only a static neighborhood model is used.
- It allows the implementation of constructive heuristics that utilize subordinate local search heuristics available in the *LocalSearch* hierarchy.

Woodruff (1997) developed a class library concerning neighborhood search metaheuristics for problems whose solutions can be represented by characteristic, permutation, or integer vectors. As in our framework **Searcher**, the solution, problem instance, and movement model concepts are partitioned. Solution representation and semantic are encapsulated in separate classes, since his framework is designed to take advantage of the common representation of different combinatorial problems.

Fink et al. (1999b) also developed a framework for the same domain. Their project is based mainly in the use of static polymorphism, with intensive use of generic classes (using C++ templates). Michel and van Hentenryck (1999) have used a very interesting approach to deal with local search procedures. They designed and implemented a domain-specific language for implementing these procedures. The language combines aspects of declarative and imperative programming, and introduces the declarative concept of *invariant*, which provides a declarative way to specify what needs to be maintained to define the neighborhood and the objective function. They present some experimental results that indicate that the language can be implemented to run with reasonable efficiency with much less development time. Their language was not able to deal with the dynamic neighborhood model of VNS.

3.7 CONCLUSIONS AND EXTENSIONS

In this work, we propose an architecture for the implementation and comparison of local search heuristics. The **Searcher** framework addresses the most important design issues, which are how to separate responsibilities and how the agents of these tasks are related. It encapsulates, in abstract classes, different aspects involved in local search heuristics, such as algorithms for the construction of initial solutions, methods for the limitation of possible neighborhoods, and criteria for the selection of movements. Encapsulation and abstraction promote unbiased comparisons between different heuristics. The classification of different aspects of heuristics simplifies their implementations and invites new extensions in the form of subclasses. The reuse of code we can accomplish, implementing common aspects of a family of heuristics, offers us a better platform for comparison, since a large part of the code is common to all members of the family.

We also describe a successful application of the **Searcher** framework to the solution of the phylogeny problem. Approximately 30% of the code generated for this application can be reused outside its domain. Besides allowing code reuse, the framework gave flexibility in programming and enabled the fast development of different algorithms, variants and combinations. Its design includes patterns, carefully chosen and often balanced against others. The use of design patterns to describe the framework automatically addresses most of its implementation issues. The choice of the C++ language for its implementation was motivated by the need to generate efficient code to manage problem instances of great size. Moreover, although C++ supports object-oriented programming, it is possible to relax the paradigm whenever one needs more speed (relaxing the data encapsulation restriction, for instance). The **Searcher** framework can be used and has particularly appropriate features for situations involving:

- Local search strategies that can use different methods for the construction of the initial solution, different neighborhood relations, or different movement selection criteria.

- Construction algorithms that utilize subordinate local search heuristics for the improvement of partially constructed solutions.

- Local search heuristics with dynamic neighborhood models.

4 HOTFRAME: A HEURISTIC OPTIMIZATION FRAMEWORK

Andreas Fink and Stefan Voß

Technische Universität Braunschweig
Institut für Wirtschaftswissenschaften
Abt-Jerusalem-Straße 7, D-38106 Braunschweig, Germany
{a.fink,stefan.voss}@tu-bs.de

Abstract: In this paper we survey the design and application of HOTFRAME, a framework that provides reusable software components in the metaheuristics domain. After a brief introduction and overview we analyze and model metaheuristics with special emphasis on commonalities and variabilities. The resulting model constitutes the basis for the framework design. The framework architecture defines the collaboration among software components (in particular with respect to the interface between generic metaheuristic components and problem-specific complements). The framework is described with respect to its architecture, included components, implementation, and application.

4.1 INTRODUCTION

There are several strong arguments in favor of reusable software components for metaheuristics. First of all, mature scientific knowledge that is aimed at solving practical problems must also be viewed from the point of view of technology transfer. If we consider the main metaheuristic concepts as sufficiently understood, we must strive to facilitate the efficient application of metaheuristics by suitable means. "No systems, no impact!" (Nievergelt (1994)) means that in practice we need easy-to-use application systems that incorporate the results of basic research. Therefore, we also have to deal with the issue of efficiently building such systems to bridge the gap between research and practice. From a research point of view, software reuse may also provide a way for a fair comparison of different heuristics within controlled and unbiased experiments, which conforms to some of the prescriptions in the literature (see, e.g., Barr et al. (1995) and Hooker (1994)).

Metaheuristics are by definition algorithmic concepts that are widely generic with respect to the type of problem. Since algorithms are generally applied in the form of software, adaptable metaheuristic software components are the natural means to incorporate respective scientific knowledge. We have built HOTFRAME, a Heuristic OpTimization FRAMEwork implemented in C++, which provides adaptable components that incorporate different metaheuristics and common problem-specific complements as well as an architectural description of the collaboration among these components (see Fink and Voß (1999b), Fink (2000)).

Others have followed or actively pursue similar research projects, which focus primarily on metaheuristics based on the local search paradigm including tabu search; see the contributions within this volume. Moreover, there is a great variety of software packages for evolutionary algorithms; see Heitkötter and Beasley (2001) as well as Chapter 10 below. On the other hand, there are some approaches to design domain-specific (modeling) languages for local search algorithm, which are partly based on constraint programming; see de Backer et al. (1999), Laburthe and Caseau (1998), Michel and van Hentenryck (1999), Michel and van Hentenryck (2001a), Nareyek (2001), as well as Chapter 9.

The adaptation of metaheuristics to a specific type of problem may concern both the static definition of problem-specific concepts such as the solution space or the neighborhood structure as well as the tuning of run-time parameters (calibration). The latter aspect of designing robust (auto-adaptive) algorithms, though being an important and only partly solved research topic, may be mostly hidden from the user. However, the general requirement to adapt metaheuristics to a specific problem is a more serious obstacle to an easy yet effective application of metaheuristics. Following the "no free lunch theorem" we assume that there can not be a general problem solver (and accordingly no universal software implementation) that is the most effective method for all types of problems (see Culberson (1998), Wolpert and Macready (1997)). This implies that one has to provide implementation mechanisms to do problem-specific adaptations to facilitate the effective application of reusable metaheuristic software components.

In this paper, we survey the design and application of HOTFRAME; for a detailed description in German we refer to Fink (2000). In the next section, we give a brief overview of the framework indicating some of its basic ideas. In Section 4.3, we analyze commonalities and variabilities of metaheuristics to the end that respective formal models provide the basis for reusable software components.

The subsequent sections are organized to follow the classical phases of software development processes. Here we assume that the reader is familiar with basic ideas of object-oriented programming and C++. The framework architecture, which specifies the collaboration among software components, is described in Section 4.4. In Section 4.5, we consider basic aspects of the implementation. In Section 4.6, we sketch the application of the framework and describe an incremental adoption path. Finally, we draw some conclusions.

4.2 A BRIEF OVERVIEW

The scope of HOTFRAME comprises metaheuristic concepts such as (iterated) local search, simulated annealing and variations, different kinds of tabu search methods (e.g., static, strict, reactive) providing the option to flexibly define tabu criteria taking into account information about the search history (attributes of performed moves and traversed solutions), evolutionary algorithms, candidate lists, neighborhood depth variations, and the pilot method.

The primary design objectives of HOTFRAME are run-time efficiency and a high degree of flexibility with respect to adaptations and extensions. The principal effectiveness of the framework regarding competitive results has been demonstrated for different types of problems; see Fink and Voß (1999a), Fink (2000), Fink et al. (2000), Fink and Voß (2001). However, claiming validity of the "no free lunch theorem", the user must generally supplement or adapt the framework at well-defined adaptation points to exploit problem-specific knowledge, as problems from practice usually embody distinctive characteristics.

The architecture of a software system specifies the collaboration among system elements. The architecture of a framework defines a reusable design of systems in the same domain. Moreover, a framework provides (adaptable) software components (typically in accordance with the object-oriented paradigm), which encapsulate common domain abstractions. Contrary to an ordinary class library, which is some kind of a toolkit with modules that are mainly usable independently from each other, a framework also specifies (some of) the control flow of a system. With respect to the variabilities of different applications in the same domain, frameworks must provide variation points for adaptation and extension. That is, to instantiate a framework for a particular application one generally has to complete certain parts of the implementation, according to some kind of interface definitions.

HOTFRAME aims for a natural representation of variation points identified in the analysis of metaheuristics. Since metaheuristics are generic (abstract) algorithms, which are variable with respect to problem-specific concepts (structures), we follow the generic programming paradigm: A generic algorithm is written by abstracting algorithms on specific types (data structures) so that they apply to arguments whose types are as general as possible (generic algorithm = code + requirements on types); see Musser and Stepanov (1994). HOTFRAME uses parameterization by type as the primary mechanism to make components adaptable. Common behavior of metaheuristics is factored out and grouped in generic templates, applying static type variation. This approach leads to generic metaheuristic components which are parameterized by (mainly problem-specific) concepts such as the solution space, the neighborhood structure, or tabu-criteria. In C++, generic components are implemented as template classes (or functions), which enables achieving abstraction without loss of efficiency. Those templates have type parameters. The architecture defines properties (an interface with syntactic and semantic requirements) to be satisfied by argument types.

For example, steepest descent (greedy local search) is generic regarding the solution space S and the neighborhood N. Accordingly, we have two corresponding type parameters, which results in the following template:

```
SteepestDescent< S, N >
```

To streamline the parameterization of metaheuristic templates, we group type parameters in so-called configuration components, which encapsulate variation points and their fixing. For example, we may define

```
struct myConfiguration
{
  typedef  MCKP_S  S;
  typedef  BV_S_N  N;
};
```

and instantiate a metaheuristic component by

```
SteepestDescent< myConfiguration >.
```

In this case, we define the solution space to be represented by a class `MCKP_S`, which encapsulates solutions for the multi-constrained 0/1-knapsack problem, and we apply a class `BV_S_N`, which represents the classical bit-flip neighborhood for binary vectors. HOTFRAME provides several reusable classes for common solution spaces (e.g., binary vectors, permutations, combined assignment and sequencing) and neighborhood structures (e.g., bit-flip, shift, or swap moves). These classes can be used unchanged (e.g., neighborhoods) or reused by deriving new classes which customize their behavior (e.g., by defining some specific objective function). In cases where one of the pre-defined problem-specific components fits the actual problem, the implementation efforts for applying various metaheuristics can be minor, since the user, essentially, only needs to implement the objective function in a special class that inherits the data structure and the behavior from some appropriate reusable class.

Metaheuristic templates can have a set of type parameters, which refer to both problem-specific concepts as well as strategy concepts. For example, local search strategies may differ regarding the rule to select neighbors (moves). We have hierarchically separated the configuration regarding problem-specific and metaheuristic concepts. Common specializations of general metaheuristic components are pre-defined by configuration components that pass through problem-specific definitions and add strategy definitions. This is exemplified in the following definition:

```
template <class C>
struct CSteepestDescent
{
  typedef  BestPositivNeighbor<C>  NeighborSelection;
  typedef  typename C::S  S;
  typedef  typename C::N  N;
};
```

Then, we may customize a general local search frame, instantiate some metaheuristic object, and use it as shown in the following example:

```
myHeuristic = new IteratedLocalSearch
                    < CSteepestDescent< myConfiguration> >;
myHeuristic->search( initialSolution );
```

Advantages of this design are that it provides a concise and declarative way of system specification, which decreases the conceptual gap between program code and domain concepts (known as achieving high intentionality) and simplifies managing many variants of a component. Moreover, the resulting code can be highly run-time efficient.

4.3 ANALYSIS

Developing a framework requires a solid understanding of the domain under consideration. That is, we must develop a common understanding of the main metaheuristic concepts and the scope of the framework. Therefore, it is important to comprehensively analyze the domain and to develop a concrete and detailed domain model. This domain model comprises the commonalities and variabilities of metaheuristics, laying the foundation for the design of reusable software components with corresponding means for adaptation. In the following, we give a semi-formal domain model of the considered metaheuristics. The concise descriptions presuppose that the reader knows about metaheuristic concepts such as (iterated) local search, simulated annealing and variations, and different kinds of tabu search methods (see, e.g., Reeves (1993), Rayward-Smith et al. (1996), Laporte and Osman (1996), Osman and Kelly (1996), Aarts and Lenstra (1997), Glover and Laguna (1997), Voß et al. (1999), and Ribeiro and Hansen (2002)).

4.3.1 Problem-Specific Concepts

The first step in analysis is the definition of the commonalities (shared features) of different types of problems by an abstract model. Such a model ("domain vocabulary"), which captures problem-specific concepts with the same external meaning, is an essential basis to define metaheuristics, i.e., algorithmic concepts which are widely generic with respect to the type of problem. Eventually, when applying metaheuristics to some specific type of problem, these abstractions have to be instantiated by the specific structures of the problem at hand.

There are different types of

$$\text{problems} \quad P$$

with

$$\text{problem instances} \quad p \in P\,.$$

For every problem, there are one or more

$$\text{solution spaces} \quad S_P(p)$$

with

$$\text{solutions} \quad s \in S_P(p)\,.$$

For ease of notation, we generally restrict to $s \in S$ (where S replaces $S_P(p)$) when there are no ambiguities; this also concerns the subsequent notation. Solutions are evaluated by an

$$\text{objective function} \quad f : S \to \mathsf{R} \ .$$

We generally formulate problems as minimization problems; i.e., we strive for minimization of f.

To efficiently manage information about solutions (in connection with some tabu search method), we may need a function

$$h : S \to S_h$$

which transforms solutions to elements of a suitable set S_h. With respect to efficiency, one mostly uses non-injective ("approximate") functions (e.g., hash-codes).

For every solution space S, there are one or more

$$\text{neighborhoods} \quad N_S \ ,$$

which define for each solution $s \in S$ an ordered set of neighboring solutions

$$N_S(s) = \{n_1(s), \ldots, n_{|N_S(s)|}(s)\} \ .$$

Such a neighborhood is sketched in Figure 4.1.

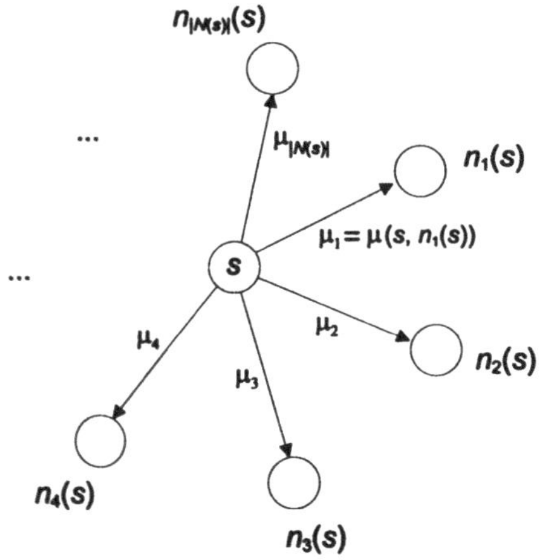

Figure 4.1 Neighborhood

From a transformation point of view every neighbor $n(s) \in N_S(s)$ of a solution $s \in S$ corresponds to a

$$\text{move} \quad \mu(s, n(s)) \ .$$

So we can also define a neighborhood as

$$N_S^\mu(s) = \{\mu_1 = \mu(s, n_1(s)), \ldots, \mu_{|N_S(s)|} = \mu(s, n_{|N_S(s)|}(s))\} \ .$$

Moves $\mu(s, n(s)) \in N_S^{\mu}(s)$ are evaluated by a

$$\text{move evaluation} \quad \hat{f}(\mu(s, n(s))) ,$$

which is often defined as $f(s) - f(n(s))$. Other kinds of move evaluations, e.g., based on measures indicating structural differences of solutions, are possible. In general, positive values should indicate "improving" moves. Move evaluations provide the basis for the guidance of the search process.

Both solutions and moves can be decomposed into

$$\text{attributes} \quad \psi \in \Psi ,$$

with Ψ representing some attribute set, which may depend on the neighborhood. A solution s corresponds to a set

$$\psi(s) = \{\psi_1(s), \ldots, \psi_{|\psi(s)|}(s)\} \text{ with } \psi_j(s) \in \Psi \; \forall j = 1, \ldots, |\psi(s)| .$$

The attributes of a move μ may be classified as plus and minus attributes, which correspond to the characteristics of a solution that are "created" or "destroyed", respectively, when the move is performed (e.g., inserted and deleted edges for a graph-based solution representation). A move μ corresponds to a set of plus and minus attributes:

$$\begin{aligned} \psi(\mu) &= \psi^+(\mu) \cup \psi^-(\mu) \\ &= \{\psi_1^+(\mu), \ldots, \psi_{|\psi^+(\mu)|}^+(\mu)\} \cup \{\psi_1^-(\mu), \ldots, \psi_{|\psi^-(\mu)|}^-(\mu)\} \\ &= \{\psi_1(\mu), \ldots, \psi_{|\psi(\mu)|}(\mu)\} . \end{aligned}$$

Finally, we denote the inverse attribute of ψ by $\overline{\psi}$.

4.3.2 *Metaheuristic Concepts*

In this section, we define some of the metaheuristic concepts included in HOTFRAME. We restrict the descriptions to (iterated) local search, simulated annealing (and variations), and tabu search, while neglecting, e.g., candidate lists, evolutionary methods and the pilot method. At places we do not give full definitions of various metaheuristics components (modules) but restrict ourselves to brief sketches. That is, the following descriptions exemplify the analysis of metaheuristics with respect to commonalities and variabilities, without providing a complete model.

There are two kinds of variabilities to be considered for metaheuristics. On the one hand, metaheuristics are generic regarding the type of problem. On the other hand, metaheuristics usually provide specific variation points regarding subordinate algorithms (aspects such as move selection rules or cooling schedules) and simple parameters (aspects such as the termination criterion). Accordingly, a configuration C of a metaheuristic H is composed of a definition of a subset C_P of the problem-specific abstractions (S, N, h, Ψ) discussed in the previous subsection, and of a configuration C_H that is specific to the metaheuristic. Given such a configuration C, a metaheuristic defines a transformation of an initial solution s to a solution s': $H_C : s \rightarrow s'$.

In the following descriptions, we use straightforward pseudo-code (with imperative and declarative constructs) and data structures such as lists or sets, without caring about, e.g., efficiency aspects. That is, such analysis models define meaning but should not necessarily prejudice any implementation aspects. In the pseudo-code description of algorithms, we generally denote a parameterization by "$< \ldots >$" to define algorithmic variation points, and we use "$(\ldots)$" to define simple value parameters (e.g., the initial solution or numeric parameters). When appropriate, we specify standard definitions, which allows using algorithms without explicitly defining all variation points. By the symbol ϕ we denote non-relevance or invalidity. To simplify the presentation, we use Ω to denote an additional termination criterion, which is implicitly assumed to be checked after each iteration of the local search process. By ω we include means to specify external termination, which is useful, e.g., in online settings. Using our notation, the interface of a metaheuristic H may be defined by $H < C > (s, T_{\max}, I_{\max}, \omega)$. Such a metaheuristic with configuration C transforms an initial solution s, given a maximum computation time $T_{\max}$, a maximum iteration number $I_{\max}$, and an external termination criterion ω.

To model the variation points of metaheuristics, we use feature diagrams, which provide a concise means to describe the variation points of concepts in a manner independent from any implementation concerns (see Czarnecki and Eisenecker (2000), Simos and Anthony (1998)).

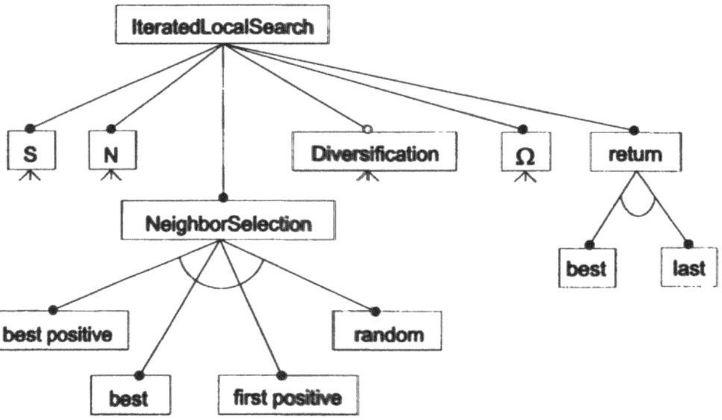

Figure 4.2 Feature Diagram for Simple Local Search Methods

4.3.2.1 Iterated Local Search. Figure 4.2 shows a feature diagram for simple local search methods (IteratedLocalSearch). In principle, all such local search procedures are variable regarding the solution space and the neighborhood structure. That is, S and N are mandatory features (denoted by the filled circles). The crowsfeet indicate that S and N are considered as abstract features, which have to be instantiated by specific definitions/implementations. At the center of any iteration of a local search procedure is the algorithm for selecting a neighbor. The diagram shows that there are four alternatives for instantiating this feature (denoted by the arc): select the best neighbor out of $N(s)$ with a positive move evaluation, select the best neighbor even if its evaluation is non-positive, select the first neighbor with a positive move

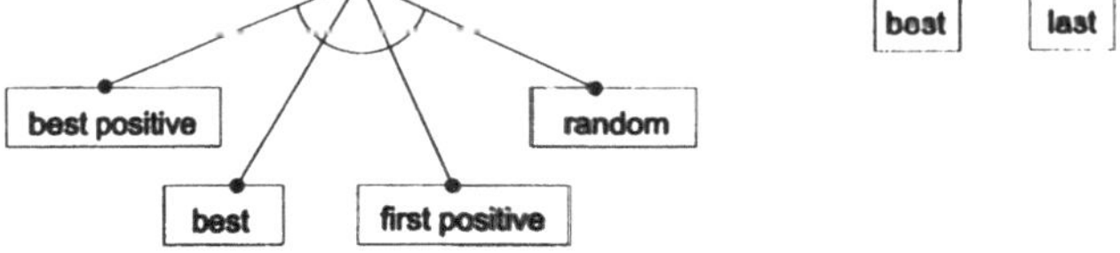

evaluation, or select a random neighbor. The diversification feature is optional (denoted by the open circle), since we do not need a diversification mechanism, e.g., when we specialize IteratedLocalSearch as a one-time application of a greedy local search procedure. The termination criterion Ω is a mandatory feature, which has to be defined in a general way. Finally, there is a feature that allows to specify whether the search procedure should return the best solution found or the last solution traversed. The latter option is useful, e.g., when we use a local search procedure with a random neighbor selection rule as a subordinate diversification component of another metaheuristic.

The feature diagram defines the basic variation points of the considered concept. On this basis, Algorithm 1 formally defines the specific meaning of IteratedLocalSearch. While not giving formal definitions of all the features, Algorithm 2 exemplifies such a sub-algorithm. By using BestPositiveNeighbor, we may instantiate IteratedLocalSearch to generate SteepestDescent (as shown in Algorithm 3). In a similar manner, we may generate a RandomWalk procedure, which may again be used to generate a more complex procedure as shown in Algorithm 4.

Algorithm 1 IteratedLocalSearch

IteratedLocalSearch
 $< S, N, \mathit{NeighborSelection}, \mathit{Diversification} >$
 $(s, T_{\max} = \infty, I_{\max} = \infty, R_{\max} = 1, \omega = \text{false}, \mathit{returnBest} = \text{true})$:

$\Omega : (t \geq T_{\max})$ **or** (ω)

$s_{\text{best}} = s$;
for $r = 1$ **to** $R_{\max}$
  if $r > 1$
    Diversification(s);
  $i = 0$;
  do
    $i = i + 1$;
    $s' = \mathit{NeighborSelection}< S, N >(s)$;
    if s' is valid
      $s = s'$;
      if $f(s) < f(s_{\text{best}})$
        $s_{\text{best}} = s$;
  while (s' is valid) **and** ($i < I_{\max}$);
if *returnBest*
  $s = s_{\text{best}}$;

Algorithm 2 BestPositiveNeighbor

BestPositiveNeighbor $< S, N >$ (s) :

$j = \text{argmax}\{\hat{f}(\mu(s, n_j(s))) \mid j = 1, \ldots, |N(s)|\};$
if $\hat{f}(\mu(s, n_j(s))) > 0$
  return $n_j(s)$;
else
  return ϕ;

Algorithm 3 SteepestDescent

SteepestDescent $< S, N >$ $(s, T_{\max}, I_{\max}, \omega)$:

IteratedLocalSearch $< S, N,$ BestPositiveNeighbor, $\phi >$
  $(s, T_{\max}, I_{\max}, 1, \omega,$ true);

Algorithm 4 IteratedSteepestDescentWithPerturbationRestarts

IteratedSteepestDescentWithPerturbationRestarts $< S, N >$
  $(s, T_{\max}, I_{\max}, R_{\max}, \textit{perturbations}, \omega)$:

IteratedLocalSearch
  $< S, N,$ BestPositiveNeighbor,
   RandomWalk $< S, N >$ $(\phi, \infty, \textit{perturbations}, \omega,$ false) $>$
  $(s, T_{\max}, I_{\max}, R_{\max}, \omega,$ true);

4.3.2.2 Simulated Annealing and Variations. Figure 4.3 shows the HOTFRAME feature diagram of simulated annealing and variations. Besides S and N, the principal features of such concepts are the acceptance criterion, the cooling schedule, and an optional reheating scheme.

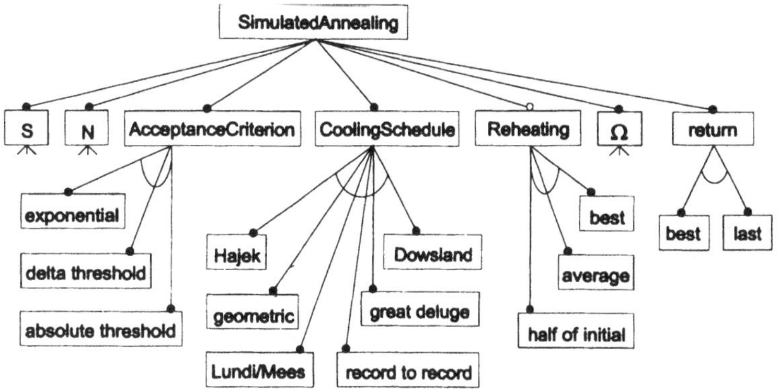

Figure 4.3 Feature Diagram for Simulated Annealing

The algorithmic commonalities of some set of simulated annealing like procedures are captured in Algorithm 5. After defining all the features (not shown here), we may generate a classic simulated annealing heuristic as shown in Algorithm 6. Some straightforward variations of this procedure are defined in Algorithms 7–9.

Nevertheless, it is not reasonable to try to capture all kinds of special simulated annealing procedures by one general simulated annealing scheme. This is exemplified by showing, in Algorithm 10, a simulated annealing procedure which strongly deviates from the general scheme of Algorithm 5. We define such a procedure separately, while we may use the general simulated annealing features as defined above. Given a parameter *initialAcceptanceFraction*, the starting temperature is set so that the initial fraction of accepted moves is approximately *initialAcceptanceFraction*. At each temperature $sizeFactor \times |N|$ move candidates are tested. The parameter *frozenAcceptanceFraction* is used to decide whether the annealing process is *frozen* and should be terminated. Every time a temperature is completed with less than *frozenAcceptanceFraction* of the candidate moves accepted, a counter is increased by one. This counter is reset every time a new best solution is found. The procedure is terminated when the counter reaches *frozenParemeter*. Then it is possible to reheat the temperature to continue the search by performing another annealing process. Algorithm 10 includes default values for the parameters according to the recommendation in Johnson et al. (1989).

Algorithm 5 GeneralSimulatedAnnealing

```
GeneralSimulatedAnnealing
      < S, N, AcceptanceCriterion, CoolingSchedule, Reheating >
      (s, T_max = ∞, I_max = ∞, ω = false, returnBest = true,
       τ_0, R_max = 1, deltaRepetitions = 1, numberOfReheatings = 0)  :

Ω : (t ≥ T_max) or (ω)

s_best = s;
for h = 1 to numberOfReheatings + 1
      if h > 1
            τ_0 = Reheating(τ_0, τ_best, τ);
      else
            τ_best = τ_0;
      τ = τ_0;
      i = 0;
      do
            for r = 1 to R_max
                  i = i + 1;
                  s' = RandomNeighbor< S, N >(s);
                  if AcceptanceCriterion(τ, f̂(μ(s, s')), f(s'))
                        τ = CoolingSchedule.moveAccepted(τ, f̂(μ(s, s')), f(s'));
                        s = s';
                        if f(s) < f(s_best)
                              s_best = s;
                              τ_best = τ;
                              τ = CoolingSchedule.newBestObjective(f(s));
                  else
                        τ = CoolingSchedule.moveRejected(τ, f̂(μ(s, s')), f(s'));
            τ = CoolingSchedule.repetitionIntervalDone(τ, τ_0, i );
            R_max = ⌊deltaRepetitions · R_max⌋;
      while i < I_max;
if returnBest
      s = s_best;
```

Algorithm 6 ClassicSimulatedAnnealing

ClassicSimpleSimulatedAnnealing $< S, N >$:
$(s, T_{\max}, I_{\max}, \omega, \tau_0, \alpha, b)$

GeneralSimulatedAnnealing $<S, N,$ ClassicExponentialAcceptanceCriterion,
GeometricCooling $< \alpha >, \phi >$
$(s, T_{\max}, I_{\max}, \omega,$ true, $\tau_0, 1, b, 0)$;

Algorithm 7 ThresholdAccepting

ThresholdAccepting $< S, N >$:
$(s, T_{\max}, I_{\max}, \omega, \tau_0, \alpha, b)$

GeneralSimulatedAnnealing $<S, N,$ ClassicThresholdAcceptanceCriterion,
GeometricCooling $< \alpha >, \phi >$
$(s, T_{\max}, I_{\max}, \omega,$ true, $\tau_0, 1, b, 0)$;

Algorithm 8 GreatDeluge

GreatDeluge $< S, N >$:
$(s, T_{\max}, I_{\max}, \omega, \tau_0, \alpha_1, \alpha_2, b)$

GeneralSimulatedAnnealing $<S, N,$ AbsoluteThresholdAcceptanceCriterion,
GreatDelugeCooling $< \alpha_1, \alpha_2 >, \phi >$
$(s, T_{\max}, I_{\max}, \omega,$ true, $\tau_0, 1, b, 0)$;

Algorithm 9 RecordToRecordTravel

RecordToRecordTravel $< S, N >$:
$(s, T_{\max}, I_{\max}, \omega, \tau_0, \alpha, b)$

GeneralSimulatedAnnealing $<S, N,$ AbsoluteThresholdAcceptanceCriterion,
RecordToRecordTravelCooling $< \alpha >, \phi >$
$(s, T_{\max}, I_{\max}, \omega,$ true, $\tau_0, 1, b, 0)$;

Algorithm 10 SimulatedAnnealingJetal

SimulatedAnnealingJetal
< $S, N, AcceptanceCriterion, CoolingSchedule, Reheating$ >
$(s, T_{\max} = \infty, I_{\max} = \infty, \omega =$ false, $returnBest =$ true,
$initialAcceptanceFraction = 0.4, frozenAcceptanceFraction = 0.02,$
$sizeFactor = 16, frozenParameter = 5, numberOfReheatings = 0)$:

$\Omega : (t \geq T_{\max})$ **or** $(i \geq I_{\max})$ **or** (ω)

perform trial annealing run to determine τ_0 such that
    initialAcceptanceFraction of the neighbors are accepted;
$s_{\text{best}} = s$;
for $h = 1$ **to** $numberOfReheatings + 1$
    if $h > 1$
        $\tau_0 = Reheating(\tau_0, \tau_{\text{best}}, \tau)$;
    else
        $\tau_{\text{best}} = \tau_0$;
    $\tau = \tau_0$;
    $frozenCounter = 0$;
    $i = 0$;
    do
        $k_1 = k_2 = 0$;
        $R_{\max} = \lfloor sizeFactor \cdot |N(s)| \rfloor$;
        for $r = 1$ **to** $R_{\max}$
            $i = i + 1$;
            s' = RandomNeighbor< S, N >(s);
            if $AcceptanceCriterion(\tau, \hat{f}(\mu(s, s')), f(s'))$
                if $\hat{f}(\mu(s, s')) > 0$
                    $k_1 = k_1 + 1$;
                else
                    $k_2 = k_2 + 1$;
                $s = s'$;
                if $f(s) < f(s_{\text{best}})$
                    $s_{\text{best}} = s$; $\tau_{\text{best}} = \tau$; $frozenCounter = 0$;
        $\tau = CoolingSchedule.repetitionIntervalDone(\tau, \tau_0, i)$;
        if $((k_1 + k_2)/R_{\max} \leq frozenAcceptanceFraction)$ **or** $(k_1 \equiv 0)$
            $frozenCounter = frozenCounter + 1$;
    while $frozenCounter < frozenParameter$;
if *returnBest*
    $s = s_{\text{best}}$;

4.3.2.3 Tabu Search. The HOTFRAME feature diagram of tabu search is shown in Figure 4.4. Besides S and N, the principal features of tabu search are the tabu criterion and the rule to select neighbors. Moreover, there may be an explicit diversification scheme. We explicitly model the strict tabu criterion (i.e., defining a move as tabu if and only if it would lead to an already traversed neighbor solution), the static tabu criterion (i.e., storing attributes of performed moves in a tabu list of a fixed size and prohibiting these attributes from being inverted), and the reactive tabu criterion according to Battiti and Tecchiolli (1994).

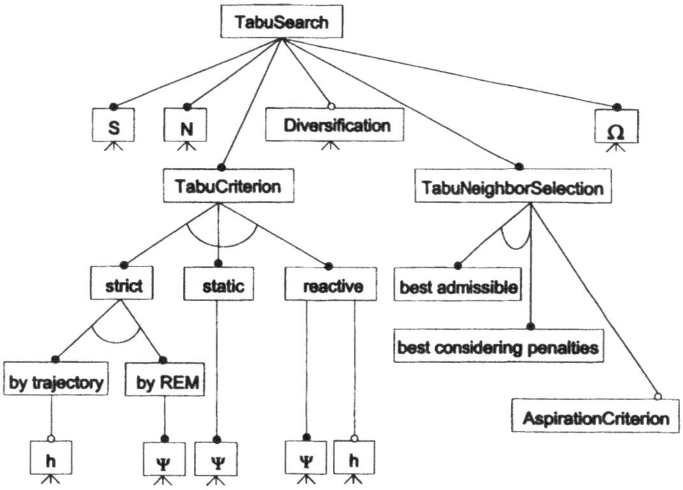

Figure 4.4 Feature Diagram for Tabu Search

The algorithmic commonalities of a tabu search metaheuristic are shown in Algorithm 11. Classic tabu search approaches control the search by dynamically classifying neighbors and corresponding moves as tabu. To implement tabu criteria, one uses information about the search history: traversed solutions and/or attributes of performed moves. Using such information, a tabu criterion defines whether neighbors and corresponding moves are classified as tabu. A move is admissible if it is not tabu or an aspiration criterion is fulfilled. That is, aspiration criteria may invalidate a tabu classification (e.g., if the considered move leads to a neighbor solution with a new best objective function value). The tabu criterion may also signal that an explicit diversification seems to be reasonable. In such a case, a diversification procedure is applied (e.g., a random walk).

The most popular approach to apply the tabu criterion as part of the neighbor selection procedure is by choosing the best admissible neighbor (Algorithm 12). Alternatively, some measure of the tabu degree of a neighbor may be used to compute a penalty value that is added to the move evaluation (Algorithm 13). With regard to the latter option, the tabu criterion provides for each move a tabu degree value (between 0 and 1). Multiplying the tabu degree with a parameter σ results in the penalty value.

The considered tabu criteria are defined in Algorithms 14–17. In each case, the tabu memory is modeled by state variables using simple container data structures such

Algorithm 11 TabuSearch

TabuSearch
  $<S, N,$ *TabuCriterion*, *TabuNeighborSelection*, *Diversification* $>$
  $(s, T_{\max} = \infty, I_{\max} = \infty, \omega = \text{false})$:

$\Omega : (t \geq T_{\max})$ **or** (ω)

$s_{\text{best}} = s$;
$i = 0$;
while $i < I_{\max}$
  $i = i + 1$;
  s' = *TabuNeighborSelection*$< S, N, TabuCriterion >(s)$;
  TabuCriterion.add$(\mu(s, s'), i)$;
  $s = s'$;
  TabuCriterion.add(s, i);
  if $f(s) < f(s_{\text{best}})$
    $s_{\text{best}} = s$;
  if *TabuCriterion.escape*()
    Diversification(s);
$s = s_{\text{best}}$;

Algorithm 12 BestAdmissibleNeighbor

BestAdmissibleNeighbor $< Aspiration >< S, N, TabuCriterion > (s)$:

if $\exists j \in \{1, \ldots, |N(s)|\}$: (**not** *TabuCriterion.tabu*$(\mu(s, n_j(s)), n_j(s))$)
    or (*Aspiration*$(\mu(s, n_j(s)), TabuCriterion)$)
  $j = \text{argmax}\{\hat{f}(\mu(s, n_j(s))) \mid j = 1, \ldots, |N(s)|,$
    (**not** *TabuCriterion.tabu*$(\mu(s, n_j(s)), n_j(s))$)
    or (*Aspiration*$(\mu(s, n_j(s)), TabuCriterion)$)$\}$;
  return $n_j(s)$;
else
  return RandomNeighbor $< S, N >(s)$;

as lists or sets, which are parameterized by the type of the respective objects. If lists are defined as having a fixed length, objects are inserted in a first-in first-out manner. Not all tabu criteria implement all functions. For instance, most of the tabu criteria do not possess means to detect and signal situations when an explicit diversification seems to be reasonable.

The strict tabu criterion can be implemented by storing information about all traversed solutions (using the function h). In Algorithm 14, we do not apply frequency

Algorithm 13 BestNeighborConsideringPenalties

BestNeighborConsideringPenalties $< Aspiration, \sigma >$
$< S, N, TabuCriterion > (s)$:

if $Aspiration(\mu(s, n_j(s)), TabuCriterion)$
  $j = \text{argmax}\{\hat{f}(\mu(s, n_j(s))) \mid j = 1, \ldots, |N(s)|\}$;
else
  $j = \text{argmax}\{\hat{f}(\mu(s, n_j(s)))$
    $+ \sigma \cdot TabuCriterion.tabuDegree(\mu(s, n_j(s)), n_j(s))$
    $\mid j = 1, \ldots, |N(s)|\}$;
return $n_j(s)$;

Algorithm 14 StrictTabuCriterionByTrajectory

StrictTabuCriterionByTrajectory $< S, h >$:

State: Set$< S_h >$ *trajectory*;

add(s, ϕ) :
  trajectory.insert$(h(s))$;

tabu(ϕ, s') :
  if $s' \in$ *trajectory*
    return true;
  else
    return false;

tabuDegree(ϕ, s') :
  if $s' \in$ *trajectory*
    return 1;
  else
    return 0;

or recency information to compute a relative tabu degree but simply use an absolute tabu classification. In principle, the strict tabu criterion is a necessary and sufficient condition to prevent cycling in the sense that it classifies exactly those moves as tabu that would lead to an already traversed neighbor. However, as one usually applies a non-injective ("approximate") function h, moves might unnecessarily be set tabu (when "collisions" occur); see Woodruff and Zemel (1993).

As an alternative implementation of the strict tabu criterion, the reverse elimination method (REM, Algorithm 15) exploits logical interdependencies among moves, their attributes and respective solutions (see Glover (1990), Dammeyer and Voß (1993), Voß (1993a), Voß (1995), Voß (1996)). A *running list* stores the sequence of the attributes of performed moves (i.e., the created and destroyed solution attributes). In every iteration, one successively computes a *residual cancellation sequence* (RCS), which includes those attributes that separate the current solution from a formerly traversed solution. Every time when the RCS exactly constitutes a move, the corresponding inverse move must be classified as tabu (for one iteration). It should be noted that it is not quite clear how to generally implement the REM tabu criterion for multi-attribute moves in an efficient way. For this reason, the REM component of HOTFRAME is restricted to single attribute moves that may be coded by natural numbers.

The strict tabu criterion is often too weak to provide a sufficient search diversification. We consider two options to strengthen the tabu criterion of the REM. The first alternative uses the parameter *tabuDuration* to define a tabu duration longer than one iteration. The second uses the parameter *rcsLengthParameter* to define a threshold for the length of the RCS, so that all possibilities to combine (subsets of) the attributes of the RCS of a corresponding maximum length as a move are classified as tabu.

The static tabu criterion is defined in Algorithm 16. The parameter Ψ represents the decomposition of moves in attributes. The parameter κ defines the capacity of the tabu list (as the number of attributes). The parameter ν defines the number of attributes of a move, for which there must be inverse correspondents in the tabu list to classify this move as tabu. Furthermore, ν is also the reference value to define a proportional tabu degree.

Algorithm 17 shows the mechanism of the tabu criterion for reactive tabu search. With regard to the adaptation of the length of the tabu list, a history stores information about traversed moves. This includes the iteration of the last traversal and the frequency. The actual tabu status/degree is defined in the same way as for the static tabu criterion using the parameter ν. The adaptation of the tabu list length is computed in dependence of the parameters δ_1 and δ_2. When a re-traversal of a solution occurs, the list is enlarged considering a maximum length κ. Depending on an exponentially smoothed average iteration number between re-traversals (using a parameter θ), the length of the tabu list is reduced if there has not been any re-traversal for some time. If there are ζ_1 solutions that each have been traversed at least ζ_2 times, the apparent need for an explicit diversification is signalled.

The parameterization of TabuSearch and of the used modules enables various possibilities to build specific tabu search heuristics. For example, Algorithm 18 (StrictTabuSearch) encodes the simplest form of strict tabu search: All traversed solutions are stored explicitly (*id* represents the identity function), which means that they are classified as tabu in the subsequent search process. Algorithm 19 (REMpen) shows the enhanced reversed elimination method in combination with the use of penalty costs. Static tabu search is shown in Algorithm 20. Algorithm 21 defines reactive tabu search in combination with the use of RandomWalk as diversification mechanism (setting most of the parameters to reasonable default values).

Algorithm 15 REMTabuCriterion

REMTabuCriterion $< S, N, \Psi >$
$(tabuDuration, rcsLengthParameter)$:

State: List$< \Psi >$ *runningList*;
Set$<$ (Set$< \Psi >$, Integer) $>$ *tabuList*;

add(μ, i) :
for $j = 1$ **to** $|\psi(\mu)|$
runningList.append$(\psi_j(\mu))$;
$RCS = \emptyset$;
for $j = |runningList|$ **downto** 1
if $\overline{runningList[j]} \in RCS$
$RCS = RCS \setminus \overline{runningList[j]}$;
else
$RCS.insert(runningList[j])$;
if $|RCS| \leq rcsLengthParameter$
$tabuList.append(\overline{RCS}, i + tabuDuration)$;

tabu(μ, ϕ) :
if $\psi(\mu) \in tabuList$ **and** μ was set tabu not longer than
tabuDuration iterations before
return true;
else
return false;

tabuDegree(μ, ϕ) :
if $\psi(\mu) \in tabuList$ **and** μ was set tabu not longer than
tabuDuration iterations before
return (remaining tabu duration of μ)$/$*tabuDuration*;
else
return 0;

Algorithm 16 StaticTabuCriterion

StaticTabuCriterion $< S, N, \Psi >$ (ν, κ) :

State: List$< \Psi, \kappa >$ *tabuList*;

add(μ, ϕ) :
    for $j = 1$ **to** $|\psi^+(\mu)|$
        tabuList.append$(\psi_j^+(\mu))$;

tabu(μ, ϕ) :
    $k = 0$;
    for $j = 1$ **to** $|\psi^-(\mu)|$
        if $\psi_j^-(\mu) \in$ *tabuList*
            $k = k + 1$;
    if $k \geq \nu$
        return true;
    else
        return false;

tabuDegree(μ, ϕ) :
    $k = 0$;
    for $j = 1$ **to** $|\psi^-(\mu)|$
        if $\psi_j^-(\mu) \in$ *tabuList*
            $k = k + 1$;
    return $\min\{k/\nu, 1\}$;

Algorithm 17 ReactiveTabuCriterion

ReactiveTabuCriterion $< S, N, \Psi, h >$ $(\nu, \delta_1, \delta_2, \kappa, \zeta_1, \zeta_2, \theta)$:

State: List$< \Psi >$ *tabuList*;
  Set$< (S_h,$ Integer, Integer$) >$ *trajectory*;
  movingAverage = *lastReaction* = 0;

add(μ, ϕ) :
  for $j = 1$ **to** $|\psi^+(\mu)|$
    tabuList.append$(\psi_j^+(\mu))$;

add(s, i) :
  if $h(s) \in$ *trajectory* (with corresponding iteration k)
    extend length l of *tabuList* to $\min\{\max\{\lceil l \cdot \delta_1 \rceil, l + \delta_2\}, \kappa\}$;
    lastReaction $= i$;
    movingAverage $= \theta \cdot (i - k) + (1 - \theta) \cdot$ *movingAverage*;
    update iteration to i and increment frequency of $h(s)$;
    if there are ζ_1 solutions in *trajectory*
      that have been traversed ζ_2 times or more
        trigger escape;
  else
    trajectory.insert$(h(s), i, 1)$;
    if $i -$ *lastReaction* $>$ *movingAverage*
      reduce length l of *tabuList* to $\max\{\min\{\lfloor l/\delta_1 \rfloor, l - \delta_2\}, \delta_2\}$;
      lastReaction $= i$;

tabu(μ, ϕ) :
  $k = 0$;
  for $j = 1$ **to** $|\psi^-(\mu)|$
    if $\psi_j^-(\mu) \in$ *tabuList*
      $k = k + 1$;
  if $k \geq \nu$ **return** true;
  else return false;

tabuDegree(μ, ϕ) :
  $k = 0$;
  for $j = 1$ **to** $|\psi^-(\mu)|$
    if $\psi_j^-(\mu) \in$ *tabuList*
      $k = k + 1$;
  return $\min\{k/\nu, 1\}$;

Algorithm 18 StrictTabuSearch

StrictTabuSearch $< S, N > (s, T_{\max}, I_{\max}, \omega)$:

TabuSearch $<S, N,$ StrictTabuCriterionByTrajectory $< S, id >,$
  BestAdmissibleNeighbor $< \phi >, \phi >$
  $(s, T_{\max}, I_{\max}, \omega);$

Algorithm 19 REMpen

REMpen $< S, N, \Psi >$
  $(s, T_{\max}, I_{\max}, \omega, \mathit{tabuDuration}, \mathit{rcsLengthParameter}, \sigma)$:

TabuSearch $<S, N,$
  REMTabuCriterion
   $< S, N, \Psi > (\mathit{tabuDuration}, \mathit{rcsLengthParameter}),$
  BestNeighborConsideringPenalties $< \phi, \sigma >, \phi >$
  $(s, T_{\max}, I_{\max}, \omega);$

Algorithm 20 StaticTabuSearch

StaticTabuSearch $< S, N, \Psi > (s, T_{\max}, I_{\max}, \omega, \nu, \kappa)$:

TabuSearch $<S, N,$ StaticTabuCriterion $< S, N, \Psi > (\nu, \kappa),$
  BestAdmissibleNeighbor $< \phi >, \phi >$
  $(s, T_{\max}, I_{\max}, \omega);$

Algorithm 21 ReactiveTabuSearch

ReactiveTabuSearch $< S, N, \Psi, h >$
  $(s, T_{\max}, I_{\max}, \omega, \nu, \kappa, \mathit{perturbations})$:

TabuSearch
  $< S, N,$ ReactiveTabuCriterion $< S, N, \Psi, h > (\nu, 1.2, 2, \kappa, 3, 3, 0.5),$
  BestAdmissibleNeighbor $< \phi >,$
  RandomWalk$<S, N>(\phi, \infty, \mathit{perturbations}, \omega,$ false$) >$
  $(s, T_{\max}, I_{\max}, \omega);$

4.4 DESIGN

Design involves modeling the principal abstractions defined in the domain analysis (Section 4.3) as artefacts, which may more or less be directly implemented as software (modules). This especially concerns the design of a framework architecture, which defines the interplay between software components by means of interface specifications. In particular, such a framework specifies (some of) the control flow of a system. Besides the architecture, a framework defines adaptable software components, which encapsulate common domain abstractions. To be adaptable with respect to the variabilities of different applications in the considered domain, a framework provides variation points, which allow modifying or completing certain parts of the implementation.

In this section, we give an overview of framework architecture. The primary design decisions are about mechanisms that define the interplay between metaheuristics and problem-specific components. These mechanisms involve advanced concepts to adapt and combine components, which requires adequate implementation mechanisms in the programming language employed.

The main constructs to implement adaptable (polymorphic) software components are object-oriented inheritance and genericity by type parameters. Inheritance allows adapting classes by deferring the specific implementation of (some of) the class operations to specialized classes, which inherit the common data structure and functionality from respective general (base) classes. Type parameterization means that methods and classes may have some kind of type "placeholders", which allow for specialization (instantiation) by fixing the type parameters by specific types/classes. In both cases, general classes serve as starting points to modify and extend the common data structures and algorithms identified in the analysis by the specific needs of the application under consideration. The widely used programming language C++ provides both kinds of adaptation constructs (inheritance as well as genericity), as well as enabling run-time efficient software implementations. For a detailed exposition and discussion of software construction, in particular object-oriented programming, inheritance, and genericity, we refer, in general, to Meyer (1997) and, in particular for C++, to Stroustrup (1997).

4.4.1 Basic Design Concepts

In the following, we describe the basic design concepts of HOTFRAME. In some cases, we simplify a little bit to explain our approach. The understanding of the basic design enables a concise description of the architecture and the components later on.

4.4.1.1 Genericity of Metaheuristics. The primary design decision of HOTFRAME is about the interplay between generic metaheuristics components and problem-specific components. The features common to metaheuristics are captured in metaheuristic algorithms (i.e., corresponding software components) as shown in Section 4.3. These algorithms operate on problem-specific structures (C_P, see Section 4.3.1), in particular the solution space and the neighborhood. The natural way to implement this kind of polymorphism is by means of type parameterization. This concept refers to the generic programming paradigm, where algorithms are variable with respect to the structures (types) on which they operate. In C++, type parameterization is implemented in a static manner, as the instantiation, the fixing of type parameters of so-called template classes, is done at compile time.

We illustrate this basic idea by means of the example of a template class SteepestDescent; see the UML class diagram shown in Figure 4.5. (In fact, the steepest descent algorithm is generated as a specialization of an iterated local search software component.) The actual search algorithm is implemented by the operation (member function) search, which transforms an initial solution (passed to the operation). We use classes (instead of singular methods) to represent metaheuristics to enable storing search parameters or the state of a (not yet completed) search process. This allows treating an algorithm as a dynamic object (an instance of a class with state information), which may be constructed, used, and stored. (Furthermore, using classes to represent algorithms allows to adapt these algorithms by means of inheritance. However, to keep the design straightforward, we do not use this feature for the methods implemented in HOTFRAME.)

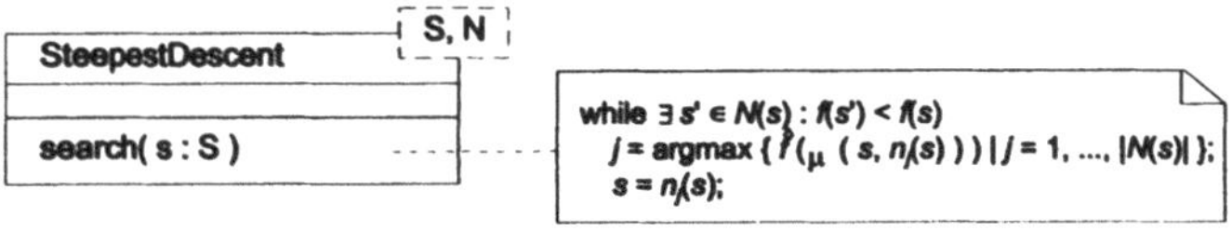

Figure 4.5 Class Diagram for SteepestDescent (Simplified)

The template parameters S and N are some sort of placeholders for the solution space and the neighborhood structure. Specific solution spaces and neighborhood structures must be implemented as classes with operations that conform to functional requirements that are derived from the analysis discussed in Section 4.3 (such interfaces are discussed below). Constructing a specialized class from a template class means defining for each template parameter a specific instantiation. This results in a class that is adapted with regard to a problem-specific configuration. That is, we have specialized a metaheuristic (template class) as a problem-specific heuristic (class). Henceforth, by calling some class constructor method (not shown in the figure) with respective parameter values (e.g., termination criteria), we can construct specific heuristic objects, which can be applied to transform an initial solution.

4.4.1.2 Variabilities of Metaheuristics. In addition to the problem-specific adaptation C_P, metaheuristics are variable with respect to the configuration C_H that is specific to the metaheuristic. For example IteratedLocalSearch (see Figure 4.2 or Algorithm 1) is variable regarding the neighbor selection rule and the diversification. These algorithmic abstractions may also be treated as template parameters of the generic metaheuristic class. On the other hand, the termination criterion concerns simple numeric parameters. That is, the termination criterion is not modeled by template parameters but by simple data parameters and corresponding class data elements. The same applies for the parameter that defines whether the algorithm should return the best or the last traversed solution. This results in the UML class diagram as shown in Figure 4.6.

Figure 4.6 Class Diagram for IteratedLocalSearch (Simplified)

Components that are used to instantiate template parameters often have type parameters by themselves. For example, the component BestPositiveNeighbor, which is used to instantiate the template parameter NeighborSelection, is parameterized by the solution space and the neighborhood structure; see Figure 4.6. To denote the partial specialization of the generic class IteratedLocalSearch as (a still generic class) SteepestDescent, we use the UML stereotype bind, which means that we fix two out of four type parameters of IteratedLocalSearch.

Since different metaheuristics certainly possess different (static as well as dynamic) parameters, we have to define and implement for each of the general metaheuristics formulated in Section 4.3.2 (IteratedLocalSearch, GeneralSimulatedAnnealing, TabuSearch) a corresponding metaheuristic component (template class). As discussed for SimulatedAnnealingJetal, there can also be distinct modifications of a metaheuristic that result in specialized components, which are not directly derived from the general metaheuristic. That is, it does not seem reasonable to follow some "one component fits all" approach, since there will always be some distinct modifications, which one has not thought about when defining and implementing the "general" metaheuristic. In particular, one should not try to capture in one large (complicated) component all kinds of variation points. HOTFRAME allows to define such new metaheuristic components, which, of course, may reuse problem-specific components and other elementary components.

4.4.1.3 Neighborhood Traversal. The iterative traversal of the neighborhood of the current solution and the selection of one of these neighbors is the core of metaheuristics that are based on the local search approach. A generic implementation of different neighbor selection rules requires flexible and efficient access to the respective neighborhood. The basic idea of generic programming – algorithms operate on abstract structures – conforms to the neighborhood traversal task and enables an efficient and flexible design.

The different neighbor selection rules imply a few basic requirements, which are illustrated in Figure 4.7 (adaptation of Figure 4.1). The traversal of the neighborhood

$N(s)$ of a solution s conforms to a sequence of moves μ that correspond to neighbor solutions $n_i(s)$, $i = 1, \ldots, |N(s)|$.

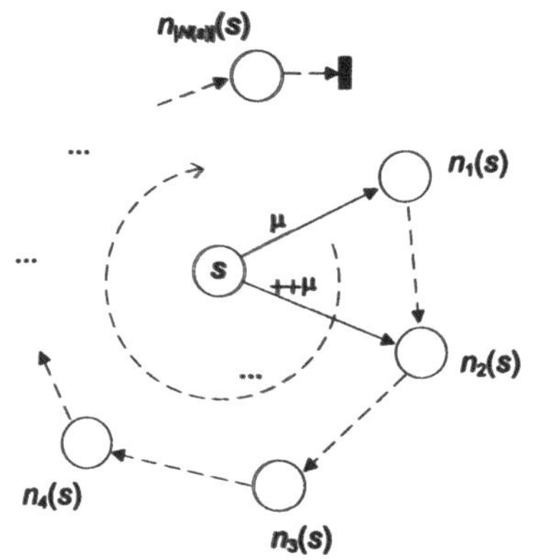

Figure 4.7 Neighborhood Traversal

The implementation of a neighbor selection rule such as BestNeighbor implies the need for the following basic functionality:

- Construction of a move to the first neighbor $n_1(s)$
- Increment of a valid move to the subsequent neighbor (in accordance with C++ represented by an increment operator "++")
- Computation of the move evaluation $\hat{f}(\mu)$
- Check for validity of a move

In principle, this functionality is also sufficient to implement the other neighbor selection rules. However, with respect to run-time efficiency we may need to directly construct a move to a random neighbor for metaheuristics such as simulated annealing. Otherwise, to construct a random neighbor, we would have to construct all moves and to select one of these by chance. This is obviously not practical for metaheuristics that require the efficient construction of a random move in each iteration. So we also require the following functionality:

- Direct construction of a move to a random neighbor

However, only suitable neighborhood structures allow an efficient implementation of the selection of a random move, if one requires to apply a uniform probability distribution. There is often a trade-off between the run-time of the random move construction and the quality of the probability distribution. So one generally may need to cope with non-uniform move selection for certain neighborhood structures.

The functionality of the neighborhood traversal largely conforms to the *iterator pattern*, which is about a sequential traversal of some (container) structure without knowing about the specific representation of this structure. That is, the iterator pattern explicitly separates structures from algorithms that operate on these structures; see Gamma et al. (1995). Accordingly, we may refer to respective neighborhood classes as neighborhood iterator classes. The design of these classes is based on the concept of the iterator classes of the *Standard Template Library*; see Musser and Saini (1996). The solution class takes the virtual role of a container class (solution space). Moves as objects of a neighborhood iterator class store a reference to a particular solution and transformational information with respect to the neighbor solution where they point to. For reasons of efficiency, it is apparently not reasonable to physically construct each neighbor solution object but only the neighbor solution that is eventually selected as the new current solution.

To illustrate the neighborhood traversal, we show a (simplified) generic method that implements the selection of the best neighbor (C++ code):

```
template <class S, class N>
N BestNeighbor( const S& s )
{
  N move = N( &s, FirstNeighbor );
  N best = move;
  if ( move.isValid() )
    ++move;
  while ( move.isValid() )
  {
    if ( *best < *move )
        best = move;
    ++move;
  }
  return best;
};
```

This template method is generic with respect to the solution space S and the neighborhood structure N, which are modeled as type parameters. For the dynamic parameter s, the first neighbor of this solution is constructed (i.e., the corresponding move). Then, in each iteration, the next move is constructed (increment operator), checked for validity, and compared with the best move obtained so far.

4.4.1.4 Commonalities and Variabilities of Problem-Specific Abstractions. The principal problem-specific abstractions (solution space, neighborhood structure, ...) are modeled by corresponding interfaces. These interfaces define the requirements that must be fulfilled by problem-specific components to be applicable as realizations of respective type parameters of metaheuristic components. Such interfaces can be modeled as classes without data elements (by using the UML stereotype interface). Figure 4.8 shows simplified interfaces for solution space template classes and neighborhood template classes. These interfaces are generic: the solution space class depends on a problem class and a neighborhood class; the neighborhood class depends on a solution space class. While the latter dependence is apparent, one may argue that a solution space class should not depend on a neighborhood class. However, for principal reasons – only the solution class knows about the objective function while the neighborhood component defines the transformations – with regard to run-time efficiency, this is indeed a sensible model. This tight relationship between solution classes and neighborhood classes will be discussed on page 116.

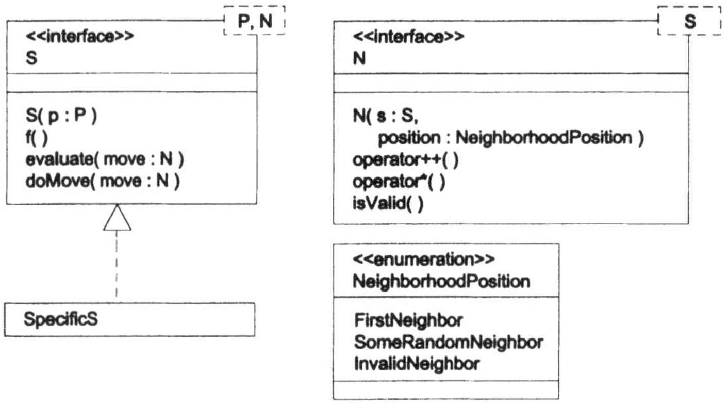

Figure 4.8 Interfaces for Solution Space Classes and Neighborhood Classes (Simplified)

The solution space interface defines the basic functionality that must be implemented by problem-specific classes: construction of a solution (given a problem instance), objective function computation, computation of the evaluation of a given move, modification of the solution according to a given move. The requirements for the neighborhood interface are in accordance with the discussion about the neighborhood traversal. That is, we need operations for the construction of a neighbor (i.e., the corresponding move) in dependence of a parameter that specifies which move is to be constructed (e.g., first versus random), for the increment to the next neighbor, for the evaluation of a move (star/dereference operator), and for the check of the validity of the move.

The relationship between the class SpecificS and the interface S, which is shown in Figure 4.8, denotes that the latter one is an implementation of the requirements set by the former one. HOTFRAME is based on such a clear separation between types (requirements defined by interfaces) and classes, which implement respective requirements.

An analysis of different types of problems shows that there are often quite similar solution spaces and neighborhood structures. So it seems reasonable to model respective commonalities with regard to data structures and algorithms by inheritance hierarchies; see Figure 4.9. This enables an (optional) reuse of the implementation of some common problem-specific structures. The objective function value of the current solution (evaluation) is obviously a common data element of the most general abstract solution space class Generic_S. There are two specializing classes (BV_S for the bit-vector solution space and Perm_S for the permutation solution space) that inherit this data element (as well as the obligation to implement the interface S) from the general class. BV_S and Perm_S, which add some specific data elements and operations, are still abstract (i.e., one cannot instantiate objects of these classes), because they lack an operation that computes the objective function.

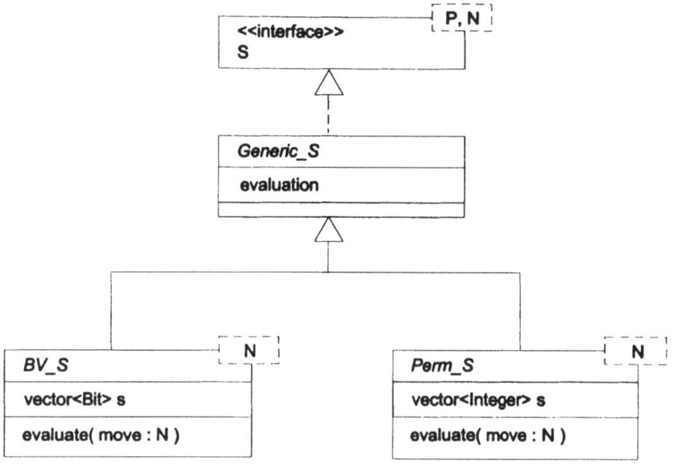

Figure 4.9 Inheritance Hierarchy of Solution Space Classes (Simplified)

4.4.2 Architecture

Building on the above discussion of the basic design ideas, in this section we describe the framework architecture (i.e., the framework components and their interplay) in more detail. For modeling variabilities we use the UML with some extensions. Unfortunately, "UML provides very little support for modelling evolvable or reusable specifications and designs" (Mens et al. (1999), p. 378), so we sporadically extend the notation if necessary.

4.4.2.1 Basic Configuration Mechanisms. Fixing the variable features of a reusable component may be called configuration. In the following, we describe the basic configuration mechanisms of HOTFRAME.

In accordance with generic programming, static configuration means fixing template parameters to define the primary behavior of reusable components (generic classes). Different metaheuristics have different options for configuration, which would lead to heterogeneous template parameter sets for different metaheuristic components. This complication can be avoided by using so-called *configuration components* that encapsulate parameter sets. That is, a configuration component defines, in accordance with respective requirements, a set of static parameters. These parameters mostly constitute type information, modeled as features of class interfaces. Since the UML provides no specific constructs to directly model requirements on such configuration components (the use of the Object Constraint Language (OCL) is not reasonable for our needs), we introduce a stereotype static interface to model respective type requirements (in a sense analogous to the common interface stereotype). By convention, we mark configuration components by a leading C, while denoting requirements by an R.

A basic kind of variability, which may concern all components, regards the numeric data types. We distinguish between a (quasi-)continuous type T (e.g., for cost data)

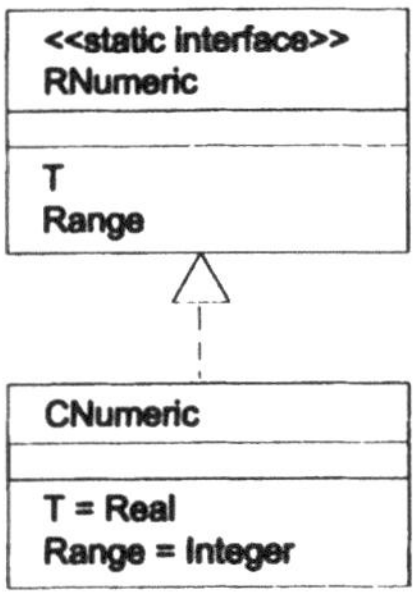

Figure 4.10 Realization of Requirements RNumeric by a Configuration Component CNumeric

and a discrete type Range (e.g., for counting objects). Both data types are required to be symmetric with respect to the representation of positive and negative values. For example, in C++, one typically defines T as float or double, and Range as int. The elementary configuration of components by T and Range is encapsulated in a configuration component with a static interface that has to conform to RNumeric as shown in Figure 4.10. CNumeric is a typical realization of such a configuration component, which may be implemented in C++ as follows:

```
struct CNumeric
{
  typedef float T;
  typedef int Range;
};
```

All problem-specific components require a configuration component that implements RNumeric. On top of this basic condition, the essential use of configuration components is to capture problem-specific type information of metaheuristic components. The specific requirements of metaheuristic components are defined in Section 4.4.2.3. In the following, we exemplify this concept for a component SteepestDescent.

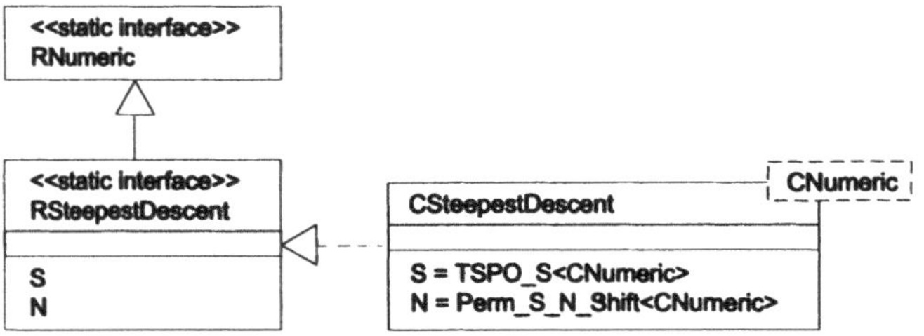

Figure 4.11 Configuration Component CSteepestDescent

Figure 4.11 illustrates modeling the requirements on the respective static interface. RSteepestDescent inherits the requirements of RNumeric, so CSteepestDescent has to define, in addition to S and N, the basic numeric types T and Range. This can be implemented in a modular way by defining CSteepestDescent as a template class where the template parameters model the numeric configuration. That is, CSteepestDescent actually deploys a hierarchical parameterization of configuration components by other configuration components. To simplify the presentation, we assume an implicit transfer of type definitions of configuration components that are used as template parameters. The components that are used in Figure 4.11 to define S and N are described in Section 4.4.2.2 and Section 4.6.2. The actual implementation of CSteepestDescent in C++ requires an explicit "transfer" of the defined types:

```
template <class CNumeric>
struct CSteepestDescent
{
  typedef typename CNumeric::T T;
  typedef typename CNumeric::Range Range;
  typedef TSPO_S<CNumeric> S;
  typedef Perm_S_N_Shift<CNumeric> N;
};
```

We illustrate the hierarchical configuration of metaheuristic components for IteratedLocalSearch (see Algorithm 1, p. 89). As described above, the configuration of metaheuristics can be decomposed in a problem-specific configuration C_P and a metaheuristic-specific configuration C_H. Accordingly, we may define a corresponding requirements hierarchy; see Figure 4.12. RpIteratedLocalSearch defines the requirements for the problem-specific configuration of IteratedLocalSearch. RIteratedLocalSearch defines the additional metaheuristic-specific requirements. The configuration component CpIteratedLocalSearch realizes RpIteratedLocalSearch in the same way as CSteepestDescent realizes RSteepestDescent (see Figure 4.11). On top of this, CIteratedLocalSearch realizes the requirements of RIteratedLocalSearch by using CpIteratedLocalSearch and additionally defining the neighbor selection rule. In Section 4.4.2.3, we shall describe the application of this design approach in more detail for different metaheuristic components.

A disadvantage of the template approach, at least when using C++, is that the template-based configuration is fixed at compile-time. Behavioral variability requirements may also be modeled by defining data elements with class-scope (in C++, such data element are called "static"). This approach is not appropriate if one wants to enable constructing objects with different state information (e.g., heuristics of the same kind with different parameter settings with respect to the termination criterion). However, class-scope is reasonable for configuration options with regard to modules of metaheuristics. Such components can indeed be parameterized by using data elements with class-scope (which are set according to respective data elements of the configuration component). This design, which is illustrated in Figure 4.13 for the cooling schedule GeometricCooling, enables the dynamic variation of respective configuration parameters at run-time. To simplify the presentation, we do not explicitly model the requirements of such components on the configuration component in the diagrams, but we assume respective requirements as implicitly defined. For example, we implic-

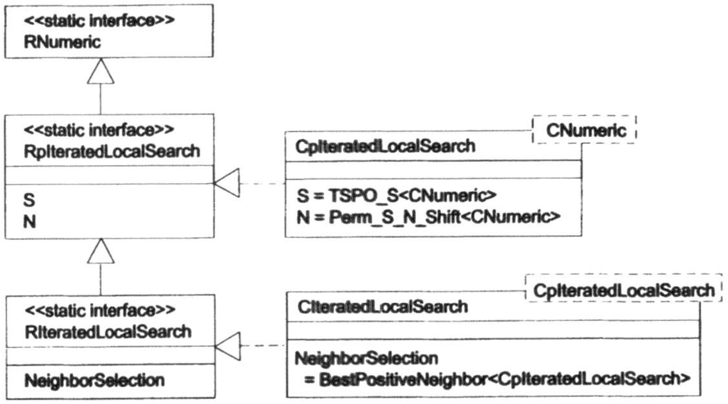

Figure 4.12 Configuration Component CIteratedLocalSearch

itly assume the configuration component CSimulatedAnnealing to be subject to the requirement of defining a numeric data element alpha.

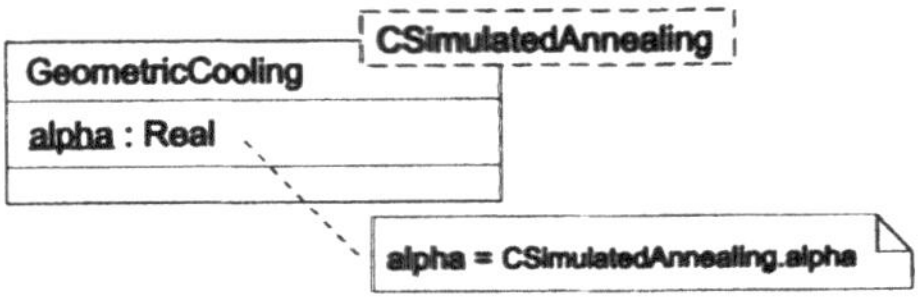

Figure 4.13 Static Parameterization of Components by Numeric Data Elements with Class-Scope

The design provides, from the application point-of-view, a flexible, efficient, and straightforward mechanism to construct components with a special functionality by a composition of adaptable elementary modules (see below).

The natural mechanism to dynamically configure objects is by plain data elements of classes, which are initialized in connection with object construction (in C++, by using constructor parameters). The specification of the termination criterion is the most obvious kind of a dynamic configuration. In accordance with the specification of the metaheuristics IteratedLocalSearch, GeneralSimulatedAnnealing, SimulatedAnnealingJetal, and TabuSearch, the dynamic parameterization conforms to the interfaces of the Algorithms 1, 5, 10, and 11, respectively.

In addition to the usual termination criteria (e.g., iteration number and run-time), we define, for each metaheuristic component, an object parameter that represents the asynchronous termination criterion ω (see p. 88). The plain abstract interface of a corresponding base class, which is called OmegaInterface, is shown in Figure 4.14. The class OmegaInterface is parameterized by the solution space. Metaheuristic components provide the interface with the current solution after each iteration (by calling the operation newSolution(s : S)). Furthermore, metaheuristics call the operation omega() after each iteration – with the metaheuristic being terminated when the operation returns true. For example, one may want to terminate the search process when an "acceptable" solution has been obtained. In general, this design enables the implementation of external termination criteria in online settings.

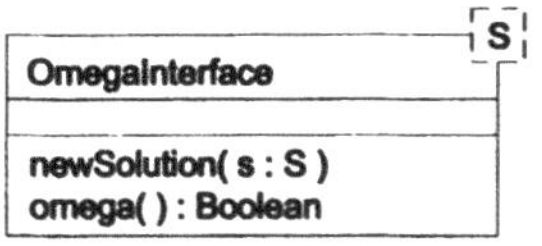

Figure 4.14 Base Class OmegaInterface

Additional parameters of metaheuristics are also modeled as data elements and corresponding dynamic parameters of constructors. This concerns both simple numeric parameters (such as the initial temperature of simulated annealing) and more complex kinds of a configuration. In particular, the tabu criterion of Algorithm 11 (see p. 96) is implemented as an object parameter, which may be explained as follows. The tabu criterion depends on actual state information (search history). The same is true for the diversification method (as a specific instance of some metaheuristic). So both variation points (TabuCriterion, Diversification) are implemented as dynamic object parameters, which allows a flexible use of such objects at run-time. (On the contrary, the static variation point *TabuNeighborSelection* is modeled as a template parameter; see p. 133.)

Variable requirements with regard to the introspection of the search process must also be modeled in a flexible and efficient way. We apply an extensible class hierarchy that represents the different kinds of introspection (e.g., a complete documentation of the move selection in every iteration versus the return of only the final solution). In Section 4.4.2.4, we describe the interface of a base class, which defines a set of operations, which are called by metaheuristic components to convey information about the search process. By deriving classes from this base class and implementing the operations in the needed form, one can model specific introspection requirements. Metaheuristics are called with instances of these classes as dynamic object parameters to implement the introspection interface.

4.4.2.2 Problem-Specific Components. In this section, we define problem-specific components in accordance with the algorithmic specifications from Section 4.3.1. In particular, we describe the component interfaces, which implicitly establishes the basic form of the interplay among components.

We do not need to define (abstract) base classes, since the problem-specific components are used to statically configure metaheuristic components by means of template parameters. Nevertheless, due to implementation reuse, it is possible to model commonalities of problem-specific components by inheritance hierarchies, which allows reuse of common data structures and algorithms (see pp. 122). Such inheritance hierarchies simplify the framework application for suitable problem types, while they do not restrain the possibilities for problem-specific adaptations (due to their optional character).

The following overview especially refers to problem-specific components and their interdependencies. Specific realizations of such components are by convention marked by a leading "X_", where X refers to the problem type that is represented. We assume that the problem-specific abstractions introduced in Section 4.3.1 are implemented by the following components:

- Problem P $\rightarrow$ Problem component X_P
- Solution space S $\rightarrow$ Solution component X_S
- (Hash-)Function h $\rightarrow$ Solution information component X_S_I
- Neighborhood N_S $\rightarrow$ Neighborhood component X_S_N
- Solution attribute out of Ψ $\rightarrow$ Attribute component X_S_A

The intended application of metaheuristics implies the need to implement a respective subset of the problem-specific components according to corresponding interfaces. While the components X_P and X_S must be implemented for all kinds of applications, one needs no neighborhood component for evolutionary methods (without local search hybridization). Components X_S_I and X_S_A are only required for particular tabu search methods.

The class diagram shown in Figure 4.15 models the typical relations between problem-specific components (neglecting data structures, operations, and template parameters). The basic dependencies may be classified as usage relationships (use-stereotype) and derivation relationships (derive-stereotype). Furthermore, there are explicit structural references that represent direct associations between objects of respective classes. The suitability of a problem-specific component as an element of a configuration component only depends on fulfilling the requirements of the interfaces that are described later on. In the class diagram, this is indicated by respective realization relationships.

The solution component X_S uses the problem component X_P to access the problem data. Accordingly, we need a reference from a solution object to the respective problem object. The information that is represented by X_S_I is derived from the data of a respective solution object. Since the state of solution objects is transient we indeed need X_S_I objects to capture such information. The same argument applies to the attribute component X_S_A. In general, attributes can be derived both from solution objects and from move (neighbor) objects. With regard to the latter, we have to distinguish between ("created") *plus attributes* and ("destroyed") *minus attributes*.

The neighborhood component X_S_N models the neighborhood traversal as described in Section 4.4.1.3. In general, the semantics of X_S_N depends on X_S, as one needs to know about the objective function (implemented as part of X_S) to be able to evaluate moves. On the contrary, X_S depends on X_S_N, because the execution of a move to a neighbor alters the solution object (X_S implements the solution data structures while X_S_N models the transformational information).

The tight relationship between, in particular, X_S and X_S_N requires to model respective interdependencies in the types of interface parameters of specific operations. For example, the solution component includes an operation that executes a move; the

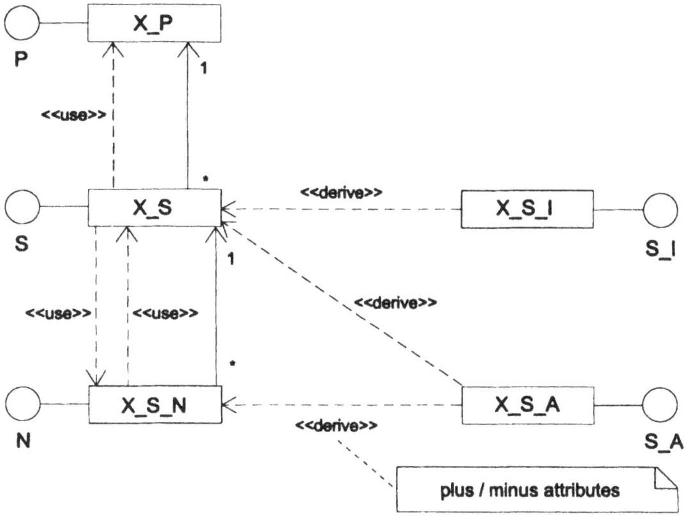

Figure 4.15 Basic Dependencies Between Problem-Specific Components

respective interface must reflect that we may have different types of neighborhoods. Because of the lack of an appropriate UML notation, we call such operations *abstract* (with italicization of respective type identifiers). (Under some restrictions, such a variability requirement could be implemented in C++ by the member template construct (i.e., member functions with separate type parameters), which would provide a modular, efficient and type-safe implementation. As shall be explained below, we actually cannot use this mechanism but have to rely on basic code supplements.)

Problem Component. A problem type P is implemented by a corresponding problem component P. In dependence on the considered type of problem, problem data is modeled by specific data structures (e.g., cost vectors). According to the object-oriented concept of encapsulation of implementation details, the component interface defines the access to the data of problem instance objects. Appropriate constructors must be defined. For example, constructor parameters may point to an input stream, which provides problem data in a specific text format. With regard to a random generation of problem instances, constructor parameters may define the respective probability distribution. The problem component may also serve as an online proxy that connects the metaheuristic to an application system or a data base that comprise the problem instance. In general, respective constructor and access operations depend on the considered application, so we only require one common interface element for all problem components: There should be a serialization of the problem data, which prints to a given output stream; see Figure 4.16.

Solution Component. The basic purposes of a solution component are the construction and representation of solutions from a solution space S, the computation of the objective function, and the modification of solutions according to a given move

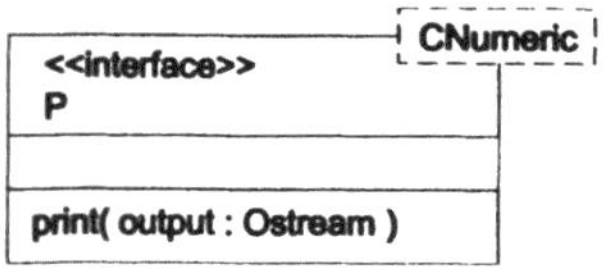

Figure 4.16 General Interface of Problem Components

to a neighbor solution. Figure 4.17 shows the general interface of solution components. (By the stereotype local search, we denote that the subsequent operations are required only if one wants to apply a local search procedure.)

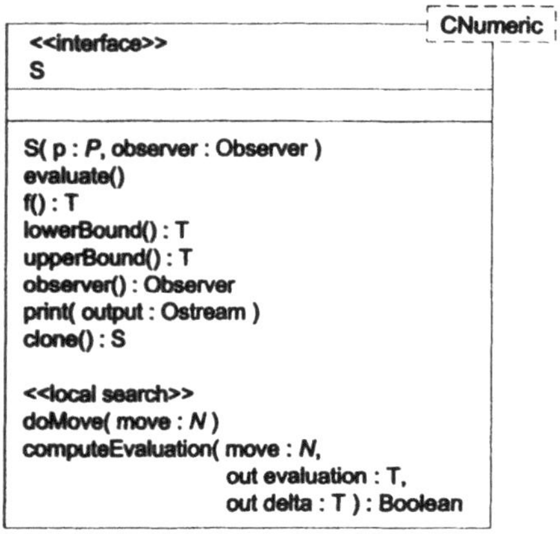

Figure 4.17 General Interface of Solution Components

Realizing the interface S by a corresponding component means implementing the following operations with "algorithmic content":

- S(p : *P*, observer : Observer)
  The constructor builds an initial solution for a given problem p. Additional parameters may be needed with respect to the actual rules of construction (e.g., random construction versus application of a simple construction algorithm).

- evaluate()
  This operation computes (and stores) the objective function value of the actual solution. Calling this operation from the outside is usually not necessary, since the operation f() returns by definition the objective function value of the current solution.

- doMove(move : *N*)
  There are two basic options to implement the modification of a solution:

  1. The neighborhood component transforms the data of the solution component, which requires the neighborhood component to know about the

respective data structures. This enables the introduction of new neighborhood structures without the need to modify solution components.

2. The solution component interprets the modification due to the move and modifies its own data accordingly. In this case, introducing new neighborhood structures requires the adaptation of an existing doMove operation (or the addition of a new doMove operation if the method is statically parameterized by the actual neighborhood type).

Due to reasons of efficiency, it is apparently not possible to dissolve this tight relationship between the solution component and the neighborhood component in a strict object-oriented manner (where each class should fully encapsulate its internal implementation).

- computeEvaluation(move : *N*, out evaluation : T, out delta : T) : Boolean
  Since only the solution component knows about the computation of the objective function, the primary responsibility to actually assess the advantageousness of a move should be assigned to the solution component. This also leads to a tight relationship between the solution component and the neighborhood component. The return parameters evaluation and delta represent the evaluation of a move and the implied change of the objective function value, respectively. These values may equal each other for many applications. In other cases, an appropriate move evaluation may require a special measurement function. In particular, this differentiation is reasonable when there is some kind of min-max objective function, where a single move usually does not directly affect the objective function. Furthermore, for some kinds of problems the exact computation of the implied change of the objective function value may be too costly as to be done for each of the considered moves. In such a case one needs to estimate the "quality" of a move in an approximate manner without computing delta; this must be indicated by returning false (otherwise the operation returns true).

  If moves are evaluated by the implied change of the objective function value, a standard implementation of this operation may be available by actually performing the move (for a copy of the solution object), computing the objective function of the resulting solution from scratch, and finally comparing the respective objective function value with the objective function value of the current solution. However, for reasons of efficiency, one usually has to implement a special, adaptive form of the move evaluation.

The following operations primarily serve for the encapsulation of the data structures of the solution component:

- f() : T
  Return of the objective function value of the solution.

- lowerBound() : T
  Return of the best known lower bound for the optimal objective function value (remember that we consider minimization problems).

- upperBound() : T
  Return of the best known upper bound for the optimal objective function value.

- observer() : Observer
  Return of the observer object.

- print(output : Ostream)
  Print the solution in some defined format to an output stream.

Furthermore, one must implement an operation clone(), which constructs a copy of the actual solution object (in the sense of a *virtual constructor*, see Stroustrup (1997), pp. 424). For some metaheuristics (in particular, evolutionary algorithms), one may need an operation that computes the "difference" between two solutions according to some measurement function (see, e.g., Woodruff (2001)). This requirement might have been specified as a member function of the solution component. However, to simplify the interface of the solution component, we use free template functions distance (with a static template parameter S and both solution objects as dynamic parameters).

Solution Information Component. A solution information component models elements of the set S_h. Respective objects are used to store the search trajectory, which means that the solution information component is only needed for tabu search methods that use such trajectory information. In accordance with this special role, the interface (and the corresponding functionality) of the solution information component is quite simple; see Figure 4.18 and the following description.

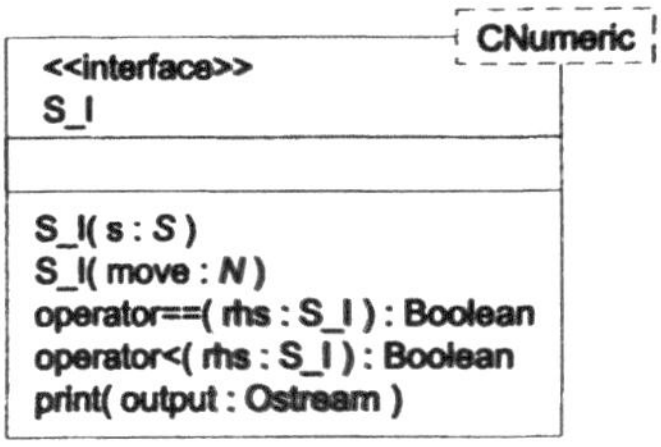

Figure 4.18 General Interface of Solution Information Components

- S_I(s : S)
  Construction of the object that corresponds to $h(s)$ by computing the transformation $h : S \to S_h$.

- S_I(move : N)
  Due to reasons of efficiency, we require a means to directly construct the solution information of the neighbor solution that corresponds to a given move. (One may also construct the solution information of a neighbor by actually constructing the solution object and deriving the solution information from this object.)

- operator==(rhs : S_I) : Boolean
  The definition of the equivalence relation is needed to implement the trajectory data structures.

- operator<(rhs : S_I) : Boolean
  Tree data structures, which may be used to store the trajectory, require the definition of an order operator.
- print(output : Ostream)
  Print the solution information in some defined format to an output stream.

Usually, the solution information component models a hash-function. In this case, the constructors include the respective computations, while the operators are simply implemented as the comparison of integer values.

Neighborhood Component. Neighborhood components (neighborhood iterators) represent neighborhood structures N_S (i.e., respective moves from N_S^{μ}) for a solution space S. Move object data generally consists of a reference to the actual solution and information about the transformation to a neighbor solution. In accordance with the discussion in Section 4.4.1.3, the basic functionality of neighborhood components is shown in Figure 4.19.

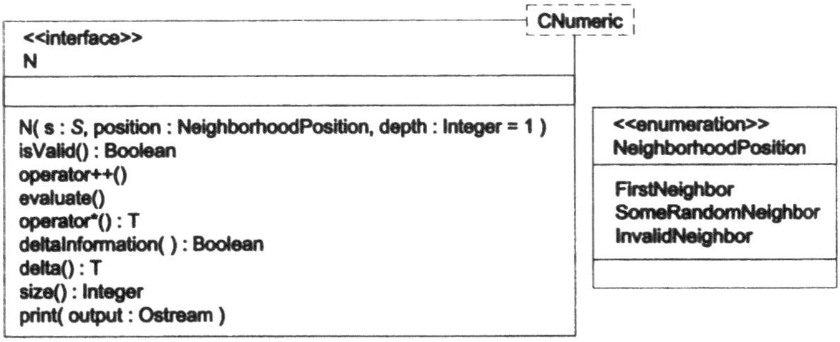

Figure 4.19 General Interface of Neighborhood Components

- N(s : S, position : NeighborhoodPosition, depth : Integer = 1)
  To construct a neighbor (i.e., the respective move) for a solution s one specifies by using the parameter position whether the first, a random, or the invalid neighbor is to be constructed. The optional parameter depth may define a neighborhood depth greater than one (i.e., a respective concatenation of elementary moves of the basic neighborhood structure).
- isValid() : Boolean
  This operation returns true if and only if the actual move is valid. The following operations may only be called for valid moves.
- operator++()
  Increment of a valid move to the next neighbor.
- evaluate()
  The evaluation of a move (according to $\hat{f}$) usually occurs when a move is constructed or incremented. The implementation of the evaluation may be delegated to the respective solution class; see the discussion in Section 4.4.2.2.

- operator*() : T
  This operation returns the move evaluation (according to $\hat{f}$).

- deltaInformation() : Boolean
  The value true is returned if and only if one may use the operation delta() to ask for the implied change of the objective function value that is due to the move; see the discussion in Section 4.4.2.2.

- delta() : T
  This operation returns the implied change of the objective function that is due to the actual moves if and only if this information is available (see the operation deltaInformation()).

- size() : Integer
  Some metaheuristics need an estimation of the neighborhood size (e.g., Algorithm 10, p. 94). For reasons of efficiency, we do not require this operation to return the exact number of neighbors of the actual solution, but only a reasonable upper bound.

- print(output : Ostream)
  Print the move in some defined format to an output stream.

The sequence diagram of Figure 4.20 illustrates the interaction between solution and neighborhood components by describing the typical neighborhood traversal of a local search approach. At the beginning, the move to the first neighbor of the actual solution is constructed and evaluated. Iteratively, the move is incremented to the next neighbor until we have reached the invalid neighbor. Finally, the best move found will be executed.

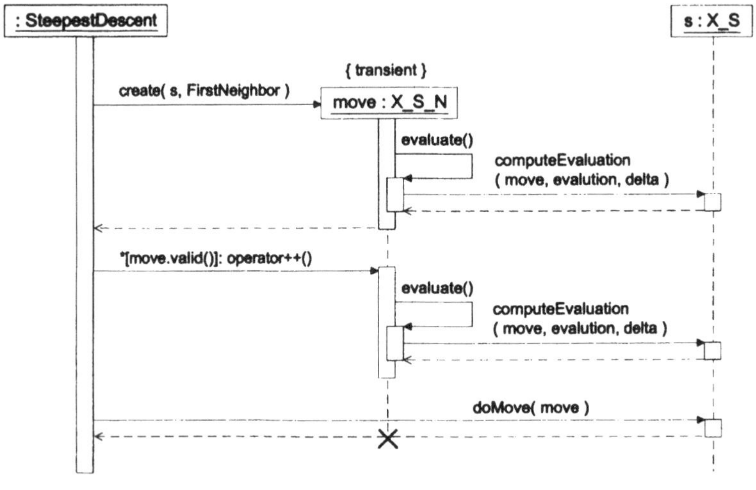

Figure 4.20 Sequence Diagram of the Interaction Between Solution and Neighborhood Components

Attribute Component. Attribute components represent elementary solution attributes. From the point of view of moves, we have to distinguish between plus and minus attributes. Such information is used by tabu search methods that apply an attribute-based memory. (The decomposition of solutions in elementary moves may be used by tabu search methods that build new solutions by combining attributes from a *vocabulary* of "promising" solution attributes.) The general interface of attribute components is shown in Figure 4.21.

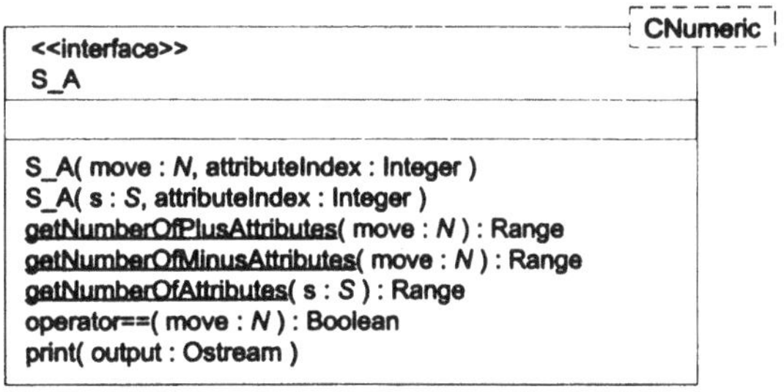

Figure 4.21 General Interface of Attribute Components

- S_A(move : *N*, attributeIndex : Integer)
  Move attributes are constructed in dependence on the parameter attributeIndex. Positive and negative values indicate plus and minus attributes, respectively. For example, the move attribute $\psi_1^-(\mu)$ is constructed when attributeIndex is set to -1.

- S_A(s : *S*, attributeIndex : Integer)
  The operation constructs the attributes of a solution.

- getNumberOfPlusAttributes(move : *N*) : Range
  This operation returns the number of plus attributes of a move ($|\psi^+(\mu)|$). The result of this and the next two operations does not depend on a specific object. (In C++, this leads to an implementation as static class methods.)

- getNumberOfMinusAttributes(move : *N*) : Range
  This operation returns the number of minus attributes of a move ($|\psi^-(\mu)|$).

- getNumberOfAttributes(s : *S*) : Range
  This operation returns the number of attributes of a solution ($|\psi(s)|$).

- operator==(move : *N*) : Boolean
  The equivalence relation between an attribute and a move reflects whether the actual attribute is "destroyed" when the considered move is executed. That is, this operation returns true if and only if the attribute corresponds to a minus attribute of the move. (For reasons of efficiency, we do not rely on the explicit construction and comparison of all minus attributes of a move, but require the implementation of this redundant operation.)

- print(output : Ostream)
  Print the attribute in some defined format to an output stream.

Standard Problem-Specific Components. As discussed in Section 4.4.1.4, different types of problems often possess some commonalities with respect to the solution space. So it seems reasonable to exploit these commonalities by defining, implementing, and (re)using problem-specific components, which provide respective data structures and algorithms. In particular, a factoring of common concepts may be modeled by inheritance hierarchies refering to the problem-specific components/interfaces that have been defined above. The goal is to provide a set of reusable classes, from which a user of the framework can derive special classes according to his application. Then, one has to implement the remaining concerns that are specific to the considered application (e.g., the objective function), while general data structures and algorithms with respect to the solution space, neighborhood structure, etc. may be reused. To summarize, applying HOTFRAME in any case means reuse of the respective architecture and metaheuristics components – if one of the standard problem-specific components that are available fits the considered problem, there is also the option to exploit implementation reuse on the problem side.

In the following, we describe standard problem-specific components for bit-vector and permutation solution spaces. The solution components BV_S and Perm_S represent solutions of the kind $x = (x_1, \ldots, x_n), x_i \in \{0,1\}, 1 \le i \le n$, and $\Pi = (\pi_1, \ldots, \pi_n)$, with Π being a permutation of $(1, \ldots, n)$, respectively. For these solution spaces, we implement commonly used neighborhood structures as respective neighborhood components. (HOTFRAME also includes components for combined assignment and sequencing problems, i.e., a solution is defined by an ordered assignment of some kind of objects to some kind of resources, which are not described in this paper.)

For BV_S, the neighborhood component BV_S_N represents neighbors (i.e., corresponding moves) that result from (one or more) bit inversions. For sequencing problems, there exist several popular neighborhood structures. So we introduce an additional layer in the class hierarchy by defining an abstract component Perm_S_N, from which three specific neighborhood components are derived, which represent three different neighborhood structures with a neighborhood size of $O(n^2)$. The component Perm_S_N_Shift represents neighbors that result from shifting one object to another position (with the option to restrict the maximum "shift distance"). Swaps of two objects are represented by the component Perm_S_N_Swap. The 2-exchange neighborhood (component Perm_S_N_2Exchange) represents those neighbors that result from replacing two predecessor-successor relations (edges when considering the sequence in a graph representation) by two other predecessor-successor relations so that a feasible solution (sequence) results; this move corresponds to an inversion of a partial sequence. There are quite a few additional neighborhood structures for sequencing problems, which may be represented by additional neighborhood components; see, e.g., Tian et al. (1999).

As an extension to Figure 4.9 (see p. 109), Figure 4.22 shows the inheritance relationship of solution components. Figure 4.23 shows the according diagram for neighborhood components. The usual way of reusing these components is by deriving a new solution component and applying some suitable neighborhood component unchanged. (The close relationship between solution components and neighborhood components ("*dual inheritance hierarchies*") leads to some subtle problems, which will be briefly discussed below.)

In the abstract base class Generic_S, we define data elements and operations that are common to all solution components. In particular, every solution component has data elements to store the objective function value, an lower and upper bound, and a reference to an observer object. For all such data elements, we define corresponding access operations. With regard to other methods, we can only define abstract interfaces, which have to be complemented by respective implementations in derived classes. (According to the UML, such operations are shown italicized in the diagram.) To simplify the diagrams, we do not show all access operations for data elements (which are implicitly assumed to exist with the identifier set according to the data element).

In the derived classes BV_S and Perm_S, one needs to implement, based on the neighborhood, the operations for the default evaluation of moves (according to the implied change of the objective function value that is due to the considered move) and the execution of moves for particular neighborhood structures/components. So we must provide respective adaptation mechanisms. Due to subtle technical reasons (see Stroustrup (1997), p. 348), C++ does not allow member template methods to be virtual, which unfortunately prevents us from using member template constructs to implement those operations for specific neighborhoods. So we must refer in the interface declarations of these operations to the most general neighborhood type (Generic_S_N). Implementing those member functions requires dynamic type-casts ("downcasts") for respective neighborhood types, applying run-time-type information (RTTI) to check for type-conformance. Adding new neighborhood structures requires changing these member functions. (See the discussion of this crucial problem of object-oriented design with regard to dual inheritance hierarchies ("codependent domains") by Martin (1997) and Coplien (1999), pp. 210–227.)

Implementing a concrete solution component with a specific objective function for a problem that may be modeled by a bit-vector or permutation solution space means deriving a new solution component from BV_S or Perm_S, respectively. To meet the requirements defined by the solution component interface S one has to implement an appropriate constructor, the clone operation, and the computation of the objective function. With regard to run-time efficiency, one may also need to implement an adaptive move evaluation.

The definition of BV_S includes a specific addition to the solution component. The member functions shown below the variable fixation stereotype enable – in combination with appropriate neighborhood components, which use these operations – restricting the neighborhood by excluding variables of the solution vector from consideration regarding bit inversions. That is, the value of some of the variables may be temporarily fixed by an appropriate implementation of the operations firstVariable, nextVariable, and randomVariable.

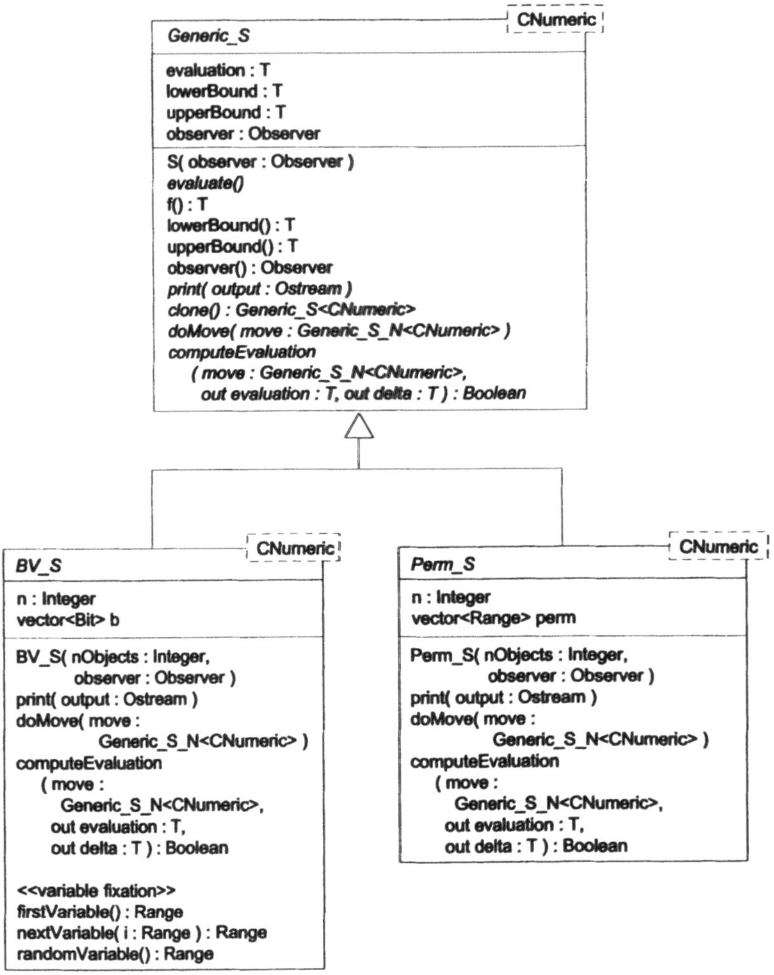

Figure 4.22 Inheritance Hierarchy for Solution Components

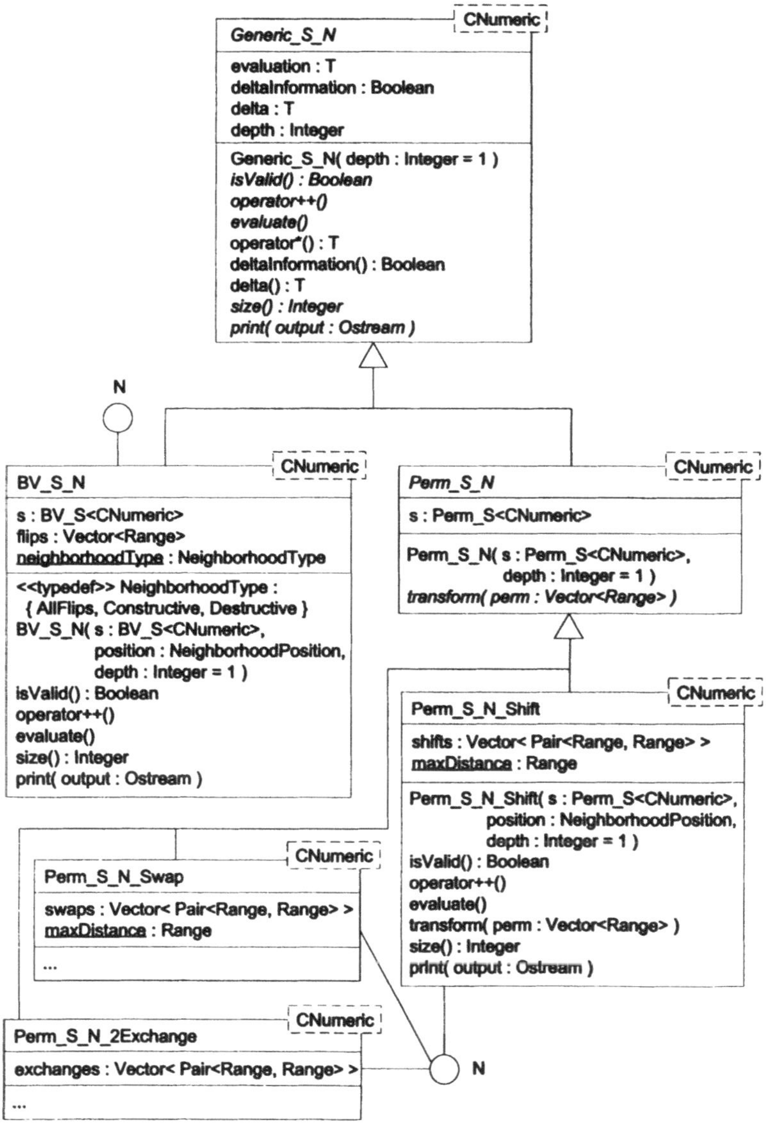

Figure 4.23 Inheritance Hierarchy for Neighborhood Components

Common data elements and respective access operations of neighborhood components are defined in the abstract base class Generic_S_N. For the bit-vector solution space, the bit inversion neighborhood is implemented by the component BV_S_N, which is directly derived from Generic_S_N. The BV_S_N component defines as data elements a reference to the respective solution object and a vector of integer numbers, which define the variables that are to be flipped. This enables modeling moves that comprise multiple bit inversions (note the depth parameter in Generic_S_N). By the static (class-scope) parameter neighborhoodType one can restrict the neighborhood to "constructive" or "destructive" moves, which means that only bit inversion from 0 to 1 or 1 to 0 are considered, respectively.

In addition to the constructors, the specific neighborhood components implement those operations that are defined in the base classes as virtual. As discussed above, the evaluation of moves is usually implemented by a delegation to the solution component.

The commonalities of the neighborhood components Perm_S_N_Shift, Perm_S_N_Swap, and Perm_S_N_2Exchange are modeled by an abstract class Perm_S_N. This component defines a reference to a solution object of Perm_S, a constructor, and the abstract member function transform, which represents the move's transformation of the solution data structure according to the implementation in the derived classes. Specialized neighborhood components include data structures that represent the move. For example, in Perm_S_N_Shift, a pair of integer numbers defines which object has to be shifted to which position. Vectors of such pairs define concatenations of moves. For Perm_S_N_Shift and Perm_S_N_Swap, the data element maxDistance can be used to restrict the neighborhood with regard to the shift or swap distance, respectively. The operations of Perm_S_N_Swap and Perm_S_N_2Exchange, which are not shown in Figure 4.23, are defined in the same way as Perm_S_N_Shift.

Solution information components and attribute components are only needed for particular tabu search methods. There is no need to define common base classes, since there are no general reference relations that refer to these components. Moreover, there are no common data structures, which may be modeled in a base class. So we define separate standard components for respective solution spaces. These components fully implement the interfaces S_I and S_A, respectively, and thus can be applied right away.

The solution information components, which are defined in Figure 4.24, represent solutions by applying hash-functions (see p. 119). The respective constructors compute hash-values for solution objects and for neighbor objects. Pre-defined implementations are available for the standard solution spaces and neighborhood structures that have been described above.

The attribute components shown in Figure 4.25 represent attributive information with regard to solutions and neighborhoods. The attributes match the elements of bit-vector and permutation solutions (elementary bit inversions and predecessor-successor-relations). Implementations for the introduced solution spaces and neighborhood structures are available.

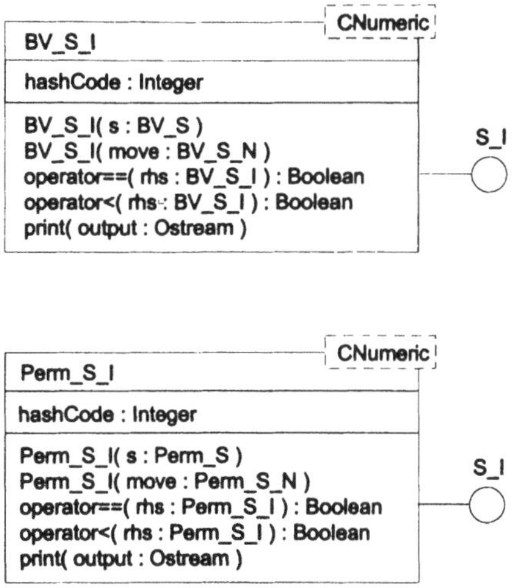

Figure 4.24 Solution Information Components

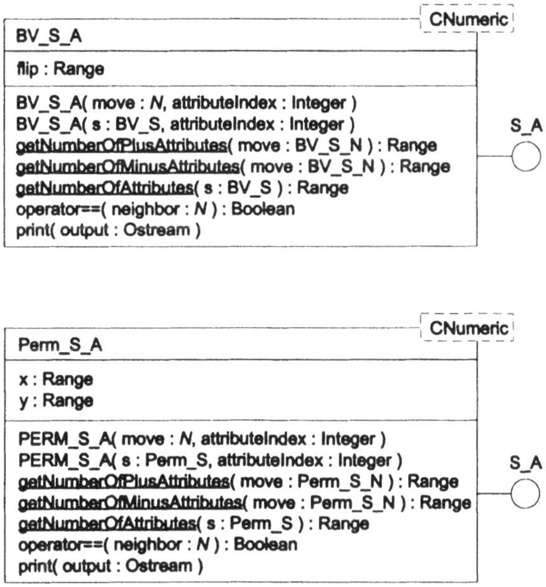

Figure 4.25 Solution Attribute Components

4.4.2.3 Metaheuristic Components. The metaheuristic concepts that have been described in Section 4.3.2 are implemented by corresponding metaheuristic components. In the following, we describe the most important operations of these components, which also includes the requirements with regard to respective configuration components; see Section 4.4.2.1. (To simplify the presentation, we do not explicitly show data elements and corresponding access operations.)

The general interface Heuristic of a metaheuristic is shown in Figure 4.26. To enable a flexible (polymorphic) application of respective algorithmic objects, specific metaheuristic classes of the framework are derived from a common base class (also named Heuristic). This base class is statically parameterized by the solution space.

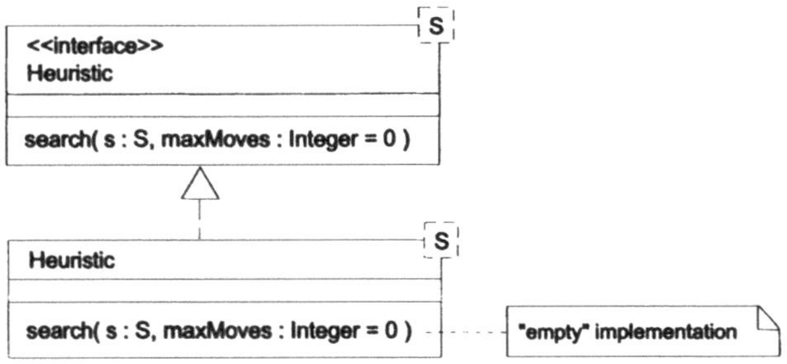

Figure 4.26 General Interface and Base Class Heuristic

The derivation of metaheuristic components from Heuristic, which is shown in Figure 4.27, follows the well-known strategy pattern; see Gamma et al. (1995). First of all, metaheuristic components provide an operation search, which transforms a solution. The optional parameter maxMoves facilitates such a modification of the termination criterion, which may be used, e.g., when triggering an adaptive diversification. The operation search of the base class Heuristic is defined as an "empty" method, so that respective objects can be used when one needs a default behavior for some feature (e.g., a non-existent diversification).

Iterated Local Search. Iterated local search and similar methods, which have been discussed in Section 4.3.2, are implemented by the component IteratedLocalSearch; see Figure 4.28. The static configuration of this component as well as the implementation of the features Diversification and Ω as dynamic object parameters have already been discussed in Section 4.4.2.1. The parameter depth determines the neighborhood depth. The parameter returnBest defines whether the algorithm should return the best solution found or the last traversed solution.

The static parameter NeighborSelection, which is an element of the configuration of IteratedLocalSearch, must conform to the interface defined in Figure 4.29, which also shows some components that realize popular neighbor selection rules.

The derivation of specific metaheuristic components is exemplified in Figure 4.30. IteratedSteepestDescent is defined by fixing the type parameter NeighborSelection accordingly.

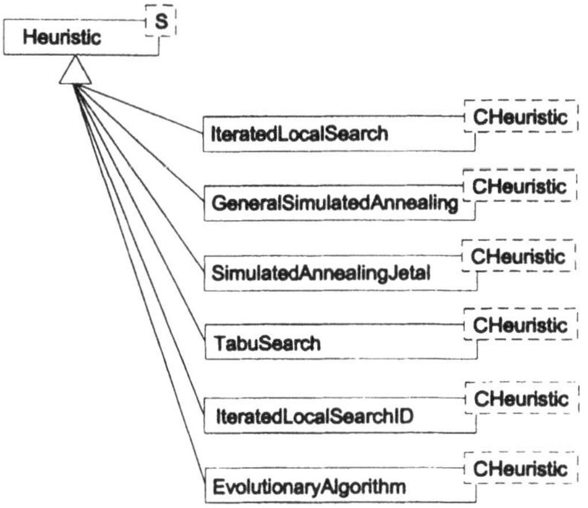

Figure 4.27 Derivation of Specific Metaheuristic Components from Heuristic

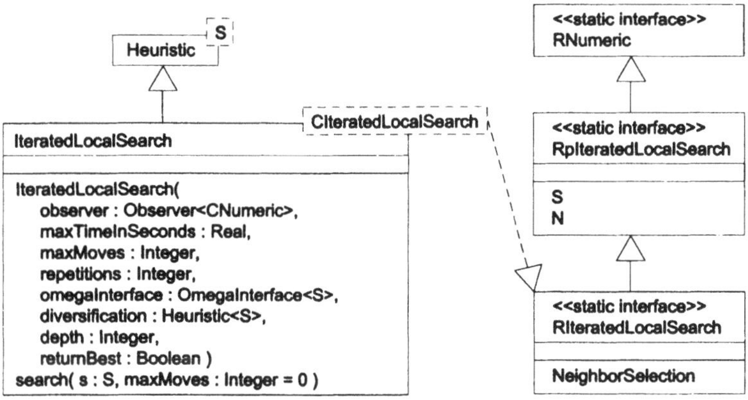

Figure 4.28 Component IteratedLocalSearch

Simulated Annealing and Variations. Simulated annealing and variations, which have been described in Section 4.3.2.2, are implemented by the components GeneralSimulatedAnnealing and SimulatedAnnealingJetal shown in Figure 4.31. With regard to the metaheuristic-specific configuration, one has to specify the features acceptance criterion, cooling schedule, and an optional reheating scheme. In the following, we describe the general interface of respective components and some common instances of these interfaces. With regard to the implementation in C++, these modules are applied by using class-scope parameters (in accordance with the discussion on p. 111).

The interface AcceptanceCriterion is defined in Figure 4.32. The acceptance criterion is implemented by the operation check. In general, the acceptance of a move depends on the current value of the temperature parameter (tau), the evaluation of

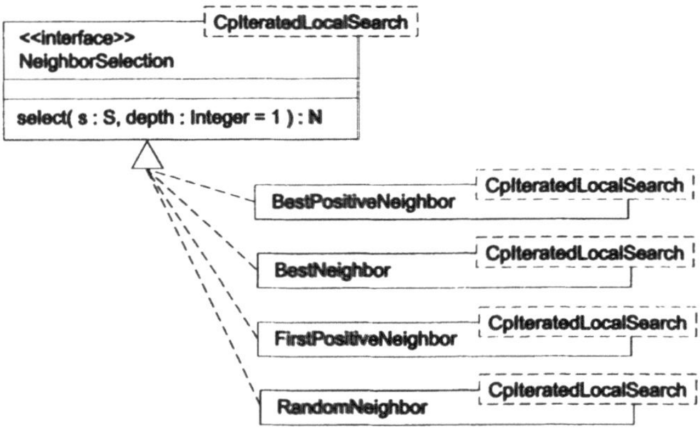

Figure 4.29 Interface NeighborSelection and Respective Components

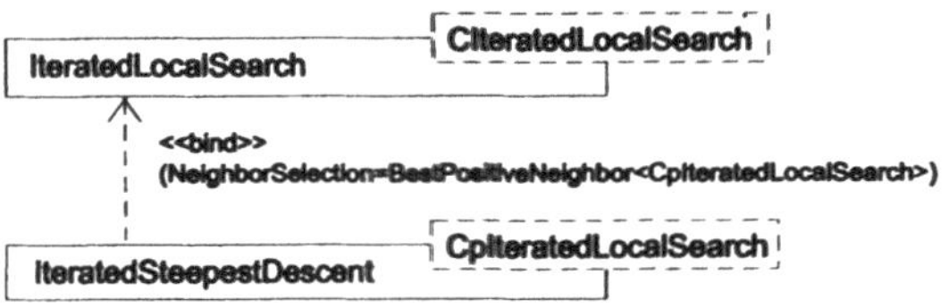

Figure 4.30 Derivation of IteratedSteepestDescent

the considered move (moveEval), and the objective function value of the respective neighbor solution (newSolutionEval).

Figure 4.33 shows the interface CoolingSchedule and some respective components, which implement popular schemes for decreasing the temperature parameter. When specific events occur, respective operations are called; see Algorithms 5 and 10. If an event is not relevant for some cooling schedule, the corresponding operation does not alter the temperature (i.e., "empty" implementation). Specific numeric parameters of the cooling schedule components are implemented as data elements with class-scope. The configuration components are implicitly required to provide corresponding definitions.

Reheating components, which have to conform to the Reheating interface shown in Figure 4.34, re-initialize the temperature parameter when called by a simulated annealing algorithm. The new initial temperature value may depend on the preceding initial value (tinitial), the temperature when the best solution was found (tbest), and the last temperature (tau). We also define an "empty" component NoReheating, which does not affect tinitial (reheating is an optional feature).

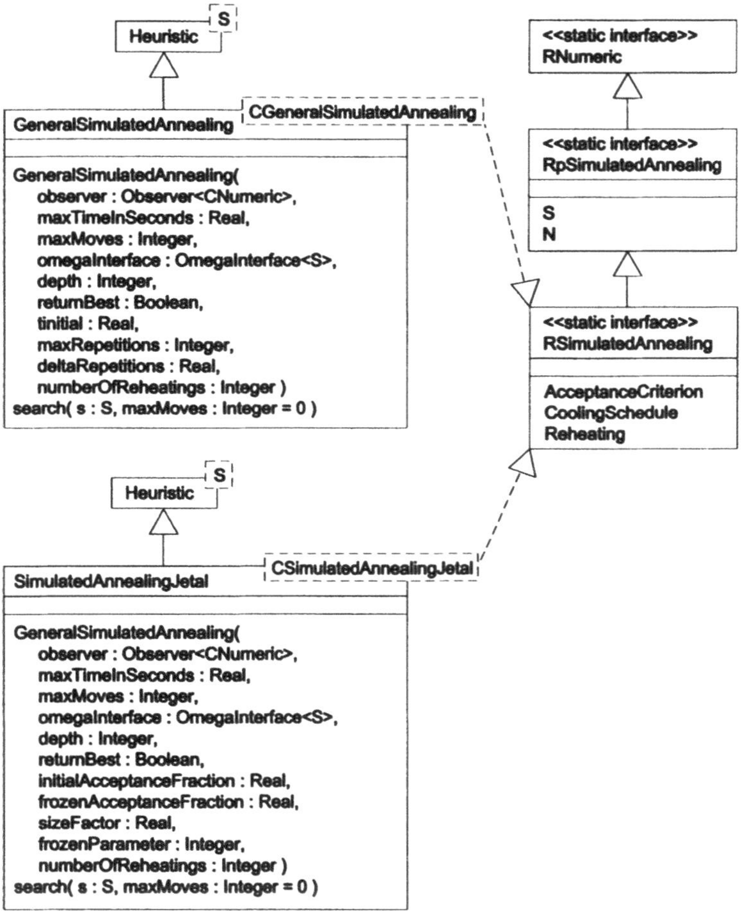

Figure 4.31 Components GeneralSimulatedAnnealing and SimulatedAnnealingJetal

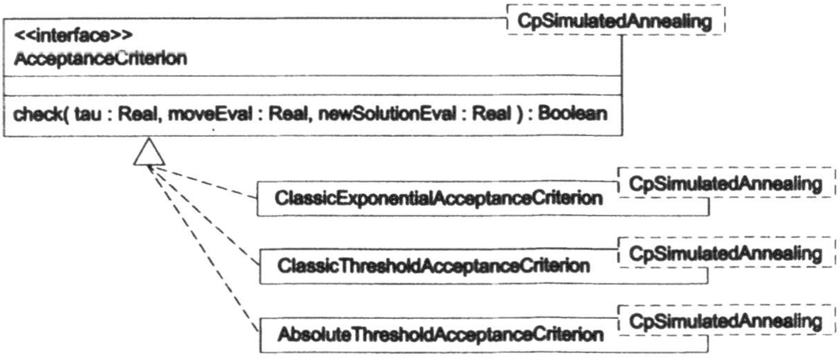

Figure 4.32 Interface AcceptanceCriterion and Respective Components

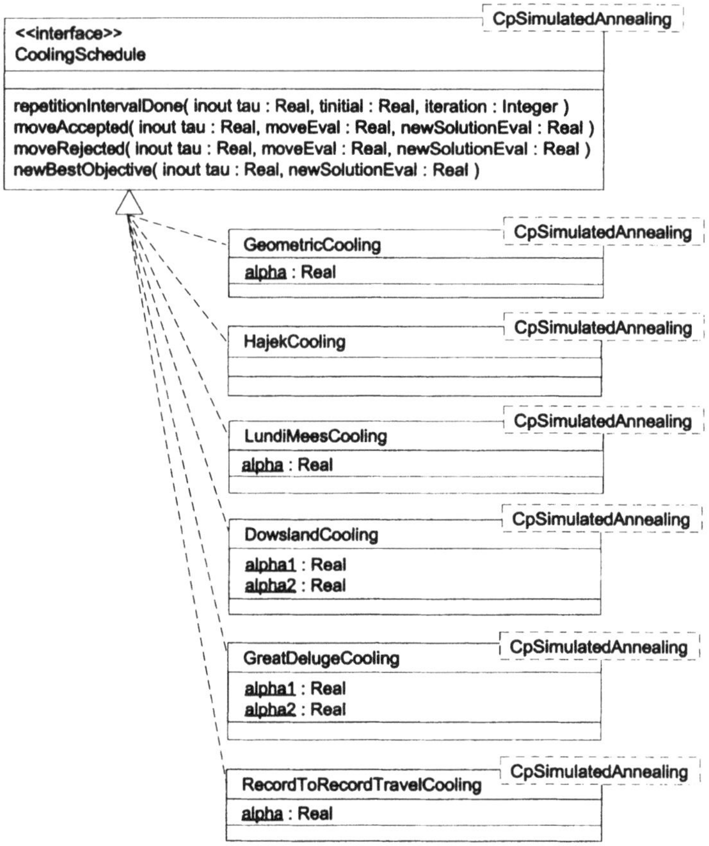

Figure 4.33 Interface CoolingSchedule and Respective Components

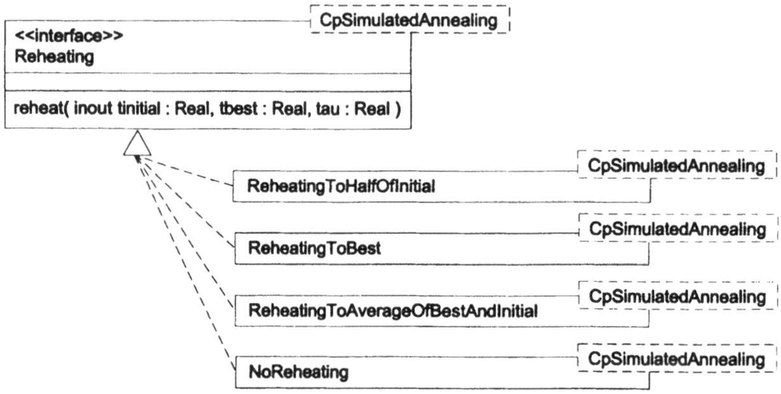

Figure 4.34 Interface Reheating and Respective Components

Tabu Search. The component TabuSearch, shown in Figure 4.35, implements the classic tabu search scheme according to Algorithm 11. A tabu search configuration component in particular defines the neighbor selection rule and an aspiration criterion. The implementation of the features TabuCriterion and Diversification as dynamic object parameters has already been discussed in Section 4.4.2.1.

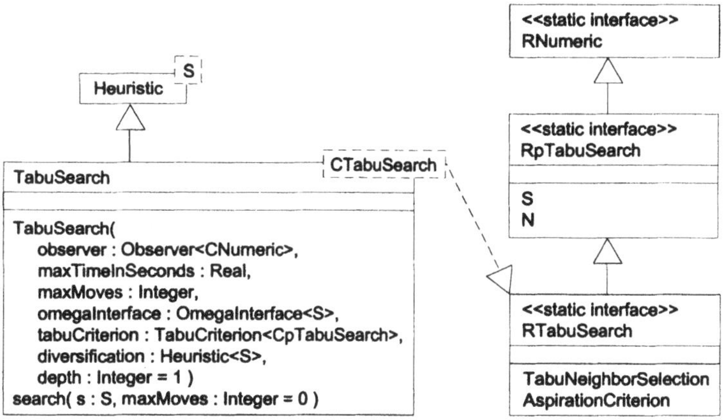

Figure 4.35 Component TabuSearch

The components for different tabu criteria (Algorithms 14–17) are derived from a common base class; see Figure 4.36. The base class defines the general interface of a tabu criterion component according to the discussion in Section 4.3.2.3. Specific requirements of the tabu criteria with regard to the need to define problem-specific components S_I as well as S_A within the configuration component CpTabuSearch conform to the feature diagram shown in Figure 4.4.

The two operations that feed the tabu criterion with applied moves and traversed solutions (addToHistory) provide a return parameter, which is used to indicate whether there is an apparent need for an explicit diversification. By returning a value k greater than zero, the tabu criterion "suggests" to apply some big "escape move" to some solution that is about k moves away from the current solution. Such an explicit diversification is usually accompanied by a call to the operation escape, which re-initializes the tabu memory in an appropriate way. The operation print writes information about the tabu memory in some format to an output stream. The dynamic configuration of the tabu criteria (i.e., the initialization of respective numeric parameter elements such as, e.g., the tabu list length) results from calling constructors, which are not shown in the diagram.

For the component REMTabuCriterion we make two restrictions with regard to an efficient implementation of Algorithm 15. Firstly, we restrict the application of this tabu criterion to single-attribute moves. Secondly, we require the definition of a free template function moveNumber, which computes for each move a unique integer number representative. We identify such free (global) functions in the diagram by using a stereotype free function; see Figure 4.36.

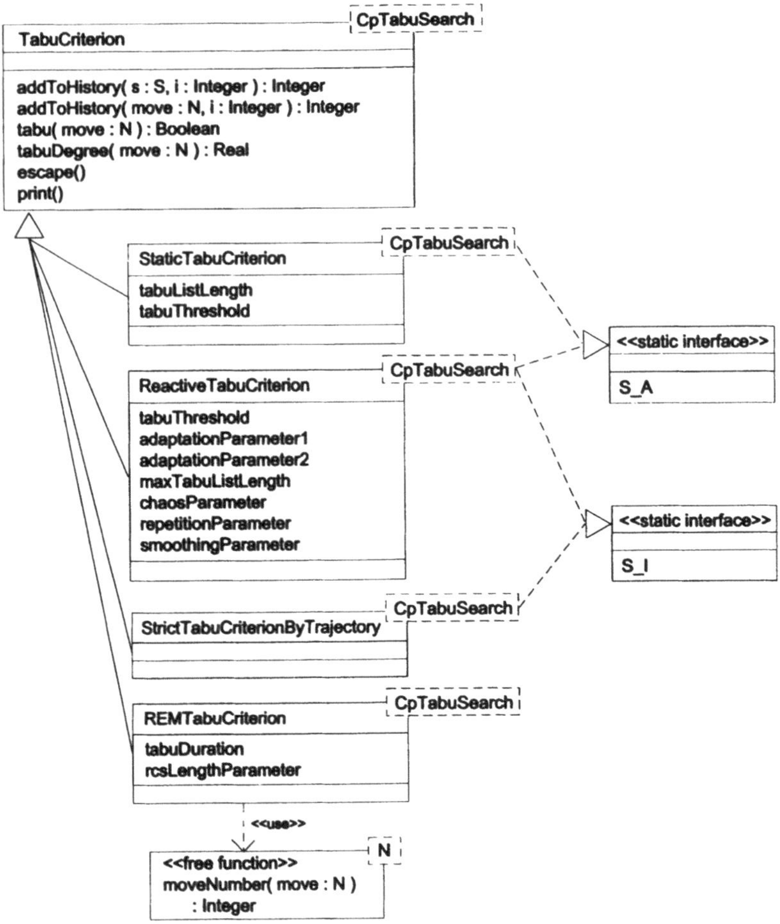

Figure 4.36 Base Class TabuCriterion and Derived Classes

The component TabuSearch requires the definition of a neighbor selection rule (i.e., a corresponding component), which itself depends on the aspiration criterion. Figure 4.37 shows the interface TabuNeighborSelection and respective components. The components BestAdmissibleNeighbor and BestNeighborConsideringPenalties use the aspiration criterion, which is configured in CTabuSearch, in accordance with Algorithms 12 and 13, respectively. The interface of the (optional) aspiration criterion is shown in Figure 4.38. The component NewBestSolutionAspirationCriterion implements the most popular aspiration criterion: A tabu-status is neglected if the move would lead to a new best solution. An efficient implementation of this criterion requires actual information about the implied change of the objective function value that is due to the considered moves (deltaInformation); see p. 119.

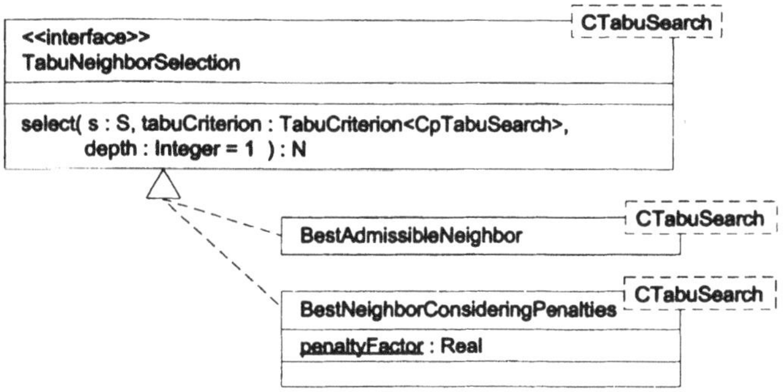

Figure 4.37 Interface TabuNeighborSelection and Respective Components

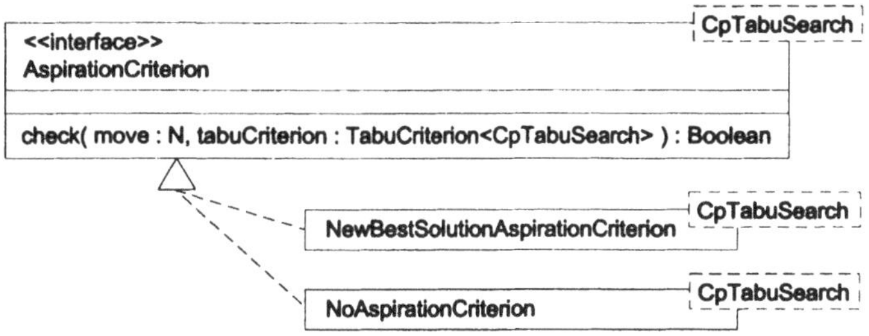

Figure 4.38 Interface AspirationCriterion and Respective Components

4.4.2.4 Introspection. In Section 4.4.2.1, we argued in favor of flexible introspection means regarding information about the search process. We apply the following design: The interface of a concrete base class Observer represents a comprehensive set of introspection elements; see Figure 4.39. The diagram shows only a subset of the interface. For example, it does not show operations that are specific to a metaheuristic (e.g., an operation that provides access to the contents of the tabu list).

Metaheuristic components and solution components are dynamically parameterized by an object of the class Observer or a derived class. Appropriate object operations are called at respective steps of the algorithms. The member functions of the class Observer are implemented as "empty" functions. So the class Observer defines the maximal introspection interface, yet actually implements a minimal introspection (nothing at all). Special introspection needs are implemented by respective member functions in derived classes, which overwrite the behavior of the base class. Figure 4.39 shows three such components, which all rely on sending data to a stream. The component StreamingObserver simply prints all received data, while the other two components only output the best objective function value and possibly the computation time.

To enable a flexible change of the degree of introspection during the search process, we provide operations to activate or deactivate an observer object. A deactivated object neglects all information sent to it. Moreover, a caller may check for the ac-

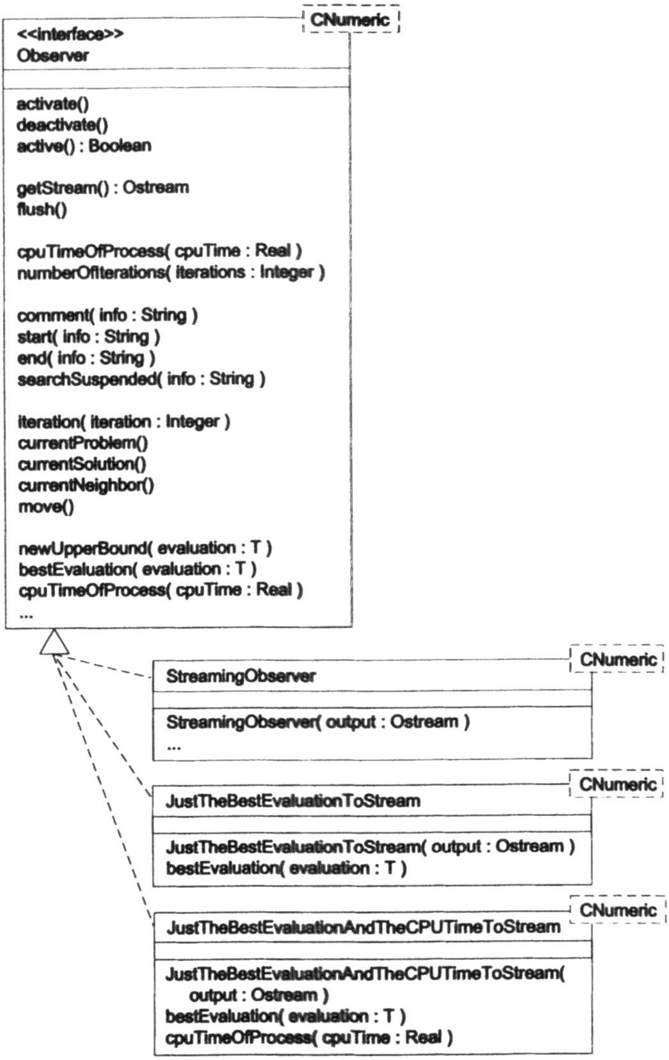

Figure 4.39 Derivation of Special Introspection Classes from the Base Class Observer

tivation status of the assigned observer object, which allows to suppress costly data preparations (which would be neglected anyway).

To simplify the transfer of non-atomic information to the observer (e.g., the solution representation in some format), the operation getStream provides the caller with a stream, where respective data may be sent to. With regard to a semantic interpretation of stream data, the caller has to call an appropriate operation before using the stream (e.g., currentSolution). After completing the transfer, the transaction must be finalized by calling the operation flush.

This design, which decouples the search process from the introspection, provides a flexible means for extension. For example, search processes may be coupled to an environment for experimental tests with regard to visualization or statistical analysis of the search process and respective results; see Jones (1994), Jones (1996). Moreover, one may integrate the search process in a decision support system; see Ball and Datta (1997).

4.5 IMPLEMENTATION

As emphasized by Nygaard, *Programming is understanding*. In this sense, implementing metaheuristics as reusable software components – and also eventually using these components – serves for the comprehension of the respective genericity of metaheuristic algorithms. In the following, we briefly describe some essential aspects of the implementation.

4.5.1 Technical Environment and Conventions

In Section 4.4, we argued that C++ is an appropriate programming language in the given context. C++ provides powerful language constructs (in particular with regard to enabling adaptation by type parameterization and inheritance), it facilitates run-time efficient implementations, and it is in wide use in practice. The main argument against C++ is its complexity; see, e.g., the discussion in Gillam (1998), p. 41. However, while this complexity indeed affects the actual developer of a framework, it is mainly hidden from the plain user of a framework.

The implementation is based on Standard C++ (ISO/IEC 14882), which should lead to wide portability. Due to deficiencies of some of the available compilers we decided not to employ member templates, covariant return types, and the exception mechanism. As primary development platform we used *Microsoft Visual C++ 6.0 (Service Pack 3 and higher)*. The code was also successfully tested with the compilers *gcc 3.0* and *MIPSpro C++ 7.3*. A few incompatibilities of these compilers are handled in the source code by relying on preprocessor variables VCC, GCC, and MIPS.

The discussion of the design in the preceding section leads to a broad use of templates. With regard to generic components, the C++ template construct provides a type-safe and run-time efficient means to implement respective adaptation requirements. (On the other hand, the unfamiliarity of most programmers with the template construct and insufficient compiler support with regard to debugging template-intense code may make debugging difficult).

To prevent name clashes all HOTFRAME components are defined in a namespace HotFrame. To apply respective components one may use explicit qualification (e.g., "HotFrame::TabuSearch<... >"), using declarations (e.g., "using HotFrame::TabuSearch<... >"), or using directives (e.g., "using HotFrame").

To streamline the implementation we follow by convention a few coding rules. Class or type identifiers start with a capital letter, while identifiers for global methods, member functions, and local variables start with a lower case letter. Member data identifiers start with an underscore ("_"); static class members start with "_s_". Composite identifiers are partitioned by using capital letters (e.g., computeEvaluation).

Member data access functions are usually named in accordance with the identifier of the member data (e.g., n() provides access to _n). For all problem-specific classes, one has to define appropriate copy constructors, assignment operators, and destructors, if the compiler-generated default versions are not adequate. Member functions that provide a dynamically allocated object for which the caller is responsible (in particular with regard to object destruction) start with create, copy, or clone. Variables that represent iteration numbers are declared as unsigned long; i.e., the maximal iteration number is typically $2^{32} \approx 4.3 \times 10^9$. With x representing an appropriate acronym that identifies the problem type, problem-specific code is usually put into two files x_p.h (problem component) and x_s.h (other components such as the solution space component etc.).

4.5.2 Problem-Specific Standard Components

Applying HOTFRAME involves appropriate problem-specific components that particularly implement the solution space and the neighborhood structure. In Section 4.4.2.3, we described respective requirements in dependence on the metaheuristics that one wants to use. In case the pre-defined problem-specific standard components do not fit for some new type of problem, one may need to implement such components from scratch. HOTFRAME includes the following problem-specific standard components (see pp. 122):

```
template <class C> class Generic_S;
template <class C> class Generic_S_N;

template <class C> class BV_S : public Generic_S<C>;
template <class C> class BV_S_N : public Generic_S_N<C>;
template <class C> class BV_S_A;
template <class C> class BV_S_I;

template <class C> class Perm_S : public Generic_S<C>;
template <class C> class Perm_S_N : public Generic_S_N<C>;
template <class C> class Perm_S_N_Shift : public Perm_S_N<C>;
template <class C> class Perm_S_N_Swap : public Perm_S_N<C>;
template <class C> class Perm_S_N_2Exchange
                              : public Perm_S_N<C>;
template <class C> class Perm_S_A;
template <class C> class Perm_S_I;
```

For the sake of completeness, we also name the components for combined assignment and sequencing problems (without description):

```
template <class C> class AS_S;
template <class C> class AS_S_N_Shift;
template <class C> class AS_S_A1;
template <class C> class AS_S_A2;
template <class C> class AS_S_I;
```

The implementation of BV_S and Perm_S illustrates the alternative strategies to implement the move interpretation as part of the member functions doMove and computeEvaluation (see the discussion on p. 116). On the one hand, BV_S directly interprets and executes the modifications to a solution that are due to a considered move. In contrast, Perm_S_N and derived classes define a member function transform, which realizes respective modifications of the solution data structures and thus can be used by Perm_S. It is important to note that respective default implementations of doMove and computeEvaluation are only valid if the derived solution class for the specific application does not define special data elements that are affected by a move. In such cases, one has to provide specialized implementations of these member functions.

The enumeration NeighborhoodPosition defines the general possibilities to construct neighbor/move objects:

```
enum NeighborhoodPosition { FirstNeighbor,SomeRandomNeighbor,
                            InvalidNeighbor };
```

For all problem-specific classes X, the following stream output operator is defined, which enables a polymorphic call of the respective print member function:

```
template <class C>
  ostream& operator<<( ostream& output, const X<C>& x );
```

4.5.3 Metaheuristic Components

We briefly describe the actual C++ interfaces of the metaheuristic components that have been described in Section 4.4.2.3. The interface of the common base class of metaheuristic components is defined as follows (see Figure 4.26):

```
template <class S>
class Heuristic
{
  public:
    virtual ~Heuristic( )
      { ; }
    virtual void search( S& s, unsigned long maxMoves = 0 )
      { ; }
};
```

4.5.3.1 Iterated Local Search. The interface of the component IteratedLocalSearch is defined as follows (see Figure 4.28):

```
template <class C>
class IteratedLocalSearch
: public Heuristic<typename C::S>
{
  public:
    typedef typename C::T T;
    typedef typename C::Range Range;
    typedef typename C::CNumeric CNumeric;
    typedef typename C::S S;
    typedef typename C::N N;
```

```
    typedef typename C::NeighborSelection NeighborSelection;

  protected:
    Observer<CNumeric> *_observer;
    float _maxTimeInSeconds;
    unsigned long _maxMoves;
    unsigned long _repetitions;
    OmegaInterface<S> *_omegaInterface;
    Heuristic<S> *_diversification;
    short _depth;
    bool _returnBest;

  public:
    IteratedLocalSearch( Observer<CNumeric> *observer = 0,
                         float maxTimeInSeconds = 0,
                         unsigned long maxMoves = 0,
                         unsigned long repetitions = 1,
                         OmegaInterface<C> *omegaInterface=0,
                         Heuristic<S> *diversification = 0,
                         short depth = 1,
                         bool returnBest = true );
    virtual void search( S& s, unsigned long maxMoves = 0 );
};
```

The type definitions conform to the requirements on the configuration components, which have been specified in Figure 4.28. The explicit re-definition of these types in the first part of the interface makes requirements explicit and simplifies the use of respective types. The data elements correspond to the method parameters as defined by the constructor interface. (To simplify the presentation, we omit type definitions and data structures from the following descriptions, because these elements can be directly deduced from the requirements formulated in Section 4.4.2.3.)

The constructor interface specifies for all parameters default values. For maxTimeInSeconds and maxMoves, the default value of 0 represents the lack of a corresponding restriction. The constructor definition consists only of the initialization of respective class data elements. The actual iterated local search algorithm is implemented in the member function search.

Components that implement a neighbor selection rule must conform to the following interface NeighborSelection (see Figure 4.29):

```
template <class C>
class NeighborSelection
{
  public:
    static N select( S& s, short depth = 1 );
}
```

The following realizations of this interface are pre-defined:

```
template <class C> class BestPositiveNeighbor;
```

```
template <class C> class BestNeighbor;
template <class C> class FirstPositiveNeighbor;
template <class C> class RandomNeighbor;
```

4.5.3.2 Simulated Annealing and Variations. Algorithms 5 and 10 are implemented by the following two components (see Section 4.4.2.3):

```
template <class C>
class GeneralSimulatedAnnealing
: public Heuristic<typename C::S>
{
  public:
    GeneralSimulatedAnnealing(
        Observer<CNumeric> *observer = 0,
        float maxTimeInSeconds = 0,
        unsigned long maxMoves = 0,
        OmegaInterface<S> *omegaInterface = 0,
        short depth = 1,
        bool returnBest = true,
        double tinitial = 100,
        unsigned long maxRepetitions = 1,
        double deltaRepetitions = 1,
        unsigned int numberOfReheatings = 0 );

    virtual void search( S& s, unsigned long maxMoves = 0 );
};

class SimulatedAnnealingJetal
: public Heuristic<typename C::S>
{
  public:
    SimulatedAnnealingJetal(
        Observer<CNumeric> *observer = 0,
        float maxTimeInSeconds = 0,
        unsigned long maxMoves = 0,
        OmegaInterface<S> *omegaInterface = 0,
        short depth = 1,
        bool returnBest = true,
        float initialAcceptanceFraction = 0.4,
        float frozenAcceptanceFraction = 0.02,
        float sizeFactor = 16,
        unsigned int frozenParameter = 5,
        unsigned int numberOfReheatings = 0 );

    virtual void search( S& s, unsigned long maxMoves = 0 );
};
```

The determination of the initial temperature for the latter component is implemented in the following way: Beginning with the initial solution, a trial run of *sizeFactor* $\cdot$ $|N(s)|$ iterations is performed, where in each iteration a neighbor solution is randomly generated and accepted if and only if the move evaluation is strictly positive. The observed evaluations of the other moves are stored and sorted. Eventually, the temperature is set so that *initialAcceptanceFraction* of the observed neighbors would have been accepted.

The interface for components that implement an acceptance criterion is defined as follows (see Figure 4.32):

```
template <class C>
class AcceptanceCriterion
{
  public:
    static bool check( double tau, double moveEval,
                       double newSolutionEval = 0 );
};
```

This interface is realized by three pre-defined generic classes:

```
template<class C> class ClassicExponentialAcceptanceCriterion;
template<class C> class ClassicThresholdAcceptanceCriterion;
template<class C> class AbsoluteThresholdAcceptanceCriterion;
```

In accordance with Figure 4.33, the interface for components that implement a cooling schedule is defined as follows:

```
template <class C>
class CoolingSchedule
{
 public:
   static void repetitionIntervalDone
    ( double& tau, double tinitial, unsigned long iteration );
   static void moveAccepted
    ( double& tau, double moveEval, double newSolutionEval );
   static void moveRejected
    ( double& tau, double moveEval, double newSolutionEval );
   static void newBestObjective
    ( double& tau, double newSolutionEval );
};
```

The following realizations define non-relevant functions by an "empty" implementation:

```
template <class C> class GeometricCooling;
  // double C::alpha;
template <class C> class HajekCooling;
template <class C> class LundiMeesCooling;
  // double C::alpha;
template <class C> class DowslandCooling;
  // double C::alpha1;
  // double C::alpha2;
template <class C> class GreatDelugeCooling;
```

```
  // double C::alpha1;
  // double C::alpha2;
template <class C> class RecordToRecordTravelCooling;
  // double C::alpha;
```

The commentaries with regard to data elements of the configuration component C indicate the respective parameterization (see Figure 4.33).

In accordance with Figure 4.34, the interface of reheating components is defined as follows:

```
template <class C>
class Reheating
{
  public:
    static void reheat( double& tinitial, double tbest = 0,
                        double tau = 0 );
};
```

This interface is realized by four pre-defined generic classes:

```
template <class C> class ReheatingToHalfOfInitial;
template <class C> class ReheatingToBest;
template <class C> class ReheatingToAverageOfBestAndInitial;
template <class C> class NoReheating;
```

4.5.3.3 Tabu Search. The interface of the fundamental tabu criteria is defined as follows (see Figure 4.35):

```
template <class C>
class TabuSearch
: public Heuristic<typename C::S>
{
  public:
    TabuSearch( Observer<CNumeric> *observer = 0,
                float maxTimeInSeconds = 0,
                unsigned long maxMoves = 0,
                OmegaInterface<S> *omegaInterface = 0,
                TabuCriterion<Cp> *tabuCriterion = 0,
                Heuristic<S> *diversification = 0,
                short depth = 1 );

    virtual void search( S& s, unsigned long maxMoves = 0 );
};
```

In accordance with Figure 4.36, the base class for the tabu criteria is defined as follows:

```
template <class C>
class TabuCriterion
{
```

```
  public:
    virtual ~TabuCriterion( )
      { ; }
    virtual unsigned long addToHistory
        ( const S& s, unsigned long iteration = 0 )
      { return 0; }
    virtual unsigned long addToHistory
        ( const N& move, unsigned long iteration = 0 )
      { return 0; }
    virtual bool tabu( const N& move ) const
      { return false; }
    virtual double tabuDegree( const N& move ) const
      { return tabu( move ); }
    virtual void escape( )
      { ; }
    virtual void print( ) const
      { ; }
};
```

Different classes are derived from TabuCriterion to implement the tabu criteria according to Algorithms 14–17. Specific requirements of these classes on the configuration component C result from Figure 4.36 (with the addition of an observer object parameter). We restrict the following descriptions of tabu criteria components to the constructor:

```
template <class C>
class StrictTabuCriterionByTrajectory
: public TabuCriterion<typename C::Cp>
{
  public:
    StrictTabuCriterionByTrajectory
        ( Observer<CNumeric> *observer = 0 );
}

template <class C>
class REMTabuCriterion
: public TabuCriterion<typename C::Cp>
{
  public:
    REMTabuCriterion(
      Observer<CNumeric> *observer = 0,
      unsigned long tabuDuration = 1,
      unsigned long rcsLengthParameter = 1,
      unsigned long maximumMoveNumber
                     = numeric_limits<unsigned long>::max());
}
```

The fourth parameter of the constructor of the REM tabu criterion results from a technical requirement of the implementation, which needs to know about a maximum move number that may occur during the search process.

```
template <class C>
class StaticTabuCriterion
: public TabuCriterion<typename C::Cp>
{
  public:
    StaticTabuCriterion(
        Observer<CNumeric> *observer = 0,
        Range tabuListLength = 7,
        Range tabuThreshold = 1 );
}

template <class C>
class ReactiveTabuCriterion
: public TabuCriterion<typename C::Cp>
{
  public:
    ReactiveTabuCriterion(
        Observer<CNumeric> *observer = 0,
        Range tabuThreshold = 1,
        float adaptationParameter1 = 1.2,
        short adaptationParameter2 = 2,
        Range maxTabuListLength = 50,
        unsigned int chaosParameter = 3,
        unsigned int repetitionParameter = 3,
        float smoothingParameter = 0.5 );
}
```

Components that implement a neighbor selection rule must conform to the following interface (see Figure 4.37):

```
template <class C>
class TabuNeighborSelection
{
  typedef typename C::AspirationCriterion AspirationCriterion;
  public:
    static N select( S& s,
                     TabuCriterion<Cp>& tabuCriterion,
                     short depth = 1 );
}
```

The following two components realize this interface in accordance with Algorithms 12 and 13:

```
template <class C>
    class BestAdmissibleNeighbor;
template <class C>
    class BestNeighborConsideringPenalties;
      // double C::penaltyFactor;
```

In Figure 4.38, the interface of aspiration criteria was specified as follows:

```
template <class C>
class AspirationCriterion
{
  public:
   static bool check( N& move,
                      TabuCriterion<Cp>& tabuCriterion );
};
```

This interface is realized by two pre-defined components:

```
template <class Cp>
    class NewBestSolutionAspirationCriterion;
template <class Cp>
    class NoAspirationCriterion;
```

The latter component represents the waiving of an aspiration criterion (i.e., **check** always returns **false**).

4.5.4 Miscellaneous Components

In Section 4.4.2.4, we described the use of observer objects to flexibly implement different kinds of introspection needs. The following components correspond to Figure 4.39:

```
template <class C>
  class Observer;
template <class C>
  class StreamingObserver : public Observer<C>;
template <class C>
  class JustTheBestEvaluationToStream : public Observer<C>;
template <class C>
  class JustTheBestEvaluationAndTheCPUTimeToStream
        : public Observer<C>;
```

As an addition to the core functionality, HOTFRAME includes some classes that provide useful functionality such as the computation of hash-codes or the representation of matrices and graphs.

4.6 APPLICATION

In this section, we provide an overview with respect to the actual application of framework components. After illustrating the requirements and procedures of the application of metaheuristics in Section 4.6.1, we discuss an incremental application process in Section 4.6.2. We mostly restrict to exemplary descriptions, which should enable to transfer a respective understanding to other application scenarios.

4.6.1 Requirements and Procedures

First of all, to apply a local search procedure one must be able to formulate the problem accordingly (solution representation, scalar objective function, neighborhood structure, etc.). That is, the problem should fit with regard to the problem-specific abstractions introduced in Section 4.3.1, which correspond to components of the framework architecture.

In conformance to the "no-free-lunch theorem" mentioned on p. 82, HOTFRAME enables the implementation of problem-specific components from scratch to fully adapt metaheuristics for the considered problem. In general, if the problem-type necessitates the use of a special solution space and neighborhood structure, the adaptation of metaheuristics may require non-trivial coding. On the contrary, if the problem fits to some typical solution space and neighborhood structure that are available as pre-defined problem-specific components, the application of HOTFRAME may reduce to a few lines of code.

To apply some metaheuristic-component one needs suitable problem-specific components according to the requirements defined in Section 4.4.2.3. Table 4.1 provides a summary of these requirements for the metaheuristics that have been described in this paper. A "+" indicates the need for the respective component; for the REMTabuCriterion one additionally needs a free function moveNumber (see p. 133). Interface requirements for S, N, S_I, and S_A have been described in Section 4.4.2.2.

	S	N	S_I	S_A
IteratedLocalSearch	+	+		
GeneralSimulatedAnnealing	+	+		
SimulatedAnnealingJetal	+	+		
Tabu Search	+	+		
- StrictTabuCriterionByTrajectory			+	
- REMTabuCriterion				
- StaticTabuCriterion				+
- ReactiveTabuCriterion			+	+

Table 4.1 Summary of the Main Problem-Specific Requirements of Metaheuristics

4.6.1.1 Iterated Local Search. To specialize the component IteratedLocalSearch regarding some common uses, there are pre-defined configuration components. In particular, the following configuration components define the neighbor selection rule to be applied:

```
template <class C> struct CSteepestDescent;
template <class C> struct CFirstDescent;
template <class C> struct CRandomWalk;
```

These simple components conform to the requirements RIteratedLocalSearch defined in Figure 4.28. In each case, a problem-specific configuration is extended by the definition of the feature NeighborSelection. For example, CSteepestDescent is implemented as follows:

```
template <class C>
struct CSteepestDescent
```

```
{
  typedef BestPositiveNeighbor< C > NeighborSelection;

  typedef typename C::T T;
  typedef typename C::Range Range;
  typedef typename C::CNumeric CNumeric;
  typedef typename C::S S;
  typedef typename C::N N;
  typedef C Cp;
};
```

Given some problem-specific configuration component Cp, which defines T, Range, CNumeric, S, and N, one can generate a steepest descent heuristic by:

```
IteratedLocalSearch< CSteepestDescent< Cp > >
```

The C++ template construct consequently enables a direct implementation of the abstract design of Figure 4.28. Thus, a typical application of some metaheuristic component is structured as a three-level hierarchical configuration: numeric base types, problem-specific abstractions solution space and neighborhood structure, and neighbor selection rule.

The actual application of a steepest descent heuristic for an initial solution s means that one has to construct a respective object and to call the member function search:

```
Heuristic< Cp::S >
  *steepestDescent
  = new IteratedLocalSearch< CSteepestDescent < Cp > >;
steepestDescent->search( s );
```

As another example, using the dynamic configuration of IteratedLocalSearch as shown in Figure 4.28, Algorithm 4 (IteratedSteepestDescentWithPerturbationRestarts) can be applied as follows:

```
Heuristic< Cp::S >
  *diversification
  = new IteratedLocalSearch< CRandomWalk< Cp > >
        ( 0, 0, 10, 1, 0, 0, 1, false );
Heuristic< Cp::S >
  *iteratedSteepestDescentWithPerturbationRestarts
  = new IteratedLocalSearch< CSteepestDescent < Cp > >
        ( 0, 0, 0, 5, 0, diversification );
iteratedSteepestDescentWithPerturbationRestarts->search( s );
```

In this example, one first constructs an object that represents the diversification (ten random moves, return of the last traversed solution). This object is applied as a parameter to the actual iterated local search procedure.

4.6.1.2 Simulated Annealing and Variations. The generation of the Algorithms 6–9, which have been defined in Section 4.3.2.2, is based on the component

GeneralSimulatedAnnealing. This component is adapted by using one of the following configuration components:

```
template <class C> struct CClassicSimulatedAnnealing;
template <class C> struct CThresholdAccepting;
template <class C> struct CGreatDeluge;
template <class C> struct CRecordToRecordTravel;
```

These configuration components fix the features cooling schedule, acceptance criterion, and reheating scheme.

The application of a typical simulated annealing procedure for 10,000 moves, using an initial temperature of 100, looks as follows:

```
Heuristic< Cp::S >
  *classicSimulatedAnnealing
  = new GeneralSimulatedAnnealing
          < CClassicSimulatedAnnealing < Cp > >
          ( 0, 0, 10000, 0, 1, true, 100 );
classicSimulatedAnnealing->search( s );
```

The main advantage of the simulated annealing algorithm according to Johnson et al. (1989) is the robustness with respect to the parameter setting. In particular, the user does not need to experiment with the initial temperature. The configuration component CSimulatedAnnealingJetal defines the classic ingredients of simulated annealing as used by Johnson et al. (1989) (exponential acceptance criterion, geometric cooling schedule, no reheating). Such an algorithm, with the default parameter setting, is applied in the following example:

```
Heuristic< Cp::S >
  *simulatedAnnealingJetal
  = new SimulatedAnnealingJetal
          < CSimulatedAnnealingJetal < Cp > >;
  simulatedAnnealingJetal->search( s );
```

4.6.1.3 Tabu Search. The peculiarity of the application of tabu search is that the main variable feature, the tabu criterion, is configured dynamically by an object parameter; see Figure 4.35. This is also the case for the optional explicit diversification procedure, while the neighbor selection rule and the aspiration criterion are defined by configuration components. The pre-defined configuration component CTabuSearchByTabuization specifies the mostly used tabu search variant: The tabu criterion is used to dynamically prohibit certain moves, while the aspiration criterion overwrites a tabu status if the move would lead to a new best solution:

```
template <class C>
struct CTabuSearchByTabuization
{
  typedef NewBestSolutionAspirationCriterion< C >
          AspirationCriterion;
  typedef BestAdmissibleNeighbor
```

```
            < CTabuSearchByTabuization< C > >
          TabuNeighborSelection;

    typedef typename C::T T;
    typedef typename C::Range Range;
    typedef typename C::CNumeric CNumeric;
    typedef typename C::S S;
    typedef typename C::N N;
    typedef typename C::S_A S_A;
    typedef typename C::S_I S_I;
    typedef C Cp;
  };
```

In the same way, the configuration component CTabuSearchByPenalties defines the neighbor selection according to Algorithm 13 (without applying an aspiration criterion):

```
  template <class C> struct CTabuSearchByPenalties
  {
    ...

    typedef NoAspirationCriterion< C > AspirationCriterion;
    typedef BestNeighborConsideringPenalties
            < CTabuSearchByPenalties< C > >
          TabuNeighborSelection;

    static double penaltyFactor;
  }
```

The tabu criterion, which is applied as an object parameter to the general tabu search component, is itself statically parameterized with regard to problem-specific aspects. Figure 4.36 and Table 4.1 summarize respective requirements for different tabu criteria; the dynamic parameterization of a tabu criterion object is described in Section 4.5.3.3. The following code example shows the construction of a tabu criterion object and its use as part of a typical application of tabu search:

```
  TabuCriterion< Cp >
    *staticTabuCriterion
    = new StaticTabuCriterion< Cp >( 0, 7, 1 );

  Heuristic< Cp::S >
    *classicTabuSearch
    = new TabuSearch< CTabuSearchByTabuization< Cp > >
                    ( 0, 0, 1000, 0, staticTabuCriterion );
```

4.6.1.4 Combination of Different Algorithms. An appropriate combination of ideas from different methods often leads to high-quality results. Having available a set of flexible metaheuristics software components greatly simplifies building and applying hybrid search strategies. This is illustrated by the following example:

- Reactive tabu search with move penalties on the basis of tabu degree information
- Neighborhood depth of 2 ("quadratic neighborhood")
- Applying the pilot method (see Duin and Voß (1999)) to evaluate neighbor solutions by, e.g., performing five steepest descent moves
- Explicit diversification by short simulated annealing runs with a different neighborhood than used for the primary search process

Eventually, after writing a few lines of code to construct a corresponding heuristic object, one may even use this object to hybridize an evolutionary algorithm (not described in this paper). That is, a framework provides the user with a powerful toolbox, which can be exploited to easily construct and apply novel algorithms.

4.6.2 Incremental Application

To fully grasp the rules and mechanisms to apply a framework one may have to manage a steep learning curve. Therefore, a framework should enable an incremental application process ("*adoption path*"); see Fink et al. (1999a). That is, the user may start with a simple scenario, which can be successively extended, if needed, after having learned about more complex application mechanisms. Such an evolutionary problem solving approach corresponds to the general tendency of a successive diffusion of knowledge about a new technology and its application; see Allard (1998) and Rogers (1995).

In the following, we describe a typical adoption path for the case that some of the problem-specific standard components are appropriate for the considered application. In this process, we quickly – after completing the first step – arrive at being able to apply several kinds of metaheuristics for the considered problem, while efficiently obtaining high-quality results may require to follow the path to a higher level.

1. **Objective Function:** After selecting an appropriate solution component, one has to derive a new class and to code the computation of the objective function. Of course, one also needs some problem component, which provides problem instance data. All other problem-specific components may be reused without change.

2. **Efficient Neighborhood Evaluation:** In most cases, the system that results from step 1 bears a significant potential with regard to improving run-time efficiency. In particular, one should implement an adaptive computation of the move evaluation (which replaces the default evaluation by computing objective function values for neighbor solutions from scratch). In this context, one may also implement some move evaluation that differs from the default one (implied change of the objective function value).

3. **Problem-Specific Adaptation:** Obtaining high-quality solutions may require the exploitation of problem-specific knowledge. This may refer to the definition (and implementation) of a new neighborhood structure or an adapted tabu criterion by specific solution information or attribute components.

4. Extension of Metaheuristics: While the preceding steps only involve problem-specific adaptations, one may eventually want to extend some metaheuristic or implement a new heuristic from scratch.

We exemplify this incremental adoption path for the "open traveling salesman problem" (TSPO): Given are n "locations" with "distances" c_{ij}, $1 \leq i, j \leq n$, between respective locations. The goal is to obtain a permutation Π that minimizes the sum of distances

$$z(\Pi) = \sum_{i=1}^{n-1} c_{\pi_i, \pi_{i+1}} .$$

By π_i we denote the location that is at position i in the sequence. This problem is obviously suited to the solution component Perm_S. So, in step 1, one can derive TSPO_S from Perm_S:

```
template <class C>
class TSPO_S : public Perm_S<C>
{
 protected:
   TSPO_P<C>& _problem;

 public:
   enum FirstSolution { Identity=0, Random, GivenSolution };
   TSPO_S( TSPO_P<C>& p,
           Observer<C> *observer = 0,
           FirstSolution firstSolution = Identity,
           vector<Range> startPermutation = vector<Range>());
   virtual void evaluate();

   ...
};
```

The constructor implementation includes the determination of the initial permutation according to alternative strategies. The computation of the objective function is implemented in the member function *evaluate*. After implementing a problem component one can immediately apply different metaheuristics by defining the following configuration component (reusing Perm_S_N_Shift, Perm_S_A, and Perm_S_I without change):

```
template <class C>
struct CpTemplateTSPO
{
  typedef typename C::T T;
  typedef typename C::Range Range;
  typedef C CNumeric;

  typedef TSPO_P<C> P;
  typedef TSPO_S<C> S;
  typedef Perm_S_N_Shift<C> N;
```

```
    typedef Perm_S_A<C> S_A;
    typedef Perm_S_I<C> S_I;
};

typedef CpTemplateTSPO<CNumeric> Cp;
```

In step 2, the neighborhood evaluation might be implemented in an efficient way in the following move evaluation operation of TSPO_S :

```
virtual bool computeEvaluation( const Generic_S_N<C>& move,
                                T& evaluation, T& delta );
```

For the considered problem, this would mean subtracting the length of the deleted edges from the sum of the length of the edges inserted, which provides the implied change of the objective function value (delta). If we use this measure to actually evaluate the advantageousness of a move, evaluation results as −delta.

In case one wants to experiment with, e.g., some new neighborhood structure (such as some kind of a 3-exchange), a respective neighborhood component might be derived from Perm_S_N and implemented (step 3). Eventually, applying ideas with regard to variable depth neighborhoods or ejection chain approaches requires to specially code respective algorithms, which is accompanied with a fluent transition to step 4.

If there are no problem-specific components available that fit for the considered problem type, one has to implement respective components in accordance with the defined requirements for the metaheuristics that one wants to apply (see Table 4.1).

4.7 CONCLUSIONS

The principal effectiveness of HOTFRAME regarding competitive results has been demonstrated for different types of problems; see, e.g., Fink and Voß (1999a), Fink (2000), Fink et al. (2000), Fink and Voß (2001). Moreover, we have used the framework for different practical scenarios in an online setting; see Böse et al. (2000) and Gutenschwager et al. (2001).

HOTFRAME may be extended in various directions. On the one hand, new problem-specific standard components may be added. Ideally, this eventually results in a large set of implemented solution spaces and corresponding components, which enables for many common problem types a straightforward and easy framework application. On the other hand, one may add new metaheuristic components.

However, with regard to the first-time use of HOTFRAME by a new user, even Step 1 of the proposed adoption path requires some easy yet crucial knowledge about the framework. For example, one needs to know which metaheuristic components do exist, how can these metaheuristic components be configured, which problem-specific components are needed, how can problem-specific components be combined with each other, which source code files must be included, and so on. To facilitate the use of the framework we experiment with a software generator with a graphical user interface. A software generator builds customized applications on the basis of high-level specifications (see, e.g., Czarnecki and Eisenecker (2000)). That is, declarative specifications are transformed to specialized code. Our generator, which is based on joint work with

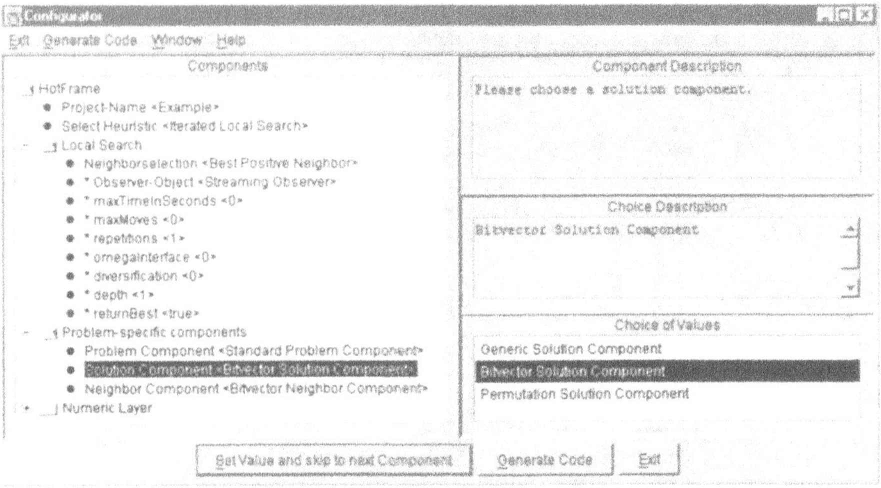

Figure 4.40 Illustration of the User Interface of the Generator

Biskup (2000), is based on a general model which allows to represent framework architectures. Specific frameworks are modeled by using a configuration language. On the basis of the design of HOTFRAME we have defined the framework architecture in this configuration language:

- Metaheuristic components and their static and dynamic parameters
- Problem-specific components and their interdependencies
- Requirements of metaheuristic components regarding problem-specific components
- Associations of components with source code templates and substitution rules regarding the actual configuration

The generator, which is implemented in Java, provides a graphical user interface; see Figure 4.40. The example shows the configuration of an iterated local search component as discussed in Section 4.4. The generator provides an intuitive interface to configure the framework regarding the intended application of some metaheuristic to some problem type. After selection of a metaheuristic, one is provided with customized options to configure dynamic search parameters, problem-specific concepts, and numeric data types. Eventually, the generator produces, in dependence on the specified configuration, customized source code with a few "holes" to be filled by the user. In simple cases, this manual programming is restricted to the coding of the objective function. In general, however, following the argumentation in Section 4.4, one may still have to do considerable parts of the implementation to exploit problem-specific knowledge. Thus, the use of HOTFRAME generally reflects the tradeoff between flexibility and ease-of-use.

5 WRITING LOCAL SEARCH ALGORITHMS USING EASYLOCAL++

Luca Di Gaspero[1] and Andrea Schaerf[2]

[1]Dipartimento di Matematica e Informatica
Università di Udine,
Via delle Scienze 208, 33100 Udine, Italy
digasper@dimi.uniud.it

[2]Dipartimento di Ingegneria Elettrica, Gestionale e Meccanica
Università di Udine,
Via delle Scienze 208, 33100 Udine, Italy
schaerf@uniud.it

Abstract: EASYLOCAL++ is an object-oriented framework that helps the user to design and implement local search algorithms in C++ for a large variety of problems.

In this paper we highlight the usability of EASYLOCAL++ by showing its contribution for the development of a solver for a real-life scheduling problem, namely the COURSE TIMETABLING problem. The COURSE TIMETABLING problem involves hard and soft constraints, and requires, in order to be solved in a satisfactory way, a non-trivial combination of different neighborhood relations. We show all steps of the implementation using EASYLOCAL++, which in our opinion is very straightforward. The resulting code is modular, small, and easy to maintain.

5.1 INTRODUCTION

Local search is a paradigm for optimization which is based on the idea of navigating the search space by iteratively stepping from one state to one of its "neighbors". The neighborhood of a state is composed by the states which are obtained by applying a simple local change to the current one.

Due to this simple schema, the developers of local search algorithms usually write their algorithms from scratch. By contrast, we believe that a clear conceptual methodology can help the user in the development of a local search application, in terms of both the quality of the software and its possible reuse. To support this claim, we designed and developed EASYLOCAL++: an object-oriented (O-O) framework to be used as a general tool for the design and the implementation of local search algorithms in C++.

A framework is a special kind of O-O library, which consists of a network of abstract and concrete classes, that are used through inheritance. The idea is that the framework provides most of the high-level structures of the program, and the user must only define suitable derived classes and implement the virtual functions of the framework. Thanks to the use of virtual functions, frameworks are characterized by the so-called *inverse control* mechanism: The functions of the framework call the user-defined ones and not the other way round.

Therefore, EASYLOCAL++ as a framework for local search, provides the full control of the invariant part of the local search algorithm, and the user classes only supply the problem-specific details and the neigborhood relations.

EASYLOCAL++ is described in details in Di Gaspero and Schaerf (2000). In this paper, we recall its main features and show a comprehensive case study of its use for the solution of a complex scheduling problem, namely the so-called COURSE TIMETABLING problem.

There are various formulations of the COURSE TIMETABLING problem (see e.g., Schaerf (1999)), which differ from each other mostly for the hard and soft constraints (or objectives) they consider. For the sake of simplicity, we describe here a basic version of the problem. Nevertheless, we have also implemented, using EASYLOCAL++, a solver for the more complex version which applies to the Faculty of Engineering of the University of Udine. This latter version is actually in use to generate the real timetable of the faculty.

In addition, we describe a new feature of EASYLOCAL++, not presented in Di Gaspero and Schaerf (2000), which allows the user to build composite neighborhoods, called *kickers*, starting from simple ones. As discussed in Section 5.5, kickers turn out to be very helpful for the solution of the COURSE TIMETABLING problem.

Both EASYLOCAL++ and the COURSE TIMETABLING solver described in this paper are available from the following web page:

`http://www.diegm.uniud.it/schaerf/projects/local++`

5.2 AN OVERVIEW OF EASYLOCAL++

Before describing the main features of EASYLOCAL++, we first briefly recall the local search paradigm. Unfortunately, though, there is not a full agreement in the research community on what the term "local search" precisely includes. We discuss here how we define it, and consequently the way it is implemented in EASYLOCAL++.

5.2.1 The Local Search Paradigm

Local search is a family of general-purpose techniques (or metaheuristics) for optimization problems. These techniques are *non-exhaustive* in the sense that they do not guarantee to find a feasible (or optimal) solution, but they search non-systematically until a specific stop criterion is satisfied.

Given an instance p of a problem P, we associate a *search space* S to it. Each element $s \in S$ corresponds to a potential solution of p, and is called a *state* of P. Local search relies on a function N (depending on the structure of P) which assigns to each $s \in S$ its *neighborhood* $N(s) \subseteq S$. Each state $s' \in N(s)$ is called a *neighbor* of s. A local search algorithm starts from an initial state s_0 and enters a loop that *navigates* the search space, stepping from one state s_i to one of its neighbors s_{i+1}. The neighborhood is usually composed by the states that are obtained by some local changes to the current one (called *moves*).

Local search metaheuristics are therefore built upon the abstract local search algorithm reported in Figure 5.1; they differ from each other according to the strategy they use to select the move in each state, to accept it, and to stop the search (encoded in the *SelectMove*, *AcceptMove* and *StopSearch* functions, respectively). In all techniques, the search is driven by a *cost function* f that estimates the quality of the state. For optimization problems, f generally accounts for the number of violated constraints and for the objective function of the problem.

procedure LocalSearch(Neighborhood N, cost function f);
begin
  s_0 := *InitialSolution*();
  $i := 0$;
  while ($\neg$*StopSearch*(s_i, i)) **do**
  begin
    m := *SelectMove*(s_i, f, N);
    if (*AcceptableMove*(m, s_i, f))
    then $s_{i+1} := s_i \oplus m$;
    else $s_{i+1} := s_i$;
    $i := i + 1$
  end
end;

Figure 5.1 The Abstract Local Search Procedure

The most common local search metaheuristics are *hill climbing* (HC), *simulated annealing* (SA), and *tabu search* (TS) (see, e.g., Aarts and Lenstra (1997) for a detailed review).

One attractive property of the local search paradigm is that different metaheuristics can be combined and alternated to give rise to complex techniques. An example of a simple mechanism for combining different techniques and/or different neighborhood relations is what we call the *token-ring* strategy: Given an initial state and a set of basic local search techniques, the token-ring search makes circularly a run of each

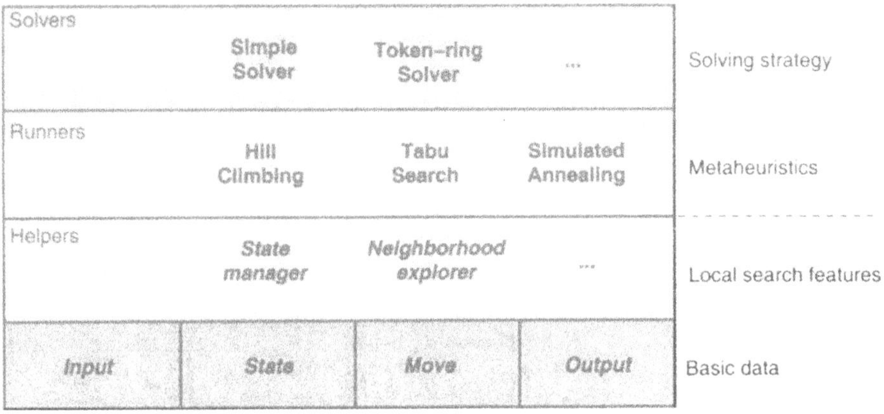

Figure 5.2 The Levels of Abstraction in EASYLOCAL++

technique, always starting from the best solution found by the previous one. The search stops when no improvement is obtained by any algorithm in a fixed number of rounds.

The effectiveness of token-ring search for two runners, called *tandem* search , has been stressed by several authors (see Glover and Laguna (1997)). In particular, when one of the two runners is not used with the aim of improving the cost function, but rather for diversifying the search region, this idea falls under the name of *iterated* local search (see, e.g., Stützle (1998)).

The ability to implement in an easy and clean way complex techniques is in our opinion one of the strengths of EASYLOCAL++.

5.2.2 *The* EASYLOCAL++ *Architecture*

The core of EASYLOCAL++ is composed by a set of cooperating classes that take care of different aspects of local search. The user's application is obtained by writing derived classes for a selected subset of the framework ones. Such user-defined classes contain only the specific problem description, but no control information for the algorithm. In fact, the relationships between classes, and their interactions by mutual method invocation, are completely dealt with by the framework.

The classes of the framework are split in four categories, and are organized in a hierarchy of abstraction levels as shown in Figure 5.2. Each layer of the hierarchy relies on the services supplied by lower levels and provides a set of more abstract operations, as we are going to describe.

5.2.2.1 Data Classes. Data classes are the lowest level of the stack. They maintain the problem representation (class `Input`), the solutions (class `Output`), the states of the search space (class `State`), and the attributes of the moves (class `Move`).

Data classes only store attributes, and have no computing capabilities. The data classes are supplied to the other classes as templates, which need to be instantiated by the user with the corresponding problem-specific types.

5.2.2.2 Helpers. The local search features are embodied in what we call *helpers*. These classes perform actions related to each specific aspect of the search. For example, the Neighborhood Explorer is responsible for everything concerning the neighborhood: selecting the move, updating the current state by executing a move, etc. Different Neighborhood Explorers may be defined in case of composite search, each one handling a specific neighborhood relation used by the algorithm.

Helpers cooperate among themselves: For example, the Neighborhood Explorer is not responsible for the computation of the cost function, and delegates this task to the State Manager which handles the attributes of each state. Helpers do not have their own internal data, but they work on the internal state of the runners, and interact with them through function parameters.

Methods of EASYLOCAL++ helper classes are split in three categories that we call *MustDef*, *MayRedef*, *NoRedef* functions:

***MustDef*:** *pure virtual* C++ functions that correspond to problem specific aspects of the algorithm; they *must* be defined by the user.

***MayRedef*:** *non-pure virtual* C++ functions that come with a tentative definition, which *may* be redefined by the user in case the definition is not satisfactory for the problem at hand. Thanks to the *late binding* mechanism for virtual functions, the program always invokes the user-defined version of the function.

***NoRedef*:** *final* (non-virtual) C++ functions and they *cannot* be redefined by the user.

5.2.2.3 Runners and Solvers. *Runners* represent the algorithmic core of the framework. They are responsible for performing a run of a local search algorithm, starting from an initial state and leading to a final one. Each runner has many data objects for representing the state of the search (current state, best state, current move, number of iterations, etc.), and it maintains links to all the helpers, which are invoked for performing problem-related tasks on its own data.

Runners may be completely abstract from the problem description, and they delegate these tasks to user-supplied classes which comply to a predefined interface.

This allows us to describe metaheuristics through incremental specification: For example, we can directly translate the abstract local search algorithm in C++ code in EASYLOCAL++. Then, we specify actual metaheuristics, by defining the strategy for move selection and acceptance (through the *SelectMove*() and *AcceptableMove*() functions, respectively), and the criterion for stopping the search (by means of the *StopSearch*() function). Three main metaheuristics have been implemented in EASYLOCAL++, namely hill climbing, simulated annealing and tabu search.

The highest abstraction level is constituted by the *solvers*, which represent the external software layer of EASYLOCAL++. Solvers control the search by generating the initial solutions, and deciding how, and in which sequence, runners have to be activated. A solver, for instance, implements the token-ring strategy described before;

other solvers implements other combinations of basic metaheuristics and/or hybrid methods.

Solvers are linked to the runners (one or more) that belong to their solution strategy. In addition, solvers communicate with the external environment, by getting the input and delivering the output.

For runners and solvers, all functions are either *MayRedef* or *NoRedef*, which means that their use requires only to define the appropriate derived class (see the case study below).

New runners and solvers can be added by the user as well. This way, EASYLOCAL++ supports also the design of new metaheuristics and the combination of already available algorithms by the user. In fact, it is possible to describe new abstract algorithms (in the sense that they are decoupled from the problem at hand) at the runner level, while, by defining new solvers, it is possible to prescribe strategies for composing pools of basic techniques.

5.2.2.4 Testers. In addition to the classes presented, the framework provides a set of *tester* classes, which act as a generic user interface of the program and support both interactive and batch runs of the algorithms. Batch runs can be programmed using a dedicated language, called EXPSPEC, which allows comparison of different algorithms and parameter settings with very little intervention of the human operator.

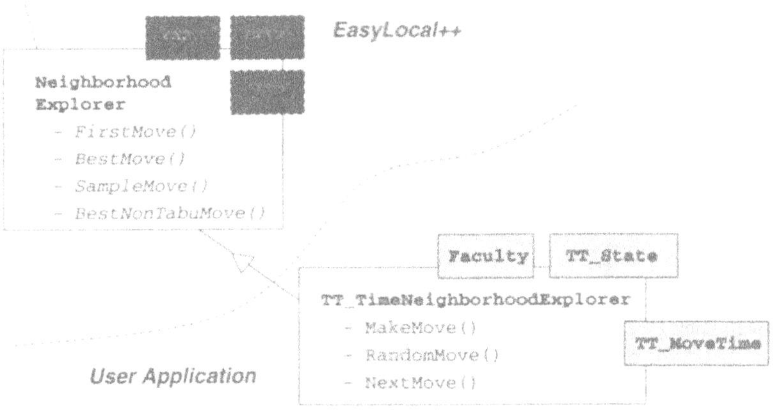

Figure 5.3 A Step of the Development Process

In order to use the framework, the user has to define the data classes and the derived helper, runner, and solver classes which encode the specific problem description. An example of a step of this process is reported in Figure 5.3 in the case of an algorithm for the COURSE TIMETABLING problem (introduced in Section 5.3). The function names drawn in the box `TT_TimeNeighborhoodExplorer` are *MustDef* functions that must be defined in the user's subclass, whereas the functions in the box `Neighbor-hoodExplorer` are *MayRedef* or *NoRedef* ones that are thus already defined in the framework. The data classes `Faculty`, `TT_State`, and `TT_MoveTime`, defined by the user, instantiate the templates `Input`, `State`, and `Move`, respectively.

5.2.2.5 Kickers. A new feature of EASYLOCAL++ (not present in Di Gaspero and Schaerf (2000)) is represented by the so-called *kickers*. A kicker is a complex neighborhood composed by chains of moves belonging to base neighborhoods. The name "kicker" comes from the metaphor of a long move as a kick given the current state in order to perturb it.

Among others, a kicker allows the user to draw a random kick, or to search for the best kick of a given length. In principle, a kicker can generate and evaluate chains of arbitrary length. However, due to the size of the neighborhood, finding the best kick is generally computationally infeasible for lengths of three or more. To reduce the computational cost, the kickers can be programmed to explore only kicks composed by certain combinations of moves. In details, a kicker searches for a chain of moves that are "related" to each other. The intuitive reason is that kickers are invoked when the search is trapped in a deep local minimum, and it is quite unlikely that a chain of unrelated moves could be effective in such situations.

The notion of "related moves" is obviously problem dependent. Therefore, the kicker classes include one *MustDef* function called `RelatedMoves()` that takes as argument two moves and return a boolean value. The user writes the complete definition for her/his specific problem. An example of a kicker, and various definitions of `RelatedMoves()` are provided in Section 5.4.4.

5.3 THE COURSE TIMETABLING PROBLEM

The university COURSE TIMETABLING problem consists in scheduling the lectures of a set of courses of a university in the given set of time periods that compose the week and using the available rooms.

There are q courses $C_1, \ldots, C_q$, p periods $1, \ldots, p$, and r rooms $R_1, \ldots, R_r$. Each course C_i consists of l_i lectures to be scheduled in distinct time periods, and it is taken by s_i students. Each room R_j has a capacity c_j, in terms of number of seats.

The output of the problem is an integer-valued $q \times p$ matrix T, such that $T_{ik} = j$ (with $1 \leq j \leq r$) means that course C_i has a lecture in room R_j at period k, and $T_{ik} = 0$ means that course C_i has no class in period k. We search for the matrix T such that the following *hard* constraints are satisfied, and the violations of the *soft* ones are minimized:

Lectures (hard): The number of lectures of each course C_i must be exactly l_i.

Room Occupation (hard): Two distinct lectures cannot take place in the same room in the same period.

Conflicts (hard): There are groups of courses that have the students in common, called *curricula*. Lectures of courses in the same curriculum must be all scheduled at different times. Similarly, lectures of courses taught by the same teacher must also be scheduled at different times.

We define a conflict matrix CM of size $q \times q$, such that $cm_{ij} = 1$ if C_i and C_j cannot be scheduled in the same period, and $cm_{ij} = 0$, otherwise.

Availabilities (hard): Teachers might not be available for some periods. We define an availability matrix A of size $q \times p$, such that $a_{ik} = 1$ if lectures of course C_i can be scheduled at period k, and $a_{ik} = 0$, otherwise.

Room Capacity (soft): The number of students that attend a course must be less or equal than the number of seats of all the rooms that host its lectures.

Minimum working days (soft): The set of periods p is split in d days of p/d periods each (assuming p divisible by d). Each period, therefore, belongs to a specific week day. The lectures of each course C_i must be spread into a minimum number of days m_i (with $m_i \leq l_i$ and $m_i \leq d$).

This concludes the (semi-formal) description of COURSE TIMETABLING. It is easy to recognize that it is a hard problem. In fact, the underlying decision problem ("does a solution satisfying the hard constraints exist?") can be easily shown to be $\mathcal{NP}$-complete through a reduction from the graph coloring problem.

5.4 SOLVING COURSE TIMETABLING USING EASYLOCAL++

We now show the solution of the COURSE TIMETABLING problem using EASYLOCAL++. We proceed in stages: We start from the data classes, then we show the helpers, the runners, and finally the solvers. The use of the testers is presented in the next section, which is devoted to the execution of the software.

For the sake of simplicity, the classes presented below are simplified with respect to the version used in the actual implementation. For example, the input and output operators (">>" and "<<") and a few auxiliary data and functions are omitted.

Nevertheless, the full software is available from the web page given in Section 5.1.

5.4.1 Data Classes

The input of the problem is represented by the class `Faculty` which stores all data about courses, rooms, periods, and constraints.

```
class Faculty
{public:
  Faculty();
  void Load(string instance); // reads an instance from file(s)

  unsigned Courses() const { return courses; }
  unsigned Rooms() const { return rooms; }
  unsigned Periods() const { return periods; }
  unsigned PeriodsPerDay() const { return periods_per_day; }
  unsigned Days() const { return periods/periods_per_day; }
```

```
  bool  Available(unsigned c, unsigned p) const
      { return availability[c][p]; }
                                   // availability matrix access
  bool Conflict(unsigned c1, unsigned c2) const
     { return conflict[c1][c2]; } // conflict matrix access
  const Course& CourseVector(int i) const
     { return course_vect[i]; }
  const Room& RoomVector(int i) const { return room_vect[i]; }
  const Period& PeriodVector(int i) const
     { return period_vect[i]; }
      // ... data and auxiliary functions (omitted for brevity)
};
```

The classes Course, Period, and Room store data on the corresponding entities (such as name, capacity, location, ...), and we omit their definitions here. The function Load(), which reads data from file(s), is a mandatory member of the input class, and EASYLOCAL++ relies on its presence.

Second, we show the output class, that stores a solution of the problem.

```
class Timetable
{public:
  Timetable(Faculty* f = NULL);
  SetInput(Faculty* f);
  unsigned operator()(unsigned i, unsigned j) const
     { return T[i][j]; }
  unsigned& operator()(unsigned i, unsigned j)
     { return T[i][j]; }
protected:
  virtual void Allocate();
  vector<vector<unsigned> > T;
  Faculty* fp;
};
```

The class stores only the matrix T itself, but no information about the input data, which are supplied through fp, a pointer to the Faculty class.

The constructor with one argument sets the value of the pointer fp, and allocates the matrix T; if no argument is supplied, the constructor leaves the object uninitialized. The function SetInput() initializes (or reinitializes) an already existing object according to the provided input. Finally, the operators "()" simply access the matrix T. As an example of the code, we show the functions SetInput() and Allocate().

```
void Timetable::SetInput(Faculty * f)
{ if (fp != f)
    { fp = f;
      Allocate();
    }
}
```

```
void Timetable::Allocate()
{ T.resize(fp->Courses(), vector<unsigned>(fp->Periods(),0)); }
```

The constructors and `SetInput()` are the mandatory members for a class that instantiates the `Output` template.

The `Input` and `Output` classes (shown above) describe the problem itself and are independent of the search technique. We now move to the classes that implement local search.

We start from the class that instantiates the `State` template. To this aim, we have first to define the search space: We decide to use the set of all possible $q \times p$ output matrices T, with the additional condition that the requirements and all availabilities are satisfied. Thus, for this problem the class `TT_State` that instantiates the template `State` is similar to the class `Timetable` that instantiates the output template. This is not always the case, because in general the search space can be an indirect representation of the output.

The difference stems from the fact that `TT_State` includes also redundant data structures used for efficiency purposes. These data store aggregate values, namely the number of lectures in a room per period, the number of lectures of a course per day, and the number of teaching days for a course. Consequently, we define as the `State` template a class derived from `Timetable`, as shown below:

```
class TT_State : public Timetable
{public:
  TT_State(Faculty * f = NULL)   : Timetable(f) {}
  unsigned RoomLectures(unsigned i, unsigned j) const
          { return room_lectures[i][j]; }
  unsigned CourseDailyLectures(unsigned i, unsigned j) const
          { return course_daily_lectures[i][j]; }
  unsigned WorkingDays(unsigned i) const
          { return working_days[i]; }
 // ... (functions that manipulates redundant data omitted)
 protected:
  void Allocate();
  // ... (redundant data omitted)
};
```

Redundant data structures make the state updates more expensive, whereas they save time for evaluations of the state and its neighbors. Their presence is justified by the fact that, as we will see later, the number of evaluations of candidate moves (neighbors) is much larger than the number of state updates.

Similarly to the `Output`, for the `State` class the default constructor, the constructor that receives a pointer to the `Input` class, and the function `SetInput()` (inherited in this case) are mandatory. Notice that in this case `SetInput()` is inherited, but it invokes the proper version of the `Allocate()` function (the one of the class `TT_State`), thanks to the late binding of virtual functions.

The fact that requirements and availabilities are always satisfied implies that they need not to be included in the cost function; on the other hand, functions that generate or modify states should be defined in such a way that these constraints are kept satisfied.

The first neighborhood relation that we consider is defined by the reschedule of a lecture in a different period (leaving the room unchanged). For implementing this kind of move, we define a class, called `TT_MoveTime`, as follows:

```
class TT_MoveTime
{public:
  unsigned course, from, to;
};
```

The effect of a move of this type is that the lecture of the course of index `course` is moved from the period `from` to the period `to`.

It is important to observe that not all possible values of a `TT_MoveTime` object represents a *feasible* move. A move is feasible if the value of `course` is included between 0 and $q-1$, and the values of `from` and `to` are between 0 and $p-1$ (remind that in C/C++ arrays start from 0); in addition, in the current state `st` there must be a lecture of course `course` in period `from` and none in period `to`, and the period `to` is available for the course.

The other move type that we consider, which is used in alternation to `TT_MoveTime`, is called `TT_MoveRoom` and its effect is to replace the room of a lecture of a given course. The corresponding data class is the following.

```
class TT_MoveRoom
{public:
  unsigned course, period, old_room, new_room;
};
```

A `TT_MoveRoom` move replaces the `old_room` with the `new_room` for the lecture of the given `course` scheduled in the given `period`. A move is feasible if `course` is between 0 and $q-1$, `period` is between 0 and $p-1$, and `old_room` and `new_room` are between 1 and r (0 represents the absence of a lecture). In addition, there must be a lecture of `course` in `period` in `old_room`.

Notice that in order to select and apply a `TT_MoveRoom` move from a given state `st` we only need the course, the period, and the new room. Nevertheless, it is necessary also to store the old room for the management of the prohibition mechanisms. In fact, the tabu list stores only the "raw" moves regardless of the states in which they have been applied.

5.4.2 Helpers

We need to define four helpers, namely the State Manager, the Output Producer, the Neighborhood Explorer, and the Prohibition Manager. Given that we deal with two move types, we have to define two Neighborhood Explorers and two Prohibition Managers, therefore, we have to define six helpers all together.

We start describing the State Manager, which is represented by the following class.

```
class TT_StateManager : public StateManager<Faculty,TT_State>
{public:
  TT_StateManager(Faculty*);
  void RandomState(TT_State&);   // mustdef
 protected:
  fvalue Violations(const TT_State& as) const;    // mayredef
  fvalue Objective(const TT_State& as) const;     // mayredef

  void UpdateRedundantStateData(TT_State& as) const; // mayredef
  void ResetState(TT_State& as);

  unsigned Conflitcs(const TT_State& as) const;
  unsigned RoomOccupation(const TT_State& as) const;
  unsigned RoomCapacity(const TT_State& as) const;
  unsigned MinWorkingDays(const TT_State& as) const;
   // ... other functions
};
```

We first describe the function `RandomState()` that assigns all lectures to random (but available) periods in random rooms.

```
void TT_StateManager::RandomState(TT_State& as)
{ ResetState(as); // make all elements of as equal to 0
  for (unsigned c = 0; c < p_in->Courses(); c++)
    { unsigned lectures = p_in->CourseVector(c).Lectures();
      for (unsigned j = 0; j < lectures; j++)
        { unsigned p;
          do // cycle until the period is free and available
            p = Random(0,p_in->Periods()-1);
          while (as(c,p) != 0 || !p_in->Available(c,p));
          as(c,p) = Random(1,p_in->Rooms());
        }
    }
  UpdateRedundantStateData(as);
}
```

The function is composed by a double cycle that assigns a (distinct) random period and a random room to each lecture. The data member `p_in` is a pointer to the input class, that is inherited from the abstract `StateManager`.

The functions `Violations()` and `Objective()` return the sum of the number of violations for each type of constraint (hard and soft). For simplicity, weights of soft constraints are assumed to be all equal to 1.

```
int TT_StateManager::Violations(const TT_State& as) const
{ return Conflitcs(as) + RoomOccupation(as); }

int TT_StateManager::Objective(const TT_State& as) const
{ return RoomCapacity(as) + MinWorkingDays(as); }
```

Among the functions that compose the cost, we show `RoomOccupation()` and `RoomCapacity()` omitting the other two that are similar.

```
unsigned TT_StateManager::RoomOccupation(const TT_State& as) const
{ unsigned cost = 0;
  for (unsigned p = 0; p < p_in->Periods(); p++)
    for (unsigned r = 1; r <= p_in->Rooms(); r++)
      if (as.RoomLectures(r,p) > 1)
        cost += as.RoomLectures(r,p) - 1;
  return cost;
}

unsigned TT_StateManager::RoomCapacity(const TT_State& as) const
{ unsigned cost = 0;
  for (unsigned c = 0; c < p_in->Courses(); c++)
    for (unsigned p = 0; p < p_in->Periods(); p++)
      { unsigned r = as(c,p);
        if (r != 0 &&
           (  p_in->RoomVector(r).Capacity()
            < p_in->CourseVector(c).Students()) )
          cost++;
      }
  return cost;
}
```

Notice that the function `RoomOccupation()` uses the redundant structure `RoomLectures`. These structures are used more intensively by the functions of the **Neighborhood Explorers** that compute the variations of the cost (see below).

We now move to the description of the first **Neighborhood Explorer**, which is represented by the class `TT_TimeNeighborhoodExplorer`, defined as follows.

```
class TT_TimeNeighborhoodExplorer
  : public NeighborhoodExplorer<Faculty,TT_State,TT_MoveTime>
{public:
  TT_TimeNeighborhoodExplorer(StateManager<Faculty,TT_State>*, Faculty*);
  void RandomMove(const TT_State&, TT_MoveTime&);          // mustdef
  bool FeasibleMove(const TT_State&, const TT_MoveTime&);  // mayredef
  void MakeMove(TT_State&,const TT_MoveTime&);             // mustdef
 protected:
  fvalue DeltaViolations(const TT_State&, const TT_MoveTime&); // mayredef
  fvalue DeltaObjective(const TT_State&, const TT_MoveTime&);  // mayredef

  int DeltaConflitcs(const TT_State& as, const TT_MoveTime& mv) const;
  int DeltaRoomOccupation(const TT_State& as, const TT_MoveTime& mv) const;
  int DeltaMinWorkingDays(const TT_State& as, const TT_MoveTime& mv) const;
  void NextMove(const TT_State&,TT_MoveTime&);    // mustdef
 private:
  void AnyNextMove(const TT_State&,TT_MoveTime&);
  void AnyRandomMove(const TT_State&, TT_MoveTime&);
};
```

We present the implementation of some of the function of the class. We start with `NextMove()` that enumerates all possible moves and is needed to explore exhaustively the neighborhood.

```
void TT_TimeNeighborhoodExplorer::NextMove(const TT_State& as,
                                           TT_MoveTime& mv)
{ do
    AnyNextMove(as,mv);
  while (!FeasibleMove(as,mv));
}
```

For modularity, we separate the generation of moves in circular lexicographic order (function `AnyNextMove()`), from the feasibility check (function `FeasibleMove()`):

```
void TT_TimeNeighborhoodExplorer::AnyNextMove(const TT_State& as,
                                               TT_MoveTime& mv)
{ if (mv.to < p_in->Periods() - 1)
    mv.to++;
  else if (mv.from < p_in->Periods() - 1)
      { mv.from++;
        mv.to = 0;
      }
  else
    { mv.course = (mv.course + 1) % p_in->Courses();
      mv.from = 0;
      mv.to = 1;
    }
}

bool TT_TimeNeighborhoodExplorer::FeasibleMove(const TT_State& as,
                                         const TT_MoveTime& mv)
{  return as(mv.course,mv.from) != 0 && as(mv.course,mv.to) == 0
          && p_in->Available(mv.course,mv.to);
}
```

Next, we show the function `MakeMove()` that performs the move by changing the state and updating the redundant data accordingly.

```
void TT_TimeNeighborhoodExplorer::MakeMove(TT_State& as,
                                     const TT_MoveTime& mv)
{      // update the state matrix
  unsigned room = as(mv.course,mv.from);
  as(mv.course,mv.to) = room;
  as(mv.course,mv.from) = 0;

     // update the redundant data
  unsigned from_day = mv.from / p_in->PeriodsPerDay();
  unsigned to_day= mv.to / p_in->PeriodsPerDay();
  as.DecRoomLectures(room,mv.from);
  as.IncRoomLectures(room,mv.to);

  if (from_day != to_day)
    {
      as.DecCourseDailyLectures(mv.course,from_day);
      as.IncCourseDailyLectures(mv.course,to_day);
      if (as.CourseDailyLectures(mv.course,from_day) == 0)
        as.DecWorkingDays(mv.course);
      if (as.CourseDailyLectures(mv.course,to_day) == 1)
        as.IncWorkingDays(mv.course);
    }
}
```

Finally, we show two of the functions that compute the difference of violations that a move would cause, namely `DeltaRoomOccupation()` and `DeltaMinWorkingDays()`.

```
int TT_TimeNeighborhoodExplorer::DeltaRoomOccupation
                (const TT_State& as, const TT_MoveTime& mv) const
{ int cost = 0;
  unsigned r = as(mv.course,mv.from);
  if (as.RoomLectures(r,mv.from) > 1) cost--;
  if (as.RoomLectures(r,mv.to) > 0) cost++;
  return cost;
}

int TT_TimeNeighborhoodExplorer::DeltaMinWorkingDays
                (const TT_State& as, const TT_MoveTime& mv) const
{
  unsigned from_day = mv.from / p_in->PeriodsPerDay();
  unsigned to_day = mv.to / p_in->PeriodsPerDay();

  if (from_day == to_day) return 0;
  if (as.WorkingDays(mv.course)
        <= p_in->CourseVector(mv.course).MinWorkingDays()
      && as.CourseDailyLectures(mv.course,from_day) == 1
      && as.CourseDailyLectures(mv.course,to_day) >= 1)
    return 1;
  if (as.WorkingDays(mv.course)
        < p_in->CourseVector(mv.course).MinWorkingDays()
      && as.CourseDailyLectures(mv.course,from_day) > 1
      && as.CourseDailyLectures(mv.course,to_day) == 0)
    return -1;
  return 0;
}
```

Notice that the above functions do not compute the values in the two states, but focus exactly on the changes. This way, thanks also to the redundant data, they have constant computational cost instead of quadratic.

The **Prohibition Manager** associated with the class `TT_TimeNeighborhood-Explorer` is represented by the class `TT_TimeTabuListManager`. The full code of the class, which consists of a constructor and of the only function that needs to be defined, `Inverse()`, is included within the class definition as shown below.

```
class TT_TimeTabuListManager : public TabuListManager<TT_MoveTime>
{
public:
  TT_TimeTabuListManager(int min = 0, int max = 0)
    : TabuListManager<TT_MoveTime>(min,max) {}
protected:
  bool Inverse(const TT_MoveTime& m1, const TT_MoveTime& m2) const
  { return m1.course == m2.course
          && (m1.from == m2.to || m2.from == m1.to); }
};
```

According to the above definition of the function `Inverse()`, we consider a move inverse of another one if it involves the same course and it has one of the two periods in common.

The **Neighborhood Explorer** based on the `TT_MoveRoom` moves is implemented by the class `TT_RoomNeighborhoodExplorer`. This class is very similar to `TT_TimeNeighborhoodExplorer`, except that the functions that contribute to

DeltaViolations() and DeltaObjective() are only DeltaSamePeriodSameRoom() and DeltaRoomCapacity(), respectively, because only these types of constraints are interested in TT_MoveRoom() moves.

We omit the definition of the class, but we show the three functions RandomMove(), MakeMove(), and DeltaRoomOccupation().

```
void TT_RoomNeighborhoodExplorer::RandomMove
                          (const TT_State& as, TT_MoveRoom& mv)
{ mv.course = Random(0,p_in->Courses() - 1);
  do
    mv.period = Random(0,p_in->Periods() - 1);
  while (as(mv.course,mv.period) == 0);
  mv.old_room = as(mv.course,mv.period);
  do
    mv.new_room = Random(1,p_in->Rooms());
  while (mv.new_room == mv.old_room);
}

void TT_RoomNeighborhoodExplorer::MakeMove
                          (TT_State& as, const TT_MoveRoom& mv)
{
  as(mv.course,mv.period) = mv.new_room;
  as.DecRoomLectures(mv.old_room,mv.period);
  as.IncRoomLectures(mv.new_room,mv.period);
}

int TT_RoomNeighborhoodExplorer::DeltaRoomOccupation
                    (const TT_State& as, const TT_MoveRoom& mv) const
{
  int cost = 0;
  if (as.RoomLectures(mv.old_room,mv.period) > 1)
    cost--;
  if (as.RoomLectures(mv.new_room,mv.period) > 0)
    cost++;
  return cost;
}
```

The corresponding Prohibition Manager is straightforward, and we do not present it here.

The Output Producer transforms a state into a solution, and vice versa. It also reads and writes solutions from file, in order to store results of current runs, and use them as starting point for future runs. In our case, given that the Timetable and TT_State classes are very similar, its functions are very simple. We show for example, the function InputState(), that passes from an output to a state.

```
void TT_OutputManager::InputState
                          (TT_State& as, const Timetable& tt) const
{ for (unsigned i = 0; i < p_in->Courses(); i++)
    for (unsigned j = 0; j < p_in->Periods(); j++)
      as(i,j) = tt(i,j);
  p_sm->UpdateRedundantStateData(as);
}
```

5.4.3 *Runners and Solvers*

We create runners which implement HC and TS for each of the two move types. No function needs to be defined for these four runners, except the constructor (which in C++ is not inherited).

The definition of the HC runner for `TT_MoveTime` move is the following.

```
class TT_TimeHillClimbing : public HillClimbing<Faculty,TT_State,TT_MoveTime>
{public:
  TT_TimeHillClimbing(StateManager<Faculty,TT_State>* psm,
      NeighborhoodExplorer<Faculty,TT_State,TT_MoveTime>* pnhe, Faculty* pin)
        : HillClimbing<Faculty,TT_State,TT_MoveTime>(psm,pnhe,pin)
            { SetName("HC-Timetabler"); }

};
```

The constructor only invokes the constructor of the base EASYLOCAL++ class, and sets the name of the runner. The name is necessary only to use the runner inside the `Tester` as described in Section 5.5. The definition of the other three runners is absolutely identical, and therefore, it is omitted.

We define now the token-ring solver, which is used for running various tandems of runners. The runners participating to the solver are selected at run-time by using the functions `AddRunner()` and `ClearRunners()`; therefore, the composition does not require any other programming effort, but, similarly to the runners, the solvers' derivation is only a template instantiation.

```
class TT_TokenRingSolver : public TokenRingSolver<Faculty,Timetable,TT_State>
{public:
  TT_TokenRingSolver(StateManager<Faculty,TT_State>* psm,
          OutputManager<Faculty,Timetable,TT_State>* pom,
          Faculty* pin, Timetable* pout)
    :  TokenRingSolver<Faculty,Timetable,TT_State>(psm,pom,pin,pout) {}
};
```

This completes the description of the software belonging to the solver itself. If the solver is embedded in a larger project, the extra code necessary to run it obviously depends on the host program. If instead the solver is stand-alone, we resort to the tester, a module of EASYLOCAL++ which allows the user to run (and debug) the solver, as explained in Section 5.5.

We describe separately the kickers, which in the current version of EASYLOCAL++ cannot be invoked by the solver, but interact only with the tester.

5.4.4 *Kickers*

In the current version of EASYLOCAL++, a kicker can be either *monomodal* or *bimodal*, which means that it has one or two atomic neighborhood components, respectively.

For our problem, we have seen that we have two move types. Therefore, we define a bimodal kicker that deals with them. The definition of the class is the following.

```
class TT_TimeRoomKicker : public BimodalKicker<Faculty,TT_State,
                                               TT_MoveTime,TT_MoveRoom>
{
public:
```

```
  TT_TimeRoomKicker(TT_TimeNeighborhoodExplorer *tnhe,
                    TT_RoomNeighborhoodExplorer *rnhe);
  bool RelatedMoves(const TT_MoveTime &mv1, const TT_MoveTime &mv2);
  bool RelatedMoves(const TT_MoveTime &mv1, const TT_MoveRoom &mv2);
  bool RelatedMoves(const TT_MoveRoom &mv1, const TT_MoveTime &mv2);
  bool RelatedMoves(const TT_MoveRoom &mv1, const TT_MoveRoom &mv2);
};
```

As already mentioned, the four functions `RelatedMoves()` are used to limit the possible compositions of the kicks (chains). There are four different definitions, rather than a single more complex one because of different parameter types. As an example, we provide two of the four definitions:

```
bool TT_TimeRoomKicker::RelatedMoves
                    (const TT_MoveTime &mv1, const TT_MoveRoom &mv2)
{ return mv1.to == mv2.period; }

bool TT_TimeRoomKicker::RelatedMoves
                    (const TT_MoveRoom &mv1, const TT_MoveRoom &mv2)
{ return mv1.period == mv2.period && mv1.new_room == mv2.old_room; }
```

The first one states that a `TT_MoveRoom` move `mv2`, in order to be related to the `TT_MoveTime` move `mv1` that precedes it, must regard the period in which `mv1` has rescheduled the lecture (i.e., `mv1.to`). Similarly, a `TT_MoveRoom` move `mv2` is related to another `TT_MoveRoom` move `mv1` if they regard the same period and `mv2` removes a lecture in the room occupied by `mv1`.

These definitions are justified intuitive by the observation that chains of moves that are not related, in the above sense, rarely give a joint improvement to the cost function.

We have concluded the code needed in order to use the kicker. In fact, the search for the best kick is completely done by an exhaustive search implemented abstractly in EASYLOCAL++, by the function `BestKick()`. It relies on the functions of the helper, like `NextMove()`, and on a vector of states that stores the intermediate states generated by the kick move.

The experiments show that the kicker takes a few minutes to find the best kick of size 3. Although this is undoubtedly a long time, quite surprisingly, a kick of size 3 is often able to improve on the best solutions found by the token-ring solver.

5.5 DEBUGGING AND RUNNING THE SOLVER

The current version of EASYLOCAL++ includes a text-based tester, which acts as a generic user interface. The tester stores pointers to all the objects involved in the solution (helpers, runners, solvers, and kickers) and activates them according to the user's inputs.

The tester can run in two modes: interactive or batch. In the interactive mode, it maintains (and displays) a current state and proposes to the user a set of possible actions upon the current state. The actions are grouped in three menus:

Move Menu: Allows the user to choose one neighborhood relation (both simple or composite), and to select and perform one move belonging to that neighborhood. The move can be either the best one, or a random one, or supplied by the user. The state obtained by the execution of the move becomes the new current one. The user can also request additional information or statistics about neighborhoods, such as their cardinality, the percentage of improving moves, etc.

Run Menu: Allow the user to select one runner, input its parameters, and run it. The best state reached by the runner becomes the new current state of the tester.

State menu: This menu includes all utilities necessary for managing the current state: The user can write it to a file, replace it with a new one read from file or generated at random, print all violations associated with it, etc.

To be precise about the State menu, what the tester writes to the file is not the state itself, but rather the solution (output). This is because, in general, the state, which can be an implicit representation of the solution, does not necessarily have an intuitive meaning. To this regard, the tester uses the Output Producer, which translates states into solutions and vice versa. A fragment of an example of a (random) output file is given in Table 5.1, where `27`, `GRT`, `DIS`, etc. are room names.

Course	(Teacher)	Monday				Tuesday				Wednesday					
		9	11	14	16	9	11	14	16	9	11	14	16	9	...
ArcTec	(Rossi)	-	-	-	GRT	E	-	27	-	D	-	-	-	50	...
TecCos	(Gialli)	-	-	36	-	30	-	-	-	DIS	-	-	-	L	...
GeoTec	(Verdi)	-	-	30	-	-	30	-	-	29	-	-	37	-	...
CosIdr	(Neri)	-	-	30	-	-	-	31	F	-	-	29	-	31	...
CalStr	(Viola)	-	-	-	-	-	-	A	-	DIS	-	-	38	-	...
IngTer	(Rossi)	-	-	-	-	-	-	27	-	-	-	-	-	-	...
IdrApp	(Marroni)	-	-	-	-	-	37	D	-	A	-	-	-	B	...
RilAmb	(Bianchi)	L	-	-	-	-	-	A	-	-	B	-	-	-	...
EcoOrg	(Verdi)	-	-	E	-	-	-	-	-	-	50	-	-	31	...

Table 5.1 Output File

The output is stored in the presented form, which is both human- and computer-readable, so that the user can interact with the tester. For example, he/she can write the current state on a file, modify it by hand, have the tester read it again, and invoke a runner on it.

In the batch mode, the tester reads a file written in the language EXPSPEC (see Di Gaspero and Schaerf (2000)), and performs all experiments specified in the file. An example of a fragment of a EXPSPEC file is the following.

```
Instance "FirstTerm"
{ Trials: 20;
  Output prefix: "TS2";
  Runner tabu search "TS-Timetabler"
  { min tabu tenure: 25;
    max tabu tenure: 40;
    max idle iteration: 1000;
    max iteration: 20000;
  }
  Runner tabu search "TS-Roomtabler"
  { min tabu tenure: 10;
    max tabu tenure: 10;
    max idle iteration: 500;
  }
}
```

The above EXPSPEC block instructs the tester to run 20 trials of the token-ring solver composed by the two TS runners on the instance called `FirstTerm`.

Various parameters are specified for each runner. For example, let us analyze the parameters of the first runner: The fact that `min tabu tenure` and `max tabu tenure` are different prescribes that the tabu list must be dynamic. Specifically, this means that when a move enters in the list at iteration i, it is assigned a random number between $i+25$ and $i+40$. This number represents the iteration in which the move will leave the tabu list. In this way, the size of the list is not fixed, but varies dynamically in the interval 25–40. The stop criterion is twofold: the search is interrupted either after 1000 iteration without an improvement of the best state, or after 20000 total iterations.

According to the specification, the best states of the trials are written into a set of files whose names start with `TS2` (i.e., `TS2-01.out`, `TS2-02.out`, ..., `TS2-20.out`). At the end of a block of experiments like this, the tester gathers statistics about results and running times, and then it moves to the next specification block.

The batch mode is especially suitable for massive night or weekend runs, in which the tester can perform all kinds of experiments in a completely unsupervised mode.

5.6 DISCUSSION AND CONCLUSIONS

The main goal of EASYLOCAL++, and similar systems, is to simplify the task of researchers and practitioners who want to implement local search algorithms. The idea is to leave only the problem specific programming details to the user. We have presented the main part of this process for a non-trivial scheduling problem.

The architecture of EASYLOCAL++ prescribes a precise methodology for the design of a local search algorithm. The user is required to identify exactly the entities of the problem at hand, which are factorized in groups of related classes in the framework: Using EASYLOCAL++ the user is forced to place each piece of code in the "right" position. We believe that this feature helps in terms of conceptual clarity, and it makes easier the reuse of the software and the overall design process.

Furthermore, we have shown that the composition of the basic entities in EASYLOCAL++ is straightforward. The user can obtain with a limited effort a set of local search algorithms, possibly based on different neighborhood relations, which can be composed in different ways.

A few other systems for local search or other techniques have been proposed in the literature, such as LOCALIZER++ (Michel and van Hentenryck (2001a)), HOTFRAME (Fink et al. (1999b)), and ABACUS (Jünger and Thienel (1997)). A comparison between EASYLOCAL++ and those systems is provided in Di Gaspero and Schaerf (2000).

Here we just want to mention that, in our opinion, the strength of EASYLOCAL++ is that it makes a balanced use of O-O features needed for the design of a framework, namely templates and virtual functions. In fact, on the one hand, data classes are provided through templates, giving a better computational efficiency and a type-safe compilation. On the other hand, algorithm's structure is implemented through virtual functions, giving the chance of incremental specification in hierarchy levels and providing the inverse control communication.

This balance does not fit perfectly for every component. For example, there is an implementation problem about kickers. Recall that we have defined monomodal and bimodal kickers, however, it would be very handful to define a generic *multimodal* kicker, which would be composed by a dynamic number of neighborhood relations. Unfortunately, due to the fact that in EASYLOCAL++ moves are supplied through templates (see Di Gaspero and Schaerf (2000) for the reasons of this design choice), it is not possible to define such a multimodal kicker without static type-checking violations.

Acknowledgments

We wish to thank Marco Cadoli for helpful suggestions. Luca Di Gaspero is partly funded by a grant from the University of Udine, "Progetto Giovani Ricercatori 2000".

6 INTEGRATING HEURISTIC SEARCH AND ONE-WAY CONSTRAINTS IN THE IOPT TOOLKIT

Christos Voudouris and Raphaël Dorne

Intelligent Complex Systems Research Group
BTexact Technologies
Adastral Park, PP12/MLB1, Martlesham Heath
Ipswich, IP5 3RE
Suffolk, United Kingdom
{chris.voudouris,raphael.dorne}@bt.com

Abstract: Heuristic search techniques are known for their efficiency and effectiveness in solving $\mathcal{NP}$-hard problems. In this chapter, we present a heuristic search framework allowing the synthesis and evaluation of a variety of algorithms and also a library of one-way constraints for problem modelling. The integration of the two is explained in the context of the iOpt toolkit. iOpt is a large software system comprising several libraries and frameworks dedicated to the development of combinatorial optimization applications based on heuristic search.

6.1 INTRODUCTION

The iOpt toolkit research project at BT Laboratories was motivated by the lack of appropriate tools to support the development of real-world applications, which are based on heuristic search methods. The goal, originally set, was to develop a set of software frameworks and libraries dedicated to heuristic search to address this problem.

Following contemporary thinking in software engineering, iOpt allows code reuse and various extensions to reduce as much as possible the fresh source code required to be written for each new application. Furthermore, application development is additive in the sense that the toolkit is enhanced by each new application, reducing further the effort in developing similar applications to those already included. iOpt is fully written in the Java programming language with all the acknowledged benefits associated with Java, including:

- easier deployment in different operating systems and environments,
- stricter object-oriented programming,
- compatibility with modern 3-tier architectures such as Enterprise Java Beans, and also
- better integration with visualization code in Web browsers and stand alone applications.

Up to recently, Java was considered too inefficient (e.g., compared to C++) for developing optimization applications. This situation has significantly improved with the introduction of new compilation technologies such as Sun's HotSpot and the ever improving performance of PCs. iOpt has taken advantage of these two developments using Java as a promising alternative to C++; at least from the software engineer's point of view who is sometimes willing to sacrifice ultimate performance for ease of use and better integration. Overall, the choice of Java in the development of iOpt has significantly reduced prototyping times and allowed for a better object orientation within the toolkit.

iOpt incorporates many libraries and frameworks such as a constraint library, a generic problem modelling framework, domain specific frameworks for particular classes of problems (e.g., scheduling), a heuristic search framework, interactive visualization facilities for constraint networks, scheduling systems, algorithm configuration and monitoring as well as a blackboard-based set of libraries for developing distributed collaborative problem solving applications. For more details on the toolkit, the reader may refer to Voudouris et al. (2001).

In the following sections, we look at the constraint library and the heuristic search framework contained in iOpt and describe how they are being used to offer a foundation for modelling and solving combinatorial optimization problems.

6.2 ONE-WAY CONSTRAINTS

The iOpt toolkit, to facilitate problem modelling, provides declarative programming capabilities within the Java programming language. The paradigm followed is similar to C++ libraries for constraint programming such as ILOG Solver (Puget (1995)), consisting of a number of built-in relations which are available to the user to state her/his problem. An efficient constraint satisfaction algorithm is utilized to transparently maintain these relations. In contrast to constraint programming tools, relations available in iOpt are based exclusively on one-way dataflow constraints (Zanden et al. (1999), Zanden et al. (1994)).

A one-way dataflow constraint is an equation (also called) formula, in which the value of the variable on the left hand side is determined by the value of the expression on the right hand side. For example, a programmer could use the equation $y = x + 10$ to constrain the value of y to be always equal to the value of x plus 10. More formally, a one-way constraint is an equation of the form (see Zanden et al. (1999)):

$$u = C(p_0, p_1, p_2, ..., p_n) \tag{6.1}$$

where each p_i is a variable that serves as a parameter to the function C. Arbitrary code can be associated with C that uses the values of the parameters to compute a value. This value is assigned to variable u. If the value of any p_i is changed during the program's execution, u's value is automatically recomputed (or incrementally updated in constant time). Note that u has no influence on any p_i as far as this constraint is concerned; hence, it is called one-way.

6.3 CONSTRAINT SATISFACTION ALGORITHMS FOR ONE-WAY CONSTRAINTS

The two most common constraint satisfaction schemes for one-way constraints are the mark/sweep strategy (Hudson (1991), Zanden et al. (1994)) and the topological ordering strategy (Hoover (1987), Zanden et al. (1994)). A mark/sweep algorithm has two phases. In the mark phase, constraints that depend on changed variables are marked out-of-date. In the sweep phase, constraints whose values are requested are evaluated and the constraints are marked as up-to-date. If constraints are only evaluated when their values are requested, then the sweep phase is called a lazy evaluator. If all the out-of-date constraints are evaluated as soon as the mark phase is complete, then the sweep phase is called an eager evaluator.

A topological ordering algorithm assigns numbers to constraints that indicate their position in topological order. Like the mark/sweep strategy, the topological ordering strategy has two phases. A numbering phase that brings the topological numbers up-to-date and a sweep phase that evaluates the constraints. The numbering phase is invoked whenever an edge in the constraint dataflow graph changes. The sweep phase can either be invoked as soon as a variable changes value or it can be delayed to allow several variables to be changed. The sweep phase uses a priority queue to keep track of the next constraint to evaluate. Initially all constraints that depend on a changed variable are added to the priority queue. The constraint solver removes the lowest numbered constraint from the queue and evaluates it. If the constraint's value changes, all constraints that depend on the variable determined by this constraint are added to the priority queue. This process continues until the priority queue is exhausted.

For the purposes of this project, we evaluated both topological ordering and mark/sweep algorithms. We opted for a mark/sweep algorithm for the following reasons also pointed out in Zanden et al. (1999):

- mark/sweep methods are more flexible supporting both eager and lazy evaluation modes,

- they are easy to implement even when extensions such as cycle handling and relations which change the constraint graph (e.g., pointer variables or conditionals; Zanden et al. (1994)) are added on top of the basic functionality,

- because of their simplicity, they can be the subject of efficient implementations something very important especially for the Java platform.

Although, theory suggests that topological algorithms should be faster at least for basic constraint types (Zanden et al. (1999)), our experience (with both approaches implemented in the Java language) showed that topological algorithms were slower, which is in agreement with the findings in Zanden et al. (1999). When cycles and dynamic relations are added then mark/sweep algorithms also become theoretically faster (Zanden et al. (1994)).

One-way constraints have been used extensively by the GUI community in building interactive user interfaces (Myers et al. (1997), Myers et al. (1990)), and also in circuits (Alpern et al. (1990)) and spreadsheet programming (e.g., MS EXCEL). LOCALIZER, a scripting language for implementing local search algorithms also uses this approach for modelling combinatorial optimization problems (Michel and van Hentenryck (1998), Michel and van Hentenryck (1999)). A specialized topological ordering algorithm is deployed there (Michel and van Hentenryck (1998)).

In LOCALIZER, one-way functional constraints are called invariants. We will also use the same term to refer to one-way constraints when used in the context of heuristic search. This would help distinguishing them from multi-way relational constraints as they are defined and used in constraint programming and also from the problem's constraints as given in its mathematical formulation.

6.4 THE INVARIANT LIBRARY OF IOPT

The *invariant library* (IL) of iOpt, as the name suggests, is solely based on invariants (i.e., one-way constraints). IL provides a number of built-in data types such as Integer, Real, Boolean, String and Object and also set versions of these types (except for Boolean). Arithmetic, logical, string, object, set and other operators are available to the user to state his/her problem (i.e., decision variables, parameters, constraints and objective function). Being a library of Java, IL brings a number of advantages such as integration with visualization components, ability for the user to extend the operators available by defining its own, facilities to work on language data structures such as Java Object and String and also easier embedding into other programs. IL incorporates many optimizations such as incremental updates for aggregate types, lazy and eager evaluation modes, constraint priorities, cycle detection facilities, propagation stopping, ability to deal with undefined parts of the dataflow graph, set constraints. Moreover, it is more geared towards computationally demanding applications compared to other Java libraries and applications. This is achieved by avoiding some built-in but inefficient data structures and also by trying to avoid the constant creation and garbage collection of objects, something very common in a strict object-oriented environment such as Java's.

Arbitrary dataflow graphs can be configured to model optimization problems by mapping the decision variables representing a solution to the objective and constraints as they are given by the problem's mathematical formulation.

The reason for selecting invariants for supporting heuristic search, over relational constraints (as used in CP), is that invariants are particularly suited for heuristic search. Heuristic search techniques require an efficient way to access the impact of incremental solution changes to the problem's constraints (i.e., constraint checks) and also the value of the objective function. Invariants are particularly adept at this task since they can incorporate specialized incremental update mechanisms for the different operators implemented, in addition to the general algorithms available for restricting the parts of the constraint graph that need to be updated after a change in the input variables (decision variables in this case).

This is in contrast to exact search methods such as branch-and-bound where the main goal is the reduction of the size of the search space to be enumerated. Relational constraints are particularly good at that since they can propagate reductions in the domain of one decision variable to the domain of other decision variables reducing quickly the number of feasible alternatives. Moreover, they can detect failures at an early stage of the solution construction process (i.e., when empty domains are encountered) preventing the algorithm from exploring large and unfruitful parts of the search tree.

Relational constraints can in principal be used to play a role similar to invariants and used to support heuristic search. Nonetheless, they carry much unnecessary functionality for the task (e.g., domains, multi-way propagation, etc.) resulting in suboptimal performances/implementations for the applications in question. Furthermore, their satisfaction algorithms such as arc consistency are usually queue based which in practice means they cannot be optimized to the degree possible with mark/sweep methods.

Small solution changes (often called moves) are the foundation of heuristic search (local search to be more precise). They are iteratively used to improve a starting solution for the problem. Devising an efficient move evaluation mechanism normally requires a person with significant expertise in the area. This hinders the widespread use of heuristic search. The invariant library addresses this problem by utilizing the generic mark/sweep algorithm mentioned above which can achieve efficient move evaluations in an automated way without any particular expertise or special programming skills required by the user. The library is generic and not dependent on the rest of the toolkit. Its applications can be as diverse as computer graphics and software modelling although we have not explored these avenues at this stage. The library also supports dynamic changes such as the addition and deletion of variables and constraints always bringing the network of invariants in a consistent state after each modification. In Figure 6.1, we provide an annotated code sample demonstrating the use of invariants and of the problem modelling framework (see Voudouris et al. (2001) for more details on this framework) to model a very simple optimization problem.

Next, we look at the iOpt's heuristic search framework, which makes use of the facilities of the invariant library to solve combinatorial optimization problems

```
/* Create a simple problem model using invariants */
Problem p = new Problem();
p.beginModelChanges();          // start changes to the problem model
RealVar x = new RealVar(10.0);
          // create a real variable x and set its initial value to 10.0
p.addDecisionVar(x);                // set x to be a decision variable
p.addObjective(Inv.plus(x, 5.0));
           // add the term x + 5.0 to the objective initially undefined
p.addConstraint(Inv.gt(x, 5.0));          // set the constraint x > 5.0
p.endModelChanges();               // end changes to the problem model

/* Change to the value of the decision variable */
/* Similar operations are performed when local search is evaluating a
move */
p.beginValueChanges();   // start changes to the values of the decision
variables
x.setValue(100.0);                 // set x to the new value of 100.00
p.endValueChanges();       // end changes to the values of the decision
variables
p.propagateValueChanges();
                  // the mark/sweep algorithm is updating the invariants
p.saveValueChanges();         // we commit the changes, we may undo them
instead
                     // in case of constraint violations or inferior cost
```

Figure 6.1 Sample Source Code Demonstrating the Use of Invariants for Problem Modelling

6.5 THE HEURISTIC SEARCH FRAMEWORK OF IOPT

The *heuristic search framework* (HSF) was created to be a generic framework for the family of optimization techniques known as heuristic search. It covers single solution methods such as local search, population-based methods such as genetic algorithms as well as hybrids combining one or more different algorithms. Heuristic search methods are known to be efficient on a large range of optimization problems but they remain difficult to design, tune, and compare. Furthermore, they tend to be problem specific often requiring re-implementation to address a new problem.

HSF proposes a new way to design a heuristic search method by identifying pertinent concepts already defined in the literature and make them available as a set of basic components. The designer has the capability to build a complete and complex algorithm using these components in a way similar to using a "Lego" kit. Three main concepts are the basis of HSF: *Heuristic Search, Heuristic Solution* and *Heuristic Problem*:

- *Heuristic Search* represents an algorithm of an optimization method. Such an algorithm is composed of several search components. A search component is the basic entity that is used to build an optimization method. Most often a search component represents a basic concept that could be encountered in a HS method. For example, the concept of Neighborhood in a Local search. A complete algorithm is a valid tree of search components (where a valid tree is a tree of search components that can be executed).

- *Heuristic Solution* is the solution representation of an optimization problem manipulated inside HSF. At the present moment, a *vector of variables* and a *set of sequences* are readily available. These representations are commonly used in modelling combinatorial optimization problems. For example, a vector of variables can model CSPs while a set of sequences can model various scheduling applications with unary resources.
- *Heuristic Problem* is only an interface between an optimization problem model implemented by IL (or by other means) and HSF. (Note that HSF is generic like the invariant library and it can be used independently from it. For example, it can be used in conjunction with a hard-coded procedural problem model.)

  This interface allows using the same algorithm for a family of optimization problems without re-implementing any separate functionality (see Figure 6.2). The following theoretical problems and real applications are already implemented using the invariant library (and also higher-level frameworks such as the problem modelling framework and the scheduling framework; Voudouris et al. (2001)) and applied by HSF using that concept: graph coloring, set partitioning, frequency assignment, job-shop scheduling, workforce scheduling, vehicle routing, car sequencing.

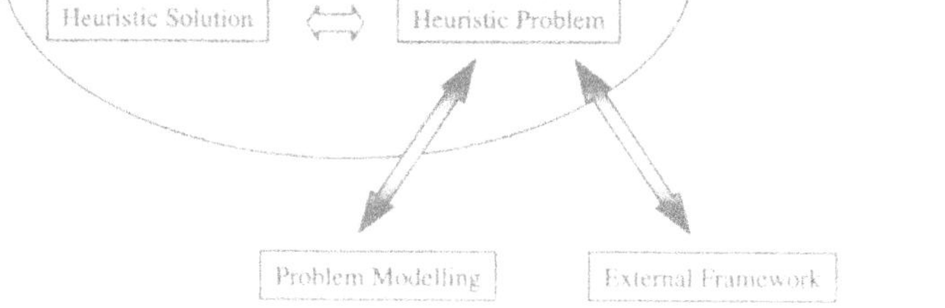

Figure 6.2 HSF Interaction with its Environment

In the current version of HSF, the search component family groups many basic concepts encountered in single solution heuristic search such as:

- Initial solution generation,

- Local search,
- Move
- Neighborhood,
- Neighborhood search,
- Aspiration criterion,
- Tabu mechanism,
- Dynamic objective function and others.

and in population-based heuristic search such as:

- Initial population generation,
- Mutation,
- Population method,
- Crossover population method,
- Selection population method,
- Crossover,
- Mutation,
- Selection,
- Restart population method, and others.

Furthermore, many popular meta-heuristics such as simulated annealing (SA), tabu search (TS), guided local search (GLS) and genetic algorithms are implemented.

Methods such as tabu search and guided local search can become observers of other search components such as neighborhood search and receive notifications for certain events, which require them to intervene (e.g., move performed, local minimum reached and others). Invariants are often used to model parts of these methods such as the acceptance criterion in SA, tabu restrictions in TS or features and their penalties in GLS.

Using the available set of search components and the other facilities explained briefly above, a heuristic search method becomes easy to build even by a non-expert. For example, a tabu search for the graph coloring problem composed of an initial generation method, a local search and a restart method is implemented by assembling the following tree of components (see Figure 6.3). (A restart method is necessary here as we consider a graph coloring problem as a sequence of k-coloring problems where k is the number of colors used to color the graph. Each time a complete coloring is found, the restart method reduces the number of available colors by one.)

Notice here that most of the components used are generic, only some components are specific to the *vector of variables* solution representation, since we use this solution representation for the graph coloring problem. However, these components are still reusable for other problems in this category.

Even the most complex hybrid methods can be modelled as the next example shows where a population-based method is composed using a local search method as the mutation, union of independent set (UIS) crossover specialized to the graph coloring (see

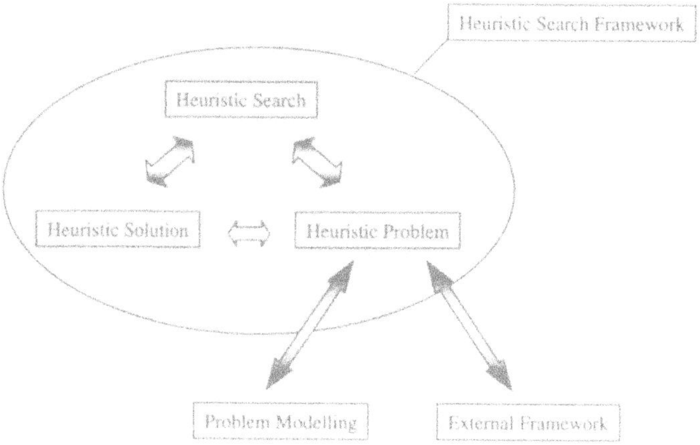

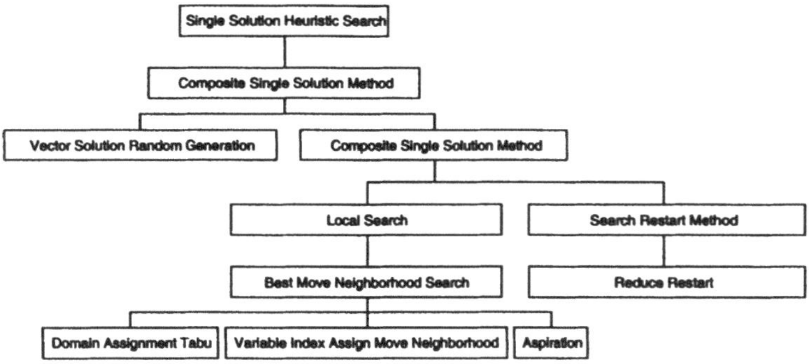

Figure 6.3 Tabu Search for a Graph Coloring Problem

Dorne and Hao (1998) for more details about this crossover), selection, and various other search components (see Figure 6.4).

As it becomes obvious, designing a heuristic search is simplified to a great extend through assembling a set of components even for a sophisticated method such as the hybrid algorithm. The framework further allows the possibility to specialize an algorithm to an optimization problem as we did above by adding a graph coloring specific crossover UIS. Thus for the development of a problem-specific algorithm, we have only to implement the specialized problem-specific components not included in HSF. These can be incorporated in the framework and possibly re-used in similar applications.

Another benefit is that we can conduct easily and fairly comparisons between different algorithms or between different components of the same category (e.g., various Neighborhood Search methods, currently fifteen of them incorporated in HSF). Since any new component can be directly plugged into the algorithmic tree replacing an old one, we can quickly determine the real benefit of different components to the search process. This is particularly useful when we are trying to determine what is the best variant of a method based on a particular philosophy (e.g., Tabu Search, Genetic Algorithms).

An additional functionality included in HSF is that any heuristic search algorithm composed by the user can be saved in an XML format (a very popular Internet format for exchanging information). An efficient algorithm is never lost, instead can be kept in a simple text file. Then you can load it again at some other time for further experiments. Moreover, a visualization tool is also available to show the tree of search components (Figures 6.3 and 6.4 have been generated using this tool), to set parameters for any of them and even watch and monitor their behaviour during the search process.

All these facilities and tools provided by HSF clearly simplify the design, understanding, tuning, and comparison of heuristic search methods. Thus HSF can be used either as a practical design tool to develop a heuristic search algorithm to solve a

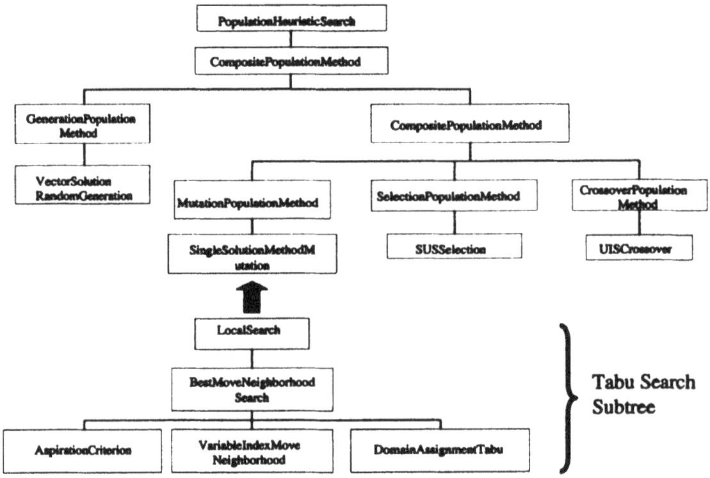

Figure 6.4 Hybrid Algorithm Using Local Search as a Mutation

real-world optimization application/problem or as a research platform to build, study, compare and tune new algorithms.

Since problem models are declarative and implemented by the invariant library, performing moves (in local search) or evaluating new solutions (in population methods) is transparent to the user. The user can instead focus on the more important task of choosing and tuning the right algorithm instead of getting involved in low-level tasks such as trying to devise hard-coded efficient move evaluation mechanisms or implementing problem-specific methods to compute a complex fitness function.

6.6 EXPERIMENTATION ON THE GRAPH COLORING AND THE VEHICLE ROUTING PROBLEM

In this section, we report results for heuristic search methods and problem models developed using the heuristic search framework and the invariant library, respectively. In particular, the *graph coloring* (GC) and the *vehicle routing problem* (VRP) are considered along with well-known meta-heuristic methods implemented using the heuristic search framework. The aim of this section is to provide a better understanding of the mark/sweep algorithm and also offer an indication of the potential of Java for developing general algorithmic frameworks as an alternative to more low-level languages such as C/C++.

All the experiments reported in this section were performed on a Pentium III 866 MHz PC running the Windows 98 operating system while the Java Virtual Machine used was the HotSpot Client VM (version 1.3.0-C) by Sun Microsystems. In cases were running times are reported, they are actual times rather than CPU times. This is to help the reader to assess more accurately the performance of methods in real-world conditions.

The problem models for graph coloring and vehicle routing are stated in a declarative form and implemented using invariants. In fact, they are both extensions of problem models already included in iOpt. Graph coloring is extending some of the foundation classes for problem modelling included in the problem modelling framework, while the vehicle routing problem is specializing the general scheduling framework, which can model various scheduling problems such as the job-shop scheduling problem and the workforce scheduling problem (Voudouris et al. (2001)).

The solution representation reused in graph coloring is that of a *vector of variables* mentioned in Section 6.5, which is suitable for CSP-type problems. The vehicle routing problem reuses the *set of sequences* representation, which is useful in modelling scheduling-type problems with unary resources. General move operators, which are applicable to these solution representations, are readily available and can be used on derived problem models such as those for the graph coloring and the vehicle routing problem. This also applies to neighborhood structures and neighborhood search mechanisms. To that extend, the user needs only to specialize the general classes available to formulate the GC and VRP problems along with selecting the search components of his/her choice to build the heuristic search methods to be evaluated.

In the experiments, we used the selection move operator in graph coloring (i.e., changes the selection of a color for a node) and the insertion move operator in vehicle routing (i.e., removes a visit from a vehicle's route and inserts it in another position in the same or another vehicle's route). Although possible using the heuristic search framework, no hard-coded move filters were applied which bypass the constraint satisfaction algorithm for efficiency purposes, nor any declarative incremental move evaluation models were incorporated which are also possible using the invariant library. Essentially, the problem objective functions where stated using invariants following the problem's mathematical formulations and any hard constraints (time windows, capacity constraints) as in the case of the VRP were defined as boolean expressions indicating the infeasibility of solutions when evaluating a move. In graph coloring only nodes in conflict were examined and the mechanism for identifying them was implemented using invariants.

We first examine the results from the point of view of the mark/sweep algorithm and its efficiency. Table 6.1 shows for the three graph coloring instances considered, the number of invariants required to model each instance and also the number of model variables modified after each move. The number of invariants influenced refers to the average number of invariants marked as out-of-date during the mark phase of the method. The number of invariants evaluated refers to the invariants updated during the sweep phase of the method. For the algorithm to be efficient, it needs to be limiting the number of invariants that need to be evaluated after each move from the overall problem model and also the number of unnecessary evaluations, which is given by the difference between the number of invariants evaluated and the number of invariants affected. This latter figure is giving the number of invariants, which actually required an update (i.e., their value changed as a result of the move).

Problem	Nodes	Density	No of Invariants	No of Variables Modified	No of Invariants Influenced	No of Invariants Evaluated	No of Invariants Affected	No of Unnecessary Evaluations
dsjc125.1.col	125	10	8553	1	100.7	39.3	28.2	11.1
dsjc125.5.col	125	50	36662	1	423.3	115.0	46.5	68.4
dsjc125.9.col	125	90	71706	1	771.5	184.7	37.4	147.2

Table 6.1 Results for the Mark/Sweep Method on Graph Coloring Instances

As we can see in this table, the algorithm is, on average, only evaluating a small percentage of invariants from the overall network after each move (from 0.25% in the best case to 0.46% in the worst case). Moreover, this number is significantly smaller compared to those that are marked out-of-date after the move (given by the number of invariants influenced). The number of unnecessary evaluations although small is sensitive to the density of the constraint graph. This is to be expected since the problem becomes harder for the algorithm, as more invariants are marked out-of-date after a move (due to the larger number of edges in the problem). This also works in conjunction with the fact that the number of invariants affected is largely independent of the density. To give an indication of the memory requirements, in our current implementation each invariant occupies on average about 100 bytes of memory.

Results for the VRP on a similar setting and for instances of different size (also called extended Solomon VRP instances; the instances are available at the following URL: http://www.fernuni-hagen.de/WINF/touren/inhalte/probinst.htm) are given in Table 6.2.

Problem	Vehicles	Visits	No of Invariants	No of Variables Modified	No of Invariants Influenced	No of Invariants Evaluated	No of Invariants Affected	No of Unnecessary Evaluations
C1_2_1	50	200	49125	10.6	266.5	43.5	34.5	8.9
C1_4_1	100	400	148225	10.2	246.9	42.4	34.3	8.0
C1_6_1	150	600	297325	10.0	225.6	41.4	34.2	7.2

Table 6.2 Results for the Mark/Sweep Method on Vehicle Routing Instances

As it is shown in this table, the mark/sweep method is very accurate in terms of evaluating only a very small percentage of invariants from the overall network after each move (from 0.01% in the best case to 0.08% in the worst case). This is also combined with a very small number of unnecessary evaluations. Another interesting result is that the algorithm remains unaffected by the size of the problem. This is because of the fact that the solutions usually contain a similar number of visits for each vehicle irrespective of the total number of visits in the problem. The memory requirements are up to 30MB of RAM for the largest problem which is in no way restrictive when using a modern PC. Next we look at the combination of the mark/sweep algorithm with different meta-heuristics.

Table 6.3 reports the results for the three graph coloring instances used above when the problem model is used by hill climbing (HC), simulated annealing (SA), and tabu search (TS).

Problem	HC		SA		TS	
	Colors	Time (sec.)	Colors	Time (sec.)	Colors	Time (sec.)
dsjc125.1.col	7	2.37	6	166.06	5	44
dsjc125.5.col	21	70.18	20	334.75	18	557
dsjc125.9.col	46	492.81	46	437.4	44	2904.75

Table 6.3 Results for Various Meta-Heuristic Methods on Graph Coloring Instances

All methods are working in conjunction with a restart strategy, which reduces the number of colors after a complete coloring is found. The exact tabu search configuration is as shown in Figure 6.3 of Section 6.5.

For those familiar with these instances, it is clear that the algorithms, especially tabu search, find high quality solutions in all three problems. The running times are certainly inferior compared to specialized C/C++ implementations. Nonetheless, the results demonstrate the viability of using the Java language for implementing generic algorithmic software which by nature is computationally demanding. Of course, this is partly possible because of the efficient implementation of the underlying constraint satisfaction method, which significantly reduces the amount of processing required to evaluate each move.

Next, we look at results for the VRP instances used above for the same three methods used in the graph coloring experiments. These results are more useful to practitioners in the area of routing/scheduling and report the number of insertion moves that can be evaluated per second for medium to large size instances of the VRP. Especially when the problem model is implemented using a general scheduling framework, which can easily accommodate variations of the problem such as various side constraints and/or complex objective functions.

As we can see in Table 6.4, depending on the metaheuristic, the overall speed of the system varies. This is because aspects of different metaheuristics are implemented using invariants (e.g., the acceptance criterion in SA, tabu restrictions), which require more or less processing on top of the time for evaluating the moves.

Problem	HC moves evaluated / second	SA moves evaluated / second	TS moves evaluated / second
C1_2_1	4628	3562	4761
C1_4_1	5174	3298	5040
C1_6_1	5147	2297	4052

Table 6.4 Evaluation Speeds for Various Meta-Heuristic Methods on Vehicle Routing Instances

In practice, these evaluation speeds can comfortably produce good quality solutions in a few minutes for medium to large size VRPs when running on modern workstations along with visualization facilities monitoring the search progress, which are also gradually becoming popular in practical applications.

Concluding, despite the several layers of software required to support advanced problem modelling/searching facilities and that in a strict object-oriented language such as Java, efficient constraint satisfaction algorithms combined with lightweight constraints can offer implementations of heuristic search, which can be very useful for:

- quickly prototyping/evaluating different heuristic search methods,
- deploying optimization algorithms on the server or client side of IT systems based on Java technologies such as Enterprise Java Beans, Java Servlets, Java Applets, etc.

As a final comment, computational results should be treated with caution when libraries/frameworks are considered for the purpose of engineering a real-world system. Since the particular facilities offered in different systems may affect other important aspects of the software development process. Performance considerations versus benefits in system development times, cost and flexibility are hard to quantify and very much depend on the particular situation.

6.7 RELATED WORK AND DISCUSSION

In terms of the invariant library, the system uses algorithms initially explored in the areas of computer graphics (Zanden et al. (1999), Zanden et al. (1994)). Related work on using one-way constraints for heuristic optimization is the LOCALIZER language described in Michel and van Hentenryck (1998), Michel and van Hentenryck (1999) also mentioned earlier in this chapter. Differences between the IL and LOCALIZER include the algorithms used for propagation (mark/sweep and topological-ordering, respectively), the ability to model dynamic problems in IL, but most important the fact that LOCALIZER is a language on its own whereas IL is using the Java programming language by providing a declarative environment within it.

In the case of the HSF most of the frameworks proposed in the literature are either local search based or genetic algorithm based. In the case of local search examples of frameworks include HOTFRAME (Fink et al. (1999b)), Templar (Jones et al. (1998)), LOCAL++ (Schaerf et al. (1999)), SCOOP (http://www.oslo.sintef.no/SCOOP/), and also the work of Andreatta et al. (1998). These tend to provide templates with the user having to define the problem model rather than use a constraint-based problem modelling paradigm. The same applies for move operators. CP toolkits such as Eclipse and ILOG Solver in their latest versions also have integrated certain basic heuristic search methods such as tabu search and simulated annealing. Finally, almost all of the related works mentioned in this section and others that we know of are heavily C/C++-based.

6.8 CONCLUSIONS

iOpt is a Java-based toolkit which integrates a number of technologies to offer a complete set of tools for developing optimization applications based on heuristic search.

At its present stage of development, it plays the role of a prototype platform for experimenting with the integration of a number of technologies such as combinatorial optimization, visualization, databases and distributed systems and utilizing the capabilities of the much promising Java language in doing so. In this chapter, we examined two aspects of the toolkit:

- a lightweight constraint library for modelling combinatorial optimization problems and
- a heuristic search framework for synthesizing local search and population-based algorithms.

We described how they could work in synergy allowing a developer to focus on the high level tasks of algorithm design, automating in an efficient way large parts of his/her work in developing a real-world optimization system. Current research is focused on including relational constraints and also exact search methods in iOpt. Relational constraints and invariants and also exact and heuristic search methods can work collaboratively in many different ways and at present we are trying to identify the most promising ones. This is analogous to the link between MP and CP. Hopefully, a successful link of HS with CP in iOpt can lead to an environment, which not only supports CP and HS but also guides the user how to exploit the best combinations of these techniques.

7 THE OPTQUEST CALLABLE LIBRARY

Manuel Laguna[1] and Rafael Martí[2]

[1]Operations and Information Systems Division
College of Business and Administration
University of Colorado
419 UCB, Boulder, CO 80309, USA
laguna@colorado.edu

[2]Departamento de Estadistica e Investigacion Operativa
Universitat de Valencia
Dr. Moliner 50, 46100 Valencia, Spain
rafael.marti@uv.es

Abstract: In this chapter we discuss the development and application of a library of functions that is the optimization engine for the OptQuest system. OptQuest is commercial software designed for optimizing complex systems, such as those formulated as simulation models. OptQuest has been integrated with several simulation packages with the goal of adding optimization capabilities. The optimization technology within OptQuest is based on the metaheuristic framework known as scatter search. In addition to describing the functionality of the OptQuest Callable Library (OCL) with an illustrative example, we apply it to a set of unconstrained nonlinear optimization problems.

7.1 INTRODUCTION

The OptQuest Callable Library (OCL), which we began developing in the fall of 1998 in collaboration with Fred Glover and James P. Kelly, is the optimization engine of the OptQuest system. (OptQuest is a registered trademark of OptTek Systems, Inc. (`www.opttek.com`). The descriptions in this chapter are based on OCL 3.0; see OptQuest (2000).) The main goal of OptQuest is to optimize complex systems, which are those that cannot be easily formulated as mathematical models and solved with classical optimization tools. Many real world optimization problems in business, engineering and science are too complex to be given tractable mathematical formulations. Multiple nonlinearities, combinatorial relationships and uncertainties often render challenging practical problems inaccessible to modeling except by resorting to more comprehensive tools (like computer simulation). Classical optimization methods encounter grave difficulties when dealing with the optimization problems that arise in the context of complex systems. In some instances, recourse has been made to itemizing a series of scenarios in the hope that at least one will give an acceptable solution. Due to the limitations of this approach, a long-standing research goal has been to create a way to guide a series of complex evaluations to produce high quality solutions, in the absence of tractable mathematical structures. (In the context of optimizing simulations, a "complex evaluation" refers to the execution of a simulation model.)

Theoretically, the issue of identifying best values for a set of decision variables falls within the realm of optimization. Until quite recently, however, the methods available for finding optimal decisions have been unable to cope with the complexities and uncertainties posed by many real world problems of the form treated by simulation. The area of stochastic optimization has attempted to deal with some of these practical problems, but the modeling framework limits the range of problems that can be tackled with such technology.

The complexities and uncertainties in complex systems are the primary reason that simulation is often chosen as a basis for handling the decision problems associated with those systems. Consequently, decision makers must deal with the dilemma that many important types of real world optimization problems can only be treated by the use of simulation models, but once these problems are submitted to simulation there are no optimization methods that can adequately cope with them.

Recent developments are changing this picture. Advances in the field of metaheuristics – the domain of optimization that augments traditional mathematics with artificial intelligence and methods based on analogs to physical, biological or evolutionary processes – have led to the creation of optimization engines that successfully guide a series of complex evaluations with the goal of finding optimal values for the decision variables. One of those engines is the search procedure embedded in OCL.

OCL is designed to search for optimal solutions to the following class of optimization problems:

$$
\begin{array}{llr l}
\text{Max or Min} & & F(x,y) & \\
\text{subject to} & & Ax \leq b & \text{(Constraints)} \\
& & g_l \leq G(x,y) \leq g_u & \text{(Requirements)} \\
& & l \leq x \leq u & \text{(Bounds)} \\
& & y = \text{ all different} &
\end{array}
$$

where x can be continuous or discrete with an arbitrary step size and y represents a permutation.

The objective $F(x, y)$ may be any mapping from a set of values x and y to a real value. The set of constraints must be linear and the coefficient matrix "A" and the right-hand-side values "b" must be known. The requirements are simple upper and/or lower bounds imposed on a function that can be linear or non-linear. The values of the bounds "g_l" and "g_u" must be known constants. All the variables must be bounded and some may be restricted to be discrete with an arbitrary step size. The set of y variables are used to represent permutations. A given optimization model may consist of any combination of continuous, discrete or "all different" (also referred to as "permutation") variables.

In a general-purpose optimizer such as OCL, it is desirable to separate the solution procedure from the complex system to be optimized. The disadvantage of this "black box" approach is that the optimization procedure is generic and has no knowledge of the process employed to perform evaluations inside of the box and, therefore, does not use any problem-specific information (Figure 7.1). The main advantage, on the other hand, is that the same optimizer can be applied to complex systems in many different settings.

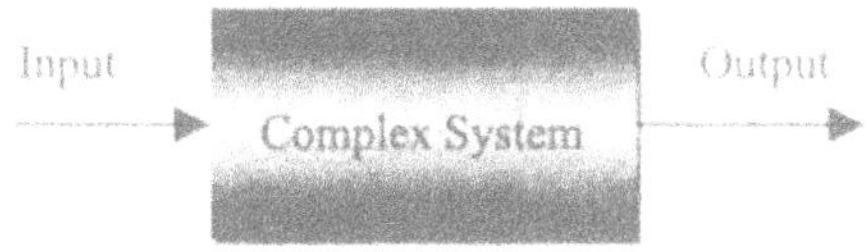

Figure 7.1 Complex System as a Black Box

OCL is a generic optimizer that overcomes the deficiency of black box systems of the type illustrated in Figure 7.1, and successfully embodies the principle of separating the method from the model. In such a context, the optimization problem is defined outside the complex system. Therefore, the evaluator can change and evolve to incorporate additional elements of the complex system, while the optimization routines remain the same. Hence, there is a complete separation between the model used to represent the system and the procedure that solves optimization problems formulated around the model.

The optimization procedure uses the outputs from the system evaluator, which measures the merit of the inputs that were fed into the model. On the basis of both current and past evaluations, the optimization procedure decides upon a new set of input values (see Figure 7.2). The optimization procedure is designed to carry out a special "strategic search," where the successively generated inputs produce varying evaluations, not all of them improving, but which over time provide a highly efficient trajectory to the best solutions. The process continues until an appropriate termination criterion is satisfied (usually based on the user's preference for the amount of time to be devoted to the search).

OCL allows the user to build applications to solve problems using the "black box" approach for evaluating an objective function and a set of requirements. Figure 7.3

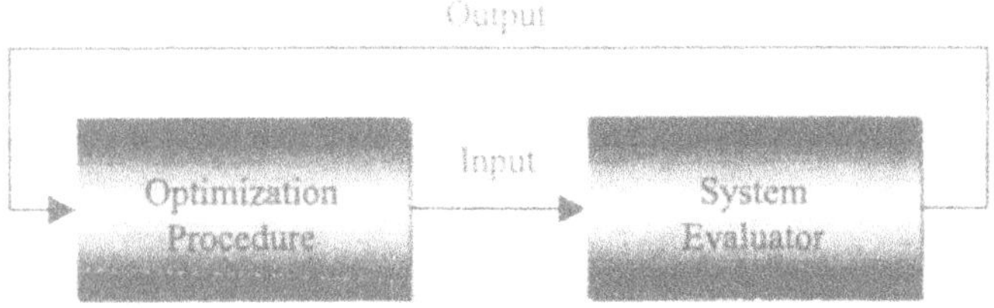

Figure 7.2 Coordination Between Optimization and System Evaluation

shows a conceptualization of how OCL can be used to search for optimal solutions to complex optimization problems.

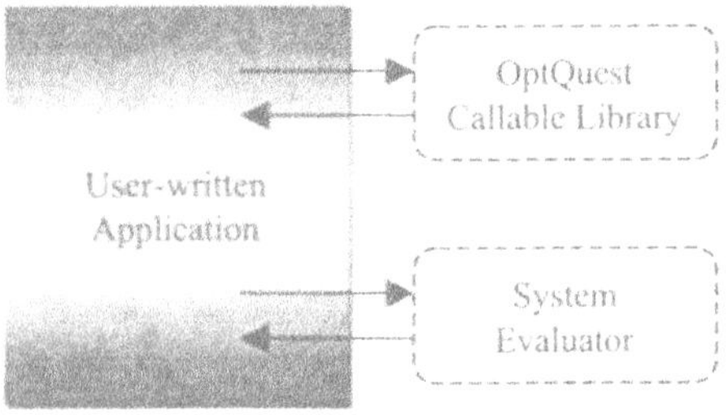

Figure 7.3 OCL Linked to a User-Written Application

Figure 7.3 assumes that the user has a system evaluator that, given a set of input values, returns a set of output values that can be used to guide a search. For example, the evaluator may have the form of a computer simulation that, given the values of a set of decision variables, returns the value of one or more performance measures (that define the objective function and possibly a set of requirements). The user-written application uses OCL functions to define an optimization problem and launch a search for the optimal values of the decision variables.

A simple example in risk analysis is to use a Monte Carlo simulator to estimate the average rate of return and the corresponding variance of a proposed investment portfolio. The average rate of return defines the objective function and the estimated variance can be used to formulate a requirement to limit the variability of the returns. The decision variables represent the allocation of funds in each investment alternative and a linear constraint can be used to limit the total amount invested. In this case, the user must provide the simulator to be coupled with OCL, while the optimization model is entirely formulated with calls to the appropriate library functions.

7.2 SCATTER SEARCH

The optimization technology embedded in OCL is the metaheuristic known as *scatter search*. Scatter search has some interesting commonalities with *genetic algorithms*

(GA), although it also has a number of quite distinct features. Several of these features have come to be incorporated into GA approaches after an intervening period of approximately a decade, while others remain largely unexplored in the GA context.

Scatter search is designed to operate on a set of points, called *reference points*, which constitute good solutions obtained from previous solution efforts. Notably, the basis for defining "good" includes special criteria such as diversity that purposefully go beyond the objective function value. The approach systematically generates combinations of the reference points to create new points, each of which is mapped into an associated feasible point. The combinations are generalized forms of linear combinations, accompanied by processes to adaptively enforce constraint-feasibility and encourage requirement-feasibility.

The scatter search process is organized to (1) capture information not contained separately in the original points, (2) take advantage of auxiliary heuristic solution methods (to evaluate the combinations produced and to actively generate new points), and (3) make dedicated use of strategy instead of randomization to carry out component steps.

Figure 7.4 sketches the scatter search approach in its original form. Extensions can be created to take advantage of memory-based designs typical of *tabu search* (Glover and Laguna (1997)). Two particular features of the scatter search proposal deserve mention. The use of clustering strategies has been suggested for selecting subsets of points in step 2, which allows different blends of intensification and diversification by generating new points "within clusters" and "across clusters." Also, the solutions generated by the combination method in step 2 are often subjected to an improvement method, which typically consists of a local search procedure. The improvement method is capable of starting from a feasible or an infeasible solution created by the combination method.

It is interesting to observe similarities and contrasts between scatter search and the original GA proposals. Both are instances of what are sometimes called "population based" or "evolutionary" approaches. Both incorporate the idea that a key aspect of producing new elements is to generate some form of combination of existing elements. However, GA approaches are predicated on the idea of choosing parents randomly to produce offspring, and further on introducing randomization to determine which components of the parents should be combined. By contrast, the scatter search approach does not emphasize randomization, particularly in the sense of being indifferent to choices among alternatives. Instead, the approach is designed to incorporate strategic responses, both deterministic and probabilistic, that take account of evaluations and history. Scatter search focuses on generating relevant outcomes without losing the ability to produce diverse solutions, due to the way the generation process is implemented. For example, the approach includes the generation of new points that are not convex combinations of the original points. The new points constitute forms of *extrapolations*, endowing them with the ability to contain information that is not contained in the original reference points.

Scatter search is an *information-driven* approach, exploiting knowledge derived from the search space, high-quality solutions found within the space, and trajectories through the space over time. The incorporation of such designs is responsible for

1. Apply a diversification generation method to build a starting set of solutions. Designate a subset of the best points (judged by quality and diversity) to be *reference points*.

while (stopping criteria are not satisfied) {

   2. Form combinations of subsets of the current reference points to create new points. The combinations are
   (a) chosen to produce points both inside and outside the convex region spanned by the reference points, and
   (b) modified by generalized mapping processes to yield feasible points according to the constraints in the problem (both linear and integrality constraints).

   3. Update the reference set by selecting points that can improve the quality and/or diversity of the set.

   if (no new combinations can be explored in the current set) {

      4. Extract a collection of the best points in the current reference set to be used as starting points for a new application of the diversification generation method.

   }

}

Figure 7.4 Scatter Search Outline

enabling OCL to efficiently search the solution space of optimization problems in complex systems.

To learn more about scatter search refer to the tutorial articles by Glover (1998b), Glover et al. (1999), Glover et al. (2000a), Laguna (2001), Laguna and Armentano (2001). Recent applications of scatter search include the linear ordering problem (see Campos et al. (1999a), Campos et al. (1999b)), permutation problems (see Campos et al. (2001)), transportation (see Corberán et al. (2000)), nonlinear optimization (see Laguna and Martí (2000)) and machine scheduling (see Laguna et al. (2000)).

7.3 THE OCL OPTIMIZER

OCL seeks to find an optimal solution to a problem defined on a vector x of bounded variables and/or a permutation y. That is, the user can define several types of optimization problems depending on the combination of variables:

- Pure continuous
- Pure discrete (including pure binary problems)
- Pure permutation ("all different" variables only)
- Mixed problems (continuous-discrete, continuous-permutation, discrete-permutation or continuous-discrete-permutation)

Also, the optimization problem may be unconstrained, include linear constraints and/or requirements. Hence, OCL can be used to formulate up to 28 different types of problems. OCL detects small pure discrete or permutation problems to trigger a complete enumeration routine that guarantees optimality of the best solution found.

In this chapter, we will describe the mechanisms that OCL employs to search for optimal solutions to problems defined with continuous and discrete variables. Similar mechanisms are used to tackle pure or mixed permutation problems and details can be found in Campos et al. (2001). The scatter search method implemented in OCL begins by generating a starting set of diverse points. This is accomplished by dividing the range of each variable into four sub-ranges of equal size. Then, a solution is constructed in two steps. First, a sub-range is randomly selected. The probability of selecting a sub-range is inversely proportional to its frequency count (which keeps track of the number of times the sub-range has been selected). Second, a value is randomly chosen from the selected sub-range. The starting set of points also includes the following solutions:

- All variables are set to the lower bound
- All variables are set to the upper bound
- All variables are set to the midpoint $x = l + (u - l)/2$
- Other solutions suggested by the user

A subset of diverse points is chosen as member of the reference set. A set of points is considered diverse if its elements are "significantly" different from one another. OCL uses a Euclidean distance measure to determine how "close" a potential new point is from the points already in the reference set, in order to decide whether the point is included or discarded.

When the optimization model includes discrete variables, a rounding procedure is used to map fractional values to discrete values. When the model includes linear constraints newly created points are subjected to a feasibility test before they are sent to the evaluator (i.e., before the objective function value $F(x)$ and the requirements $G(x)$ are evaluated). Note that the evaluation of the objective function may entail the execution of a simulation, and, therefore, it is important to be sure to evaluate only those solutions that are feasible with respect to the set of constraints. For ease of notation, we represent the set of constraints as $Ax \leq b$, although equality constraints are also allowed. The feasibility test consists of checking (one by one) whether the linear constraints are satisfied. If the solution is infeasible with respect to one or more constraints, OCL formulates and solves a linear programming (LP) problem. The LP (or mixed-integer program, when x contains discrete variables) has the goal of finding a feasible solution x^* that minimizes a deviation between x and x^*. Conceptually, the problem can be formulated as:

$$\begin{array}{lll} \text{Minimize} & d^- + d^+ & \\ \text{subject to} & Ax^* & \leq b \end{array}$$

$$\begin{aligned} x - x^* - d^- + d^+ &= 0 \\ l \leq x^* &\leq u \\ d^-, d^+ &\geq 0 \end{aligned}$$

where d^- and d^+ are, respectively, negative and positive deviations of x^* from the infeasible point x. The implementation of this mechanism within OCL includes a scaling procedure to account for the relative magnitude of the variables and adds a term to the objective function to penalize maximum deviation. Also, OCL treats pure binary problems differently, penalizing deviations without adding deviation variables or constraints. When the optimization problem does not include constraints, infeasible points are made feasible by simply adjusting variable values to their closest bound and rounding when appropriate. That is, if $x > u$ then $x^* = u$ and if $x < l$ then $x^* = l$.

Once the reference set has been created, a combination method is applied to initiate the search for optimal solutions. The method consists of finding linear combinations of reference solutions. The combinations are based on the following three types, which assume that the reference solutions are x' and x'':

$$\begin{aligned} x &= x' - d \\ x &= x' + d \\ x &= x'' - d \end{aligned}$$

where $d = r\frac{x'' - x'}{2}$ and r is a random number in the range (0,1). Because a different value of r is used for each element in x, the combination method can be viewed as a sampling procedure in a rectangle instead of a line in a two dimensional space (see Ugray et al. (2001)). The number of solutions created from the linear combination of two reference solutions depends on the quality of the solutions being combined. Specifically, when the best two reference solutions are combined, they generate up to five new solutions, while when the worst two solutions are combined they generate only one.

In the process of searching for a global optimum, the combination method may not be able to generate solutions of enough quality to become members of the reference set. If the reference set does not change and all the combinations of solutions have been explored, a diversification step is triggered (see step 4 in Figure 7.4). This step consists of rebuilding the reference set to create a balance between solution quality and diversity. To preserve quality, a small set of the best (*elite*) solutions in the current reference set is used to seed the new reference set. The remaining solutions are eliminated from the reference set. Then, the diversification generation method is used to repopulate the reference set with solutions that are diverse with respect to the elite set. This reference set is used as the starting point for a new round of combinations.

7.3.1 Constraints Versus Requirements

So far, we have assumed that the complex system to be optimized can be treated by OCL as a "black box" that takes x as an input to produce $F(x)$ as an output. We have also assumed that for x to be feasible, the point must be within a given set of bounds and, when applicable, also satisfy a set of linear constraints. We assume that both the bounds and the coefficient matrix are known. However, there are situations where the feasibility of x is not known prior to performing the process that evaluates $F(x)$, i.e., prior to executing the "black box" system evaluator. In other words, the feasibility test for x cannot be performed in the input side of the black box but instead has to be performed within the black box and communicated as one of the outputs. This situation is depicted in Figure 7.5.

This figure shows that when constraints are included in the optimization model, the evaluation process starts with the mapping $x \to x^*$. If the only constraints in the model are integrality restrictions, the mapping is achieved with a rounding mechanism that transforms fractional values into integer values for the discrete variables. If the constraints are linear, then the mapping consists of formulating and solving the above mentioned linear programming problem. Finally, if the constraints are linear and the model also includes discrete variables, then the linear programming formulation becomes a mixed-integer programming problem that is solved accordingly. Clearly, if the optimization model has no constraints or discrete variables then $x^* = x$.

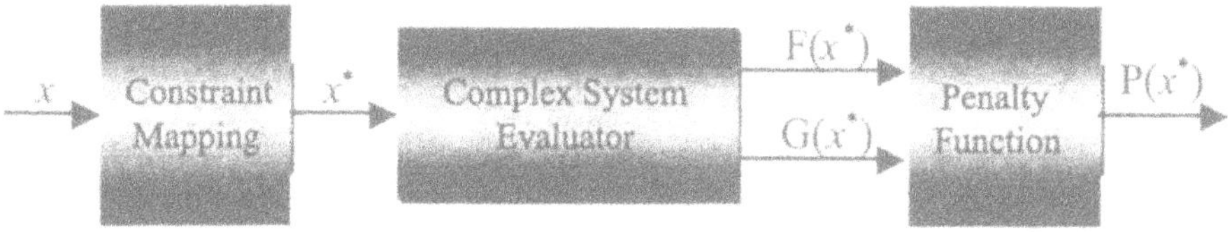

Figure 7.5 Solution Evaluation

The complex system evaluator processes the mapped solution to obtain a set of performance measures (i.e., the output of the evaluation). One of these measures is used as the *objective function* value $F(x)$ and provides the means for the search to distinguish high-quality from inferior solutions. Other measures $G(x)$ associated with the performance of the system can be used to define a set of *requirements*. A requirement is expressed as a bound on the value of a performance measure $G(x)$. Thus, a requirement may be defined as an upper or a lower bound on an output of the complex system evaluator. Instead of discarding requirement-infeasible solutions, OCL handles them with a composite function $P(x)$ that penalizes the requirement violations. The penalty is proportional to the degree of the violation and is not static throughout the search. OCL assumes that the user is interested in finding a requirement-feasible solution if one exists. Therefore, requirement-infeasible solutions are penalized more heavily when no requirement-feasible solution has been found during the search than when one is already available. Also, requirement-feasible solutions are always considered superior to requirement-infeasible solutions.

To illustrate the evaluation process in the context of a simulated system, let us revisit the investment problem briefly introduced at the end of Section 7.1. In this problem, x represents the allocation of funds to a set of investment instruments. The objective is to maximize the expected return. Assume that a Monte Carlo simulation is performed to estimate the expected return $F(x)$ for a given fund allocation. Hence, in this case, the complex system evaluator consists of a Monte Carlo simulator.

Restrictions on the fund allocations, which establish relationships among the x variables, are handled within the linear programming formulation that maps infeasible solutions into feasible ones. Thus, a restriction of the type "the combined investment in instruments 2 and 3 should not exceed the total investment in instrument 7," results in the linear constraint $x_2 + x_3 \leq x_7$. On the other hand, a restriction that limits the variability of the returns (as measured by the standard deviation) to be no more than a critical value c cannot be enforced in the input side of the Monte Carlo simulator. Clearly, the simulation must be executed first in order to estimate the variability of the returns. Suppose that the standard deviation of the returns is represented by $G(x)$, then the requirement in this illustrative situation is expressed as $G(x) \leq c$.

Note that the constraint-mapping mechanism within OCL does not handle nonlinear constraints. However, nonlinear constraints can be modeled as requirements and incorporated within the penalty function $P(x)$. For example, suppose that an optimization model must include the following nonlinear constraint:

$$x_1 x_2 - x_1^2 \leq 120$$

Then, the complex system evaluator calculates, for a given solution x, the left-hand side of the nonlinear constraint and communicates the result as one of the outputs. OCL uses this output and compares it to the right-hand side value of 120 to determine the feasibility of the current solution. If the solution is not feasible a penalty term is added to the value of the objective function $F(x)$.

7.4 OCL FUNCTIONALITY

The OptQuest Callable Library consists of a set of 43 functions that are classified into seven categories:

- General
- Variables, Constraints and Requirements
- Solutions
- Parameter Setting
- Event Handling
- Neural Network
- Linear Programming

The functions are classified according to their purpose and the library is available for both C and Visual Basic applications. Table 7.1 shows the complete set of OCL functions. A problem can be formulated and optimized with as few as five functions, which are indicated with an asterisk in Table 7.1.

Additional functions (not marked with an asterisk) in the library are used to change parameter settings or perform advanced operations such as monitoring and changing the composition of the reference set. The library also allows the user to define and

Category	Function	Brief Description
General	OCLGoodBye*	Frees OCL's memory
	OCLInitPop*	Generates an initial set of diverse solutions
	OCLRestart	Restarts the search
	OCLSetup*	Defines the optimization problem and parameters
	OCLGetNull	Retrieves the OCLNULL value
	OCLGetPopStats	Retrieves some search statistics
Variables, Constraints and Require-ments	OCLDefineVar*	Defines a decision variable. (This function is not needed for "all different" variables.)
	OCLDefineReq	Defines a requirement
	OCLConsCoeff	Changes the value of a constraint coefficient
	OCLConsRhs	Changes the values of the right-hand-side of a constraint
	OCLConsType	Changes the type of constraint
Solutions	OCLGenerateAllSolutions	Generates all solutions to a pure discrete problem
	OCLGetBest	Retrieves the best solution currently in OCL's memory
	OCLGetPopSolution	Retrieves a solution from the current reference set
	OCLGetSolution*	Retrieves a trial solution for evaluation
	OCLPutPopSolution	Replaces one of the solutions that is currently in the reference set
	OCLPutSolution*	Places the evaluation of a trial solution in OCL's memory
	OCLSugPopSolution	Places a suggested solution in a temporary memory to be added to the reference set when the set is rebuilt
	OCLSugSolution	Places a suggested solution in OCL's database
Parameter Setting	OCLSetAutoStop	Sets the automatic stop option
	OCLSetCheckDup	Activates or deactivates the use of the data-base to check duplicated solutions
	OCLSetObjPrecision	Sets the precision for the objective function
	OCLSetVarPrecision	Sets the precision for the decision variables
	OCLSetDatabaseSize	Sets the maximum size of the database
	OCLSetFileSolutions	Activates a log file of solutions
	OCLSetNNFilter	Activates the Neural Net Filter
	OCLSetNull	Sets the OCLNULL value
	OCLSetPopSize	Sets the population size
	OCLSetRandomSeed	Sets the seed for the random number generator
	OCLSetSearchStrategies	Sets parameter values for the search strategy to be used
	OCLSetSolutions	Sets the total number of solutions generated during the search
	OCLSetTaguchi	Activates the use of Taguchi solutions
Event Handling	OCLCallback	Identifies the callback function
Neural Network	OCLNNActivation	Defines the activation function
	OCLNNCallback	Identifies the callback function
	OCLNNGoodBye	Frees neural net memory
	OCLNNPrediction	Retrieves the net's output for a given input
	OCLNNSearchLevel	Sets the level of training
	OCLNNSetup	Defines the neural net
	OCLNNTrain	Trains the neural net
	OCLNNTrainData	Defines the data for training
Linear Programming	OCLDefineObj	Defines the objective function
	OCLSolveLP	Solves the linear program

*Required function.

Table 7.1 OCL Functions

train a neural network as well as to define and solve a linear (or mixed integer) programming problem. Regardless of the complexity of the application that uses OCL as its optimization engine, the following general structure is generally followed:

- Allocate memory for the optimization model by indicating the number of variables, constraints and requirements in the problem, as well as defining the direction of the optimization as minimize or maximize (OCLSetup).
- Define continuous and/or discrete decision variables (OCLDefineVar).
- Initialize the reference set (OCLInitPop) or generate all solutions in the case of small pure discrete problems (OCLGenerateAllSolutions).
- Iterate by retrieving a solution from OCL's database (OCLGetSolution), evaluating the solution (user-provided system evaluator) and placing the evaluated solution back into OCL's database (OCLPutSolution).

Suppose that we would like to use the C version of OCL to search for the optimal solution to the following unconstrained nonlinear optimization problem:

$$\begin{aligned}\text{Minimize}\quad & 100\left(x_2 - x_1^2\right)^2 + \left(1 - x_1\right)^2 + 90\left(x_4 - x_3^2\right)^2 + \left(1 - x_3\right)^2 + \\ & 10.1\left(\left(x_2 - 1\right)^2 + \left(x_4 - 1\right)^2\right) + 19.8\left(x_2 - 1\right)\left(x_4 - 1\right)\end{aligned}$$

$$\text{subject to}\quad -10 \leq x_i \leq 10 \quad \text{for } i = 1, \ldots, 4$$

According to the general structure of OCL, we need to start by allocating memory and indicating the direction of the optimization. To do this, we use the OCLSetup function, which has the following prototype:

```
long OCLSetup(long nvar, long nperm, long ncons,
              long req, char *direc, long lic);
```

`nvar`	An integer indicating the number of continuous and discrete decision variables in the problem.
`nperm`	An integer indicating the number of permutation ("all different") decision variables in the problem.
`ncons`	An integer indicating the number of constraints in the problem.
`req`	An integer indicating the number of requirements in the problem.
`direc`	An array of characters with the word "`MAX`" to indicate maximization or "`MIN`" to indicate minimization.
`lic`	A valid license number.

Therefore, the OCLSetup function call for our example would look like this:

```
nprob = OCLSetup(4, 0, 0 , 0, ''MIN'', ??????);
```

where `nprob` is a positive integer that indicates a unique problem number within OCL's memory. If OCLSetup returns a negative value, then the setup operation has failed. (Note that in an actual code "??????" must be replaced with a valid license number.) After setting up the problem, we need to define the decision variables using the OCLDefineVar function that has the following prototype:

```
long OCLDefineVar(long nprob, long var, double low,
                  double sug, double high, char *type,
                  double step);
```

`nprob`	A unique number that identifies an optimization problem within OCL's memory. This is the identifier returned by OCLSetup.
`var`	An integer indicating the variable number that corresponds to the current definition.
`low`	A double indicating the minimum value for the corresponding variable.
`sug`	A double indicating the suggested value for the corresponding variable. The suggested value is typically included in the initial reference set, unless the value results in an infeasible solution. The `OCLNULL` value can be used when no suggested value is available. The `OCLNULL` value is typically used to ignore an argument of an OCL function and is obtained with a call to OCLGetNull.
`high`	A double indicating the maximum value for the corresponding variable.
`type`	An array of characters with the word "`CON`" to define a continuous variable or "`DIS`" to define a discrete variable.
`step`	A double indicating the step size for a discrete variable. Step sizes may be integer or fractional and must be strictly greater than zero. Step sizes for continuous variables are ignored.

The function call to define the variables in our example can be programmed as follows:

```
for (i = 1; i <= 4; ++i)
   OCLDefineVar(nprob,i,-10,OCLGetNull(nprob),
                10,''CON'',1);
```

Note that although we use a "1" as the last argument of the function, this value is ignored because all the variables are defined as continuous.

We are now ready to build the starting reference set. This step is performed with a call to the OCLInitPop function. The "Pop" in the function name refers to "population", which is the terminology preferred in the GA community, although the term "reference set" is more common in scatter search implementations. The prototype of this function consists of a single argument containing the unique problem identifier returned by OCLSetup. The function call looks as follows, which assumes that `nprob` is the identifier returned by OCLSetup:

```
OCLInitPop(nprob);
```

It is important to point out that all the functions in OCL return an integer value. If the return value is positive, the function call was successful. Otherwise, if the return value is negative, the function call failed and the return value is the error code.

After a successful initialization of the reference set, the search can begin. The search is performed with a series of calls to three functions: OCLGetSolution, a user-provided system evaluator and OCLPutSolution. The first function retrieves a solution from OCL's database, the second evaluates the solution and the third places the evaluated solution back into OCL's database. The prototype for OCLGetSolution is:

```
long OCLGetSolution(long nprob, double *sol);
```

`sol` A pointer to an array of doubles where OCL places the solution. The array should have enough space to hold the variable values in positions 1 to `nvar`, as defined in OCLSetup.

As before, `nprob` is the unique problem identifier returned by OCLSetup. If the call to OCLGetSolution is successful, the function returns a unique solution identifier `nsol`. The prototype for OCLPutSolution is:

```
long OCLPutSolution(long nprob, long nsol, double
                    *objval, double *sol);}
```

`nsol` A solution identifier returned by OCLGetSolution.

`objval` An array of doubles with the values of the objective function and the requirements. The array should have a size of at least `req+1` positions, as defined in OCLSetup. The objective function value should be `objval[0]` and the ith requirement should be `objval[i]`. Note that if no requirements are defined, then `objval` can be dimensioned as a simple double variable. OCLNULL may be used to instruct OCL to discard the solution.

`sol` An array of doubles with the values of the decision variables. If the solution values are the same as the ones retrieved with a call to OCLGetSolution, a pointer to the OCLNULL variable may be used.

The evaluation function is outside the scope of OCL and is the responsibility of the user. For our example, we can code the evaluator simply as:

```
double evaluate(double *x)
{
    return(100*pow(x[2]-pow(x[1],2),2)+pow(1-x[1],2)
          +90*pow(x[4]-pow(x[3],2),2)+pow(1-x[3],2)
          +10.1*((pow(x[2]-1,2)+pow(x[4]-1,2))
          +19.8*(x[2]-1)*(x[4]-1));
}
```

Assuming that we want to search for the optimal solution to this problem allowing OCL to perform a maximum of 10000 function evaluations, the code to perform such a search has the following form:

```
for (i = 1; i <= 10000; i++)
{
    nsol = OCLGetSolution(nprob, x);
    if (nsol < 0) {
        printf(''OCLGetSolution error code %d\n'', nsol);
        exit(1);
    }
    objval = evaluate(x);
    status = OCLPutSolution(nprob, nsol, &objval, x);
    if (status < 0) {
        printf(''OCLPutSolution error code %d\n'',
        status);
        exit(1);
    }
}
```

This code retrieves, evaluates and returns the objective function value of 10000 solutions. It also checks for possible error codes from calls to OCLGetSolution and OCLPutSolution. Note that the partial code above does not keep track of the best solution found. This can be done by adding an "if" statement that compares the objective function value of the current solution with the best objective function value found during the search. Alternatively, the OCLGetBest function can be called at any time during the search to retrieve the values associated with the best solution, which OCL automatically monitors. The function is called with the following arguments:

```
OCLGetBest(nprob, x, &objval);
```

where `nprob` is the unique problem identifier, x is the array where the variable values are stored and `objval` is the variable where the objective function value is returned. The entire C code for this example is shown in Appendix A at the end of this chapter.

7.4.1 Constraints and Requirements

The illustration in the previous section does not include the OCL functions for constraints and requirements. In this section, we briefly describe how constraints and requirements can be defined with OCL. Assume that we would like to add the following linear constraint to the optimization model for our four-variable example problem:

$$x_1 + 8x_2 - 3x_4 \leq 5$$

Then after a call to OCLSetup and before a call to OCLInitPop, we add calls to the constraint-related functions: OCLConsCoeff (to change the coefficient of a constraint), OCLConsRhs (to change the right-hand-side of a constraint) and OCLConsType (to change the constraint type). The calls to these functions can be made in any order, as long as they are made before the reference set is initialized and certainly before the search begins through calls to OCLGetSolution and OCLPutSolution. The prototypes for the constraint-related functions are:

```
long OCLConsCoeff(long nprob, long cons, long var,
                  double coeffval);
long OCLConsRhs(long nprob, long cons, double rhsval);
long OCLConsType(long nprob, long cons, long type);
```

In these prototypes, `nprob` is the unique problem identifier returned by OCLSetup, `cons` is the constraint number, `var` is the variable number, `coeffval` is the coefficient value, `rhsval` is the right-hand-side value and `type` is the constraint type (`OCLLE` = less-than-or-equal, `OCLGE` = greater-than-or-equal, and `OCLEQ` = equal). The following function calls can be used to define the constraint in our example:

```
    OCLConsCoeff(nprob, 1, 1, 1);
    OCLConsCoeff(nprob, 1, 2, 8);
    OCLConsCoeff(nprob, 1, 4, -3);
    OCLConsRhs(nprob, 1, 5);
    OCLConsType(nprob, 1, 1, OCLLE);
```

These function calls assume that the constraint is the first one in the model. Also, the variable numbers match the ones used when OCLDefineVar was called.

The definition of requirements is slightly different from the definition of constraints. As mentioned before, a requirement is basically a bound on an output value of the system evaluator. Suppose that we would like to define a requirement to impose the following nonlinear restriction to our illustrative example:

$$10.8x_1^2 - 2.4x_3x_4 \geq 15.9$$

We use a call to OCLDefineReq to define the requirement. This function has the following prototype:

```
long OCLDefineReq(long nprob, long req, double low,
                  double high);
```

Where `nprob` is the unique problem identifier returned by OCLSetup, `req` is the requirement number, `low` is the lower bound for the requirement, and `high` is the upper bound for the requirement. The variable `OCLNULL` can be used to leave either `low` or `high` undefined. The function call for our example is:

```
        OCLDefineReq(nprob, 1, 15.9, OCLGetNull(nprob));
```

In addition to this definition, we need to modify the system evaluator. The evaluator must return the value of the objective function in `objval[0]` and the value of the requirement in `objval[1]`. The new evaluator looks like this:

```
void evaluate(double *x, double *objval)
```

```
{
    objval[0] = 100*pow(x[2]-pow(x[1],2),2)+pow(1-x[1],2)
                +90*pow(x[4]-pow(x[3],2),2)+pow(1-x[3],2)
                +10.1*((pow(x[2]-1,2)+pow(x[4]-1,2))
                +19.8*(x[2]-1)*(x[4]-1));
    objval[1] = 10.8*pow(x[1],2)-2.4*x[3]*x[4];
}
```

Other small changes are necessary to make OCL work with the requirement that we have defined. For example, the declaration of `objval` must be changed from a single double to an array of doubles and the `evaluate` function must be changed from double to void. The changes are reflected in the following partial code:

```
void evaluate(double *, double *);
.
.
    double objval[3];}
.
.
    for (i = 1; i <= TOT_ITER; i++) {
        nsol = OCLGetSolution(nprob, x);
        evaluate(x, objval);
        OCLPutSolution(nprob, nsol, objval, x);
    }
```

When requirements are included in an optimization model, OCLGetBest indicates the feasibility of the best solution. The return values for OCLGetBest can be either 0 if the best solution is requirement-feasible, 1 if the best solution is requirement-infeasible or a negative value representing an error code. Recall that while all the solutions generated by OCL are constraint-feasible (if the constraint space is not empty), some may be requirement-infeasible and this is why the OCLGetBest is designed to provide information regarding the requirement-feasibility of the best solution found during the search.

The discussion about the functionality of OCL in which we have engaged is not meant to be exhaustive. OCL has additional functions that can be used to modify the way the search is performed. For example, OCL includes functions to change the default parameter settings. A function such as OCLSetPopSize can be used to change the number of solutions that are carried out in the reference set throughout the search. In some applications, changing this parameter may improve the quality of the best solution found. Other functions, such as OCLSetBoundFreq, are meant for the advanced user, who has a thorough understanding of both the problem context to solve and the way the search method works.

7.4.2 Boundary Search Strategy

The OCLSetBoundFreq controls the application of the boundary search strategy. This strategy is a mechanism to generate non-convex combinations of two solutions. In particular, the set of points generated on the line defined by x' and x'', but beyond the segment $[x', x'']$ "outside" of x'' is given by:

$$x = x'' + (x'' - x')\, t \qquad \text{for } t > 0$$

While the "standard" combination method in OCL defines $t = r/2$, where r is a random number between 0 and 1, the boundary strategy considers three strategies to generate non-convex combinations in the (x',x'') line:

- *Strategy* 1: Computes the maximum value of t that yields a feasible solution x when considering both the bounds and the constraints in the model.

- *Strategy* 2: Considers the fact that variables may hit bounds before leaving the feasible region relative to other constraints. The first departure variable from the feasible region may happen because some variable hits a bound, which is followed by the others, before any of the linear constraints is violated. In such a case, the departing variable is fixed at its bound when it hits it, and the exploration continues with this variable held constant. OCL does this with each variable that encounters a bound before other constraints are violated. The process finishes when the boundary defined by the other constraints is reached.

- *Strategy* 3: Considers that the exploration hits a boundary that may be defined by either bounds or any of the linear constraints. When this happens, one or more constraints are binding and the corresponding $t$value cannot be increased without causing the violation of at least one constraint. At this point, OCL chooses a variable to make a substitution that geometrically corresponds to a projection that makes the search continue on a line that has the same direction, relative to the constraint that was reached, as it did upon approaching the hyperplane defined by the constraint. The process continues until the last unfixed variable hits a constraint. At this point, the value of all the previously fixed variables is computed.

Each of these three boundary strategies generates a "boundary solution" x^b outside x''. OCL also generates the solution in the midpoint between x^b and x''. Interchanging the role of x' and x'' gives the extension outside the "other end" of the line segment. The mechanism results in a total of four non-convex solutions out of a single combination of a pair of reference points. The user establishes, by setting the appropriate value in the OCLBoundFreq function, the percentage of combinations in which the boundary strategy is applied. The default value calls for applying the boundary strategy 50% of the time. Although this default value is fairly robust across a variety of problem classes, for problems where the best values are suspected to lie near the boundary of the feasible region, improved solutions may be found when the frequency value is increased. Similarly, if the best solutions to an optimization problem are not near a boundary, improved outcomes may result from decreasing the frequency for applying the boundary strategy.

7.5 OCL APPLICATION

In this section, we apply OCL to a set of hard nonlinear and unconstrained optimization problems. We have built an application based on OCL that allows us to test the performance of the library as compared to a competing generic optimizer based on genetic algorithms.

Table 7.2 shows the set of 30 test problems that we have gathered from the following web pages:

```
http://www.maths.adelaide.edu.au/Applied/llazausk/alife/realfopt.htm
http://solon.cma.univie.ac.at/~neum/glopt/my_problems.html
```

The number between parentheses associated with some of the function names are the parameter values for the corresponding objective function. A typical parameter refers to the number of variables in the function, since many of these functions expand to an arbitrary number of variables. Although the objective functions are built in a way that the optimal solutions are known, the optimization problems cannot be trivially solved by search procedures that do not exploit the special structure that characterizes each function. A detailed description of the objective functions is provided in Appendix B.

Our OCL-based code has a "main" function with two arguments: the problem number and the total number of function evaluations. The problem number is used to select the correct objective function from the catalog of functions inside the system evaluator. The code also uses this number to allocate the appropriate memory using OCLSetup and to define the bounds for each variable calling OCLDefineVar.

We compare the performance of OCL with the results obtained from running Genocop III (`http://www.coe.uncc.edu/~zbyszek/gchome.html`) on the same set of test problems. Genocop III is the third version of a genetic algorithm designed to search for optimal solutions to optimization problems with continuous variables and linear and nonlinear constraints. The description of the first version of Genocop appears in the book by Michalewicz (1999). While this GA optimizer does not handle discrete variables, it does provide a way for explicitly defining nonlinear constraints. The performance of the Genocop depends on a set of twelve parameters (without counting the frequency distribution for the application of each operator). We did not attempt to find the best parameter setting for the set of problems on hand and instead we used the default values that the system recommends. Similarly, we used the default values for the OCL parameters, such as the size of the reference set and the frequency of application of the boundary strategy. A summary of our comparison is shown in Table 7.3.

The results in Table 7.3 were obtained allowing each procedure to perform 10000 function evaluations. The values in bold indicate which procedure yields the solution with the better objective function value for each problem. The following observations can be made from the results in this table.

- OCL finds solutions that on the average are better than those found by Genocop, within the scope of the search (i.e., 10000 evaluations).

Problem number	Name and parameter values	Optimal value
1	Colville	0
2	Perm(4,0.5)	0
3	Perm(9,10^8)	0
4	Perm0(4,10)	0
5	Perm0(10,100)	0
6	PowerSum(8,18,44,114)	0
7	PowerSumZ(z_1,...,z_8,1,15) z_1=1, $z_i = 2 + z_{i-1}$	0
8	PowerSumZ(z_1,...,z_{10},0,2) $z_i = 1/i$	0
9	PowerSumZ(z_1,...,z_6,-6,6) $z_i = i - 1$	0
10	Trid(6)	-50
11	Trid(10)	-210
12	Rosenbrock(8)	0
13	Rosenbrock(20)	0
14	Rosenbrock2(2)	0
15	Rosenbrock2(6)	0
16	SixHumpCamelBack	-1.0316285
17	Levy(30)	0
18	Beale	0
19	Booth	0
20	Matyas	0
21	Powell(24)	0
22	Griewank(20)	0
23	Rvan(25)	0
24	Rastrigin(10)	0
25	Rastrigin(20)	0
26	Schwefel1(10)	0
27	Schwefel1(20)	0
28	Ackley(50)	0
29	Ackley(100)	0
30	Sphere(100)	0

Table 7.2 Test Problems

- OCL finds better solutions than Genocop more frequently (20 times for OCL versus 3 times for Genocop).
- Genocop is on average four times faster than OCL.
- Both OCL and Genocop struggled with Problems 3 and 7.

In addition to comparing the performance of both systems when considering the final solutions and the total computational time to find them, it is important to assess how quickly each method reaches the best solutions. Reaching good solutions quickly becomes more critical when the complexity of the system evaluator increases. Consider, for example, an application for which a single evaluation of the objective function consists of the execution of a computer simulation that requires two CPU minutes. Clearly, in this context, the time to generate each solution becomes negligible. At the same time, it is not feasible to search for the best solution employing 10000 function evaluations, unless one is willing to wait for two weeks to obtain a rel-

		Objective function value			CPU seconds	
Prob.	Var.	Optimal	OCL	Genocop	OCL	Genocop
1	4	0	**0.000361**	0.729826	3.8	0.3
2	4	0	**0.130655**	1.794891	4.0	0.9
3	9	0	**4.5839E+15**	2.12E+16	4.8	3.7
4	4	0	**0.073611**	0.79538	4.3	0.8
5	10	0	**74.020773**	276.9857	5.4	3.7
6	4	0	**0.00883**	0.02553	151.6	0.6
7	8	0	**8.30E+08**	1.23E+14	3.5	2.9
8	10	0	0	0	5.9	3.4
9	6	0	**147.688802**	1452.561	6.8	1.8
10	6	-50	**-49.958165**	-46.481	33.6	0.4
11	10	-210	**-127.15042**	-31.5584	3.8	0.9
12	8	0	**0.483572**	5.601258	3.2	0.8
13	20	0	**5.599877**	7.684962	6.9	2.8
14	2	0	**1.990099**	36.3949	4.8	0.3
15	6	0	**5.949696**	273.3088	6.3	0.8
16	2	-1.03163	-1.03163	-1.03163	5.8	0.3
17	30	0	0.69632	**0.300885**	74.3	5.5
18	2	0	0	0.00075	4.1	0.2
19	2	0	0	0	2.8	0.2
20	2	0	0	0	4.8	0.2
21	24	0	**0.085342**	2.740024	8.7	3.2
22	20	0	**0**	1.076349	3.8	2.6
23	25	0	**0.679836**	37.37577	42.9	3.9
24	10	0	**0**	1.025595	4.5	0.9
25	20	0	**0**	10.5079	6.3	2.5
26	10	0	844.068704	**1.387158**	1.8	0.8
27	20	0	1506.06706	**134.4911**	2.4	2.1
28	50	0	0	0	16.8	13.4
29	100	0	0	0	103.6	46.6
30	100	0	**2.418512**	1114.451	60.3	43.7

Table 7.3 Comparison of OCL and Genocop III

atively good answer. A more reasonable approach is to limit the search to 500 function evaluations, whose execution will require somewhat less than 17 hours to execute.

Figure 7.6 depicts the trajectory followed by OCL and Genocop when tracking the average objective function value in a set of 28 problems that excludes problems 3 and 7. Note that the graph is drawn on a logarithm scale to be able to accommodate the large average objective function value of 8.8E+12 yielded by Genocop after 100 evaluations.

The purpose of constructing the performance graph depicted in Figure 7.6 is to assess the aggressiveness of each procedure as measure by the speed in which the search is capable of finding reasonably good solutions. Based on our testing, OCL seems to be more aggressive than Genocop. The aggressiveness in OCL, however, does not compromise the quality of the final solution. In other words, OCL aggressively attempts to improve upon the best solution during the early stages of the search and at the same time it makes use of diversifying mechanisms to be able to sustain an

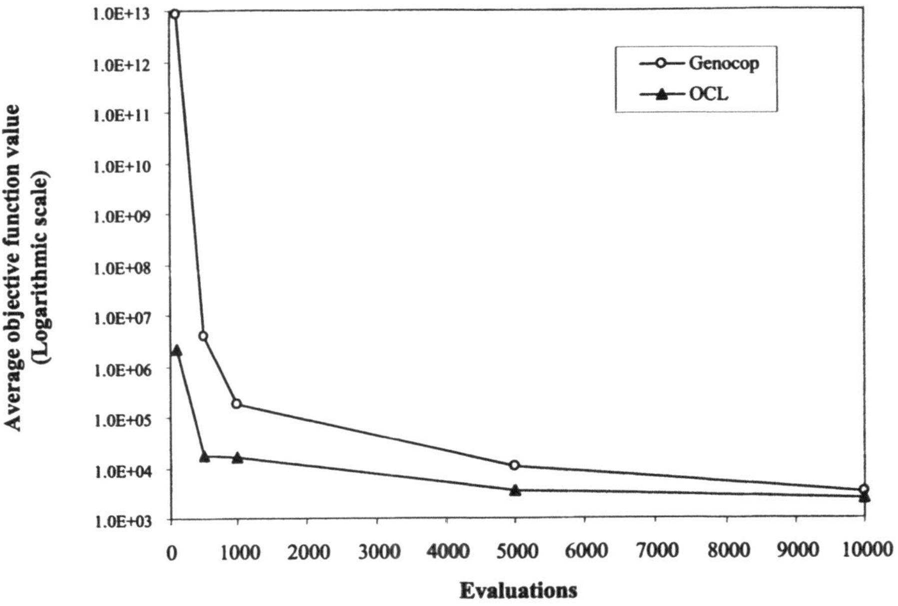

Figure 7.6 Performance Graph for OCL and Genocop III

improving trajectory when allowed to extend the search beyond a limited number of objective function evaluations.

The aggressiveness of OCL as compared to Genocop can be measured by the gap between the average objective function values found by each method. After 500 evaluations, Genocop average is more than 200 times larger than OCL average. OCL average remains lower than Genocop values throughout the search with the final average for Genocop being 36% higher than OCL's, as shown in Table 7.4.

Evaluation number	Percentage deviation of the Genocop solutions from OCL solutions
100	417266887%
500	21639%
1000	983%
5000	202%
10000	36%

Table 7.4 Genocop Solutions Compared to OCL Solutions

The performance graph in Figure 7.6 and the percentage deviation values in Table 7.4 seem to substantiate that OCL would be the preferred solution method for optimizing complex systems, for which the evaluation of the objective function determines the computational time of the search. These exhibits also show a general trend that seems to reveal that the quality of the solutions found by both methods becomes more alike as the number of evaluations increases.

The outcomes obtained with OCL for this class of problems (i.e., unconstrained nonlinear optimization problems) can be improved by adding a local optimizer. This is evident in the results reported in Ugray et al. (2001), where OCL is combined with an implementation of a generalized reduced gradient (GRG) procedure.

7.6 CONCLUSIONS

In this chapter, we discussed the notion of optimizing a complex system when a "black box" evaluator can be used to estimate performance based on a set of input values. In many business and engineering problems, the black box evaluator has the form of a simulation model capable of mapping inputs into outputs, where the inputs are values for a set of decision variables and the outputs include the objective function value. We also addressed the development and application of a function library that uses scatter search in the context of optimizing complex systems.

The functionality of the library was illustrated with an example in the context of nonlinear optimization. Our illustration resulted in a sample code that can be modified and expanded to apply OCL in other situations. Finally, we tested OCL by comparing its performance with Genocop III, a third-generation genetic algorithm. Our experiments with 30 nonlinear optimization problems show that OCL is a search method that is both aggressive and robust. It is aggressive because it finds high-quality solutions early in the search. It is robust because it continues to improve upon the best solution when allowed to search longer. These characteristics make OCL an ideal solution method for applications in which the evaluation of the objective function requires a non-trivial computational effort.

Acknowledgments

Manuel Laguna was partially supported by the visiting professor fellowship program of the University of Valencia (Grant Ref. No. 42743).

Appendix A

The following is the complete C code to search for the optimal solution of the example problem in Section 7.4 using OCL. For simplicity we have left out the error-code checking. In actual applications, however, it is recommended to check for the return values of the OCL functions in order to detect errors during the search.

```
#include ''ocl.h'' #include <stdio.h> #include <math.h>

#define NUM_VAR 4
#define TOT_ITER 10000

double evaluate(double *);

void main(void)
{
```

```
    double x[NUM_VAR+1], objval;
    long nprob, nsol;
    int i;
    /* Allocating memory */
    nprob = OCLSetup(NUM_VAR,0,0,0,''MIN'',??????);

    /* Defining variables */
    for (i = 1; i <= NUM_VAR; i++)
        status = OCLDefineVar(nprob,i,-10,OCLGetNull(nprob),
                              10,''CON'',1);

    /* Initializing the reference set */
    OCLInitPop(nprob);

    /* Generate and evaluate TOT_ITER solutions */
    for (i = 1; i <= TOT_ITER; i++) {
        nsol = OCLGetSolution(nprob,x);
        objval = evaluate(x);
        OCLPutSolution(nprob,nsol,&objval,(double*)OCLGetNull(nprob));
    }

    /* Display the best solution found */
    status = OCLGetBest(nprob,x,&objval);
    printf(''Best solution value is %9.6f\n'', i, objval);
    for(i = 1; i <= NUM_VAR; i++)
        printf(''x[%2d] = %9.6lf\n'', i, x[i]);

    /* Free OCL memory */
    status = OCLGoodBye(nprob);
}

/* Evaluation function */
double evaluate(double *x)
{
    return(100*pow(x[2]-pow(x[1],2),2)+pow(1-x[1],2)
          +90*pow(x[4]-pow(x[3],2),2)+pow(1-x[3],2)+
          10.1*((pow(x[2]-1,2)+pow(x[4]-1,2))+19.8*(x[2]-1)*(x[4]-1));
}
```

Appendix B

This appendix contains the description of the set of test functions in Table 7.2. The description consists of the objective function and the bounds for each variable. (A C code of the functions in Appendix B is available from the authors.)

Colville	
Minimize	$f(x) = 100(x_2 - x_1^2)^2 + (1 - x_1)^2 + 90(x_4 - x_3^2)^2 + (1 - x_3)^2 +$ $10.1\left((x_2 - 1)^2 + (x_4 - 1)^2\right) + 19.8(x_2 - 1)(x_4 - 1)$
subject to	$-10 \leq x_i \leq 10$ for $i = 1, \ldots, 4$

Perm(n, β)	
Minimize	$f(x) = \sum_{k=1}^{n} \left(\sum_{i=1}^{n} (i^k + \beta) \left(\left(\frac{x_i}{i}\right)^k - 1 \right) \right)^2$
subject to	$-n \leq x_i \leq n$ for $i = 1, \ldots, n$

Perm0(n, β)

Minimize $f(x) = \sum_{k=1}^{n} \left(\sum_{i=1}^{n} (i+\beta) \left(x_i^k - \left(\frac{1}{i}\right)^k \right) \right)^2$

subject to $-n \leq x_i \leq n$ for $i = 1, \ldots, n$

PowerSum($b_1, \ldots, b_n$))

Minimize $f(x) = \sum_{k=1}^{n} \left(\left(\sum_{i=1}^{n} x_i^k \right) - b_k \right)^2$

subject to $0 \leq x_i \leq n$ for $i = 1, \ldots, n$

PowerSumZ($z_1, \ldots, z_n, l, u$)

Minimize $f(x) = \sum_{k=1}^{n} \left(\left(\sum_{i=1}^{n} x_i^k \right) - b_k \right)^2$

subject to $b_k = \sum_{i=1}^{n} z_i^k \quad k = 1, \ldots, n$

$l \leq x_i \leq u$ for $i = 1, \ldots, n$

Trid(n)

Minimize $f(x) = \left(\sum_{i=1}^{n} (x_i - 1)^2 \right) - \sum_{i=2}^{n} x_i x_{i-1}$

subject to $-n^2 \leq x_i \leq n^2$ for $i = 1, \ldots, n$

Rosenbrock(n)

Minimize $f(x) = \sum_{i=1}^{n/2} 100(x_{2i} - x_{2i-1}^2)^2 + (1 - x_{2i-1})^2$

subject to $-10 \leq x_i \leq 10$ for $i = 1, \ldots, n$

Rosenbrock2(n)

Minimize $f(x) = \sum_{i=1}^{n} 100(x_i - x_{i-1}^2)^2 + (1 - x_{i-1})^2$

subject to $-10 \leq x_i \leq 10$ for $i = 1, \ldots, n$

SixHumpCamelBack

Minimize $f(x) = 4x_1^2 - 2.1x_1^4 + \frac{1}{3}x_1^6 + x_1 x_2 - 4x_2^2 + 4x_2^4$

subject to $-10 \leq x_1, x_2 \leq 10$

Levy(n)

Minimize $f(x) = \sin^2(\pi y_1) +$

$\sum_{i=1}^{k-1} (y_i - 1)^2 \left(1 + 10\sin^2(\pi y_i + 1)\right) + (y_k - 1)^2(1 + \sin^2(2\pi x_k))$

subject to $y_i = 1 + \frac{x_i - 1}{4}$ for $i = 1, \ldots, n$

$-10 \leq x_i \leq 10$ for $i = 1, \ldots, n$

Beale

Minimize $f(x) = (1.5 - x_1 + x_1 x_2)^2 + (2.25 - x_1 + x_1 x_2^2)^2 +$

$(2.625 - x_1 + x_1 x_2^3)^2$

subject to $-4.5 \leq x_1, x_2 \leq 4.5$

Booth	
Minimize	$f(x) = (x_1 + 2x_2 - 7)^2 + (2x_1 + x_2 - 5)^2$
subject to	$-10 \leq x_1, x_2 \leq 10$

Matyas	
Minimize	$f(x) = 0.26(x_1^2 + x_2^2) - 0.48x_1x_2$
subject to	$-10 \leq x_1, x_2 \leq 10$

Powell(n)	
Minimize	$f(x) = \sum_{j=1}^{n/4} (x_{4j-3} + 10x_{4j-2})^2 + 5(x_{4j-1} - x_{4j})^2 + (x_{4j-2} - 2x_{4j-1})^4 + 10(x_{4j-3} - x_{4j})^4$
subject to	$-4 \leq x_i \leq 5$ for $i = 1, \ldots, n$

Griewank(n)	
Minimize	$f(x) = \sum_{i=1}^{n} \frac{x_i^2}{4000} - \prod_{i=1}^{n} \cos\left(\frac{x_i}{\sqrt{i}}\right) + 1$
subject to	$-600 \leq x_i \leq 600$ for $i = 1, \ldots, n$

Rvan(n)	
Minimize	$f(x) = \sum_{i=1}^{n} i(2x_i^2 - x_{i-1})^2 + (x_1 - 1)^2$
subject to	$-10 \leq x_i \leq 10$ for $i = 1, \ldots, n$

Rastrigin(n)	
Minimize	$f(x) = 10n + \sum_{i=1}^{n} x_i^2 - 10\cos(2\pi x_i)$
subject to	$-5.12 \leq x_i \leq 5.12$ for $i = 1, \ldots, n$

Schwefel1(n)	
Minimize	$f(x) = 418.9829n + \sum_{i=1}^{n} -x_i \sin\sqrt{\lvert x_i \rvert}$
subject to	$-500 \leq x_i \leq 500$ for $i = 1, \ldots, n$

Ackley(n)	
Minimize	$f(x) = 20 + e - 20e^{-0.2\sqrt{\frac{1}{n}\sum_{i=1}^{n} x_i^2}} - e^{\frac{1}{n}\sum_{i=1}^{n}\cos(2\pi x_i)}$
subject to	$-30 \leq x_i \leq 30$ for $i = 1, \ldots, n$

Sphere(n)	
Minimize	$f(x) = \sum_{i=1}^{n} x_i^2$
subject to	$0 \leq x_i \leq 100$ for $i = 1, \ldots, n$

8 A CONSTRAINT PROGRAMMING TOOLKIT FOR LOCAL SEARCH

Paul Shaw[1], Vincent Furnon[2] and Bruno De Backer[2]

[1]ILOG S.A.
Les Taissounieres HB2, 1681, route des Dolines
06560 Valbonne, FRANCE
pshaw@ilog.fr

[2]ILOG S.A.
9, rue de Verdun, BP 85
94253 Gentilly Cedex, FRANCE
{vfurnon,bdebacker}@ilog.fr

Abstract: Constraint programming is both a modeling and problem solving technology applicable to a wide range of optimization problems. Its inherent flexibility means that it is one of the leading technologies used in practical applications – aided largely by the existence of *constraint programming systems*.

ILOG Solver is a commercial constraint programming system taking the form of a C++ component library that users can tightly integrate with their own applications. In this chapter we concentrate on one aspect of ILOG Solver – the local search facilities added in version 5.0. The chapter is tutorial in nature, with local search features being elaborated by examples.

8.1 INTRODUCTION

Constraint programming (Rossi (2000), Tsang (1993), van Hentenryck (1989)) is a technology which is becoming of growing importance in the field of optimization, its main advantage over other techniques being that one can directly describe and solve general models using it. The practical success of constraint programming in recent years has been greatly aided by both academics and, most importantly, commercial software vendors who have produced *constraint programming systems*. These systems form a basis for the production of optimization tools, which are becoming increasingly industrially important. Examples of academic systems are Claire (Caseau and Laburthe (1995)), OZ (Henz et al. (1993), Henz et al. (1995)), clp(fd) (Condognet and Diaz (1996)), and CHR (Frühwirth (1998)). The most prolific commercial libraries are ILOG Solver (ILOG (2000b)), ECLiPSe (IC-PARC (2001)) and CHIP (Dincbas et al. (1988), Simonis et al. (1995)). In this chapter, we describe the evolution of one of these constraint programming systems, ILOG Solver, with respect to the local search support which has recently been introduced.

Our main objective in the addition of local search support to Solver was to provide concepts and objects which were as natural as possible for the local search practitioner, or indeed beginner. Therefore, the concepts should follow closely what already exists in the literature: see, e.g., Aarts and Lenstra (1997). For example, because the concepts of solutions and neighborhood structures are fundamental, these should be important objects in ILOG Solver's local search. Second, we wished to provide a level of local search support which would be flexible and extensible. Thus, we follow the already proven approach already taken in Solver, which is to provide support for common, but not all conceivable, demands. To cope with less common or esoteric requests, Solver is *extensible*; users are at liberty to extend the important parts of the constraint programming system. This means that typical users are not burdened with learning a huge system, and yet such a system can still meet a wide variety of needs. In the same spirit, Solver's local search support provides objects for constructing the most common types of local search such as hill climbing, simulated annealing, and tabu search. More esoteric methods, which commonly have specific application in any case, can be written by users to exactly suit their needs.

This chapter is set out as follows. Section 8.2 gives an overview of constraint programming for those unfamiliar with the technology. Section 8.3 describes Solver's local search mechanisms in detail, discussing the main components, and how they interact. Examples are given of the use of the main components, and code of a complete local search example is presented. Finally, we discuss the advantages and disadvantages of embedding a local search toolkit in a constraint programming system. Section 8.4 then presents a complete real-world example of facility location, solved by three different search methods. Of particular interest is a local/complete search hybrid. Section 8.5 demonstrates how the local search features of Solver can be extended and an example is given of how to code a new neighborhood. Section 8.6 gives a more complete demonstration of how Solver can be extended on all fronts, resulting in a constraint programming system for vehicle routing, ILOG Dispatcher. Dispatcher's main concepts are introduced through an example. We concentrate on the extension of Solver's generic local search facilities, which have been targeted towards vehicle routing. Section 8.7 describes related work both in the constraint programming domain, and in respect of systems for local search. We conclude in Section 8.8.

8.2 CONSTRAINT PROGRAMMING PRELIMINARIES

This chapter concentrates on the local search features of ILOG Solver. As such, we introduce other notions of constraint programming only when relevant. However, we gain a certain economy if basic ideas are presented together. This section then – which readers familiar with constraint programming can skip – gives an overview of constraint programming and its basic notions.

Fundamentally, a constraint programming problem is defined by a set of unknowns or variables, and a set of constraints. Each variable has a finite set of possible values (a domain), and each constraint involves a subset of the variables. Each such constraint indicates which partial assignments, involving only the variables incident on the constraint, will violate (or, alternatively, satisfy) the constraint. The problem is then one of finding an assignment (single element of the domain) for each variable such that no constraints are violated. This problem is $\mathcal{NP}$-complete (Garey and Johnson (1979)). The optimization variant of this problem is one where we must find the minimal value for one of the variables such that no constraints are violated.

8.2.1 Tree Search

In constraint programming systems, solutions to problems are generally found using *tree search* and, in the simplest case, the tree search mechanism known as depth-first search. The advantage of depth-first search is that it uses memory economically while being complete: it guarantees to find a solution if one exists, or reports that there is no solution if the problem is not solvable. Tree search in the worst case examines all combinations of values for all variables, but this worst case is rarely attained. Most commonly, significant parts of the search tree are pruned by the action of constraints.

We demonstrate how depth-first search operates by means of a small example. Consider a problem with three variables a, b and c which have domains $\{0, 2\}$, $\{0, 1\}$ and $\{0, 1\}$, respectively. We shall also consider three constraints: $a \neq b$, $a \geq c$ and $b \leq c$. By inspection, the solutions to this problem are: $a = 2, b = 0, c = 0$, $a = 2, b = 0, c = 1$ and $a = 2, b = 1, c = 1$. The other five possible combinations of values for a, b and c violate at least one constraint and so are not admissible. We will perform a depth-first search, choosing to instantiate variables in lexicographical order: first a, then b, then c. We shall choose possible values for variables in ascending order. The specification of the order that variables and values are chosen, and more generally the description of the growth of the search tree, is called a constraint programming *goal*. The resulting search tree for this goal is depicted in Figure 8.1. Nodes in the tree represent states of the assignment of variables. When a node is crossed, i.e. $\otimes$, this state is one in which at least one constraint is violated. Arcs are transitions between states and are labelled with the assignment that takes place at the state transition. Finally, non-crossed leaves in the tree are *solution states*, where all variables are assigned and no constraint is violated.

The depth-first search proceeds as follows. To begin, we take variable a and assign it its smallest value, 0, followed by an assignment of variable b to 0, which is its smallest value. Although we do not have a complete solution yet, we can check the validity of the constraint $a \neq b$ as all the variables involved in the constraint are

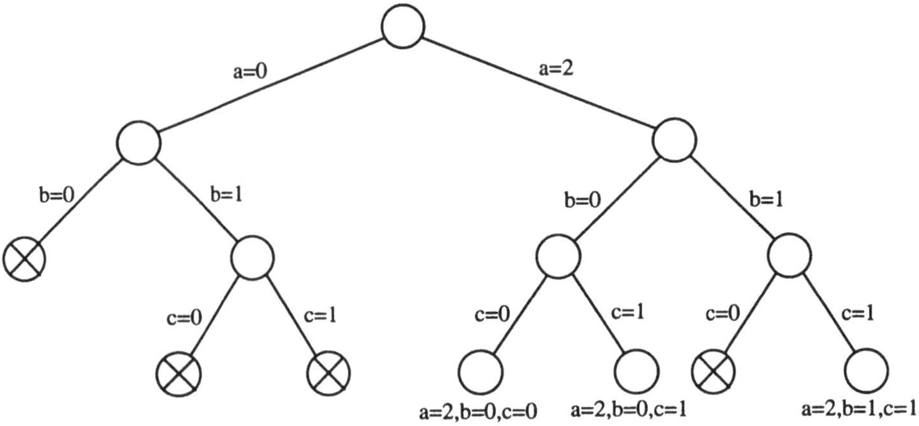

Figure 8.1 Operation of Depth-First Search

instantiated. That is, no assignments we make after this point can change the value – violated or satisfied – of the constraint. In this case we see that the constraint $a \neq b$ is violated: we call this state a *failure* (marked by $\otimes$). One of the assignments $a = 0$ or $b = 0$ needs to be changed: we undo assignments in a "depth-first" manner by undoing the most recent assignment first. We move the search back to the state before assigning b, a movement known as *backtracking*, and try the next value for b which is 1. This time, the constraint $a \neq b$ is not violated, and so we continue, by instantiating c. In fact, we find that neither value for c is satisfactory as 0 violates constraint $b \leq c$, whereas 1 violates constraint $a \geq c$. We have now proved that there exist no solutions for which $a = 0$ as we have explored all combinations of values for b and c with $a = 0$. We therefore backtrack to the top of the tree and assign $a = 2$. Consequently, we find that instantiating both b and c to 0 results in no constraints being violated: a solution. If we were interested in any solution to the problem we could simply stop. However, we can also continue to find more solutions via backtracking. When we backtrack we can change the assignment of c to 1 to find a second solution. Assigning $b = 1$ followed by $c = 0$ results in a failure (violates $b \leq c$), but assigning $c = 1$ on backtrack produces the final solution.

8.2.2 *Constraint Propagation*

The technique of constraint checking described above is known as backward checking as constraints are checked retrospectively after all variables involved in the constraint have been assigned. This is an improvement over waiting until a leaf node is reached before checking the validity of all constraints (known as *generate and test*). Nevertheless, it is normally too weak a technique to solve all but the smallest of problems. Constraint programming systems normally use a more powerful approach to pruning known as *constraint propagation*. Constraint propagation is a much more active technique than backward checking. Instead of checking the validity of constraints, future possibilities are filtered out using an appropriate filtering algorithm. This nearly al-

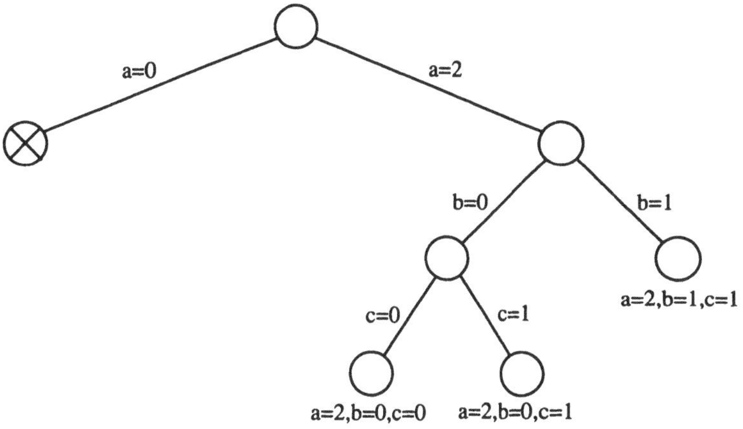

Figure 8.2 Operation of Depth-First Search with Propagation

ways results in less failures occurring – that is, dead-ends in the search tree are noticed higher up, removing the need to explicitly explore the sub-tree.

In backward checking, each variable was either assigned to a single value, or had not yet been assigned. Constraint propagation, on the other hand, maintains the *current domains* of each variable. The current domain begins equal to the initial domain, and is filtered as search progresses down the tree. If a value can be shown not to be possible for a variable (as it would violate at least one constraint), it is removed from the current domain. Such removed values are no longer considered as branching possibilities, and so the search tree is more effectively pruned than it would be for a backward checking algorithm. The algorithms that prove the illegality of domain values and filter them are typically polynomial in complexity; general methods (Bessiere and Regin (1997), Mackworth (1977)) and algorithms specific to the constraint type exist (Beldiceanu and Contjean (1994), Régin (1994), Régin (1996)).

Filtering of domain values is performed *independently* for each constraint; the only manner in which the constraints communicate is through changes to the current domains of variables. Roughly speaking, the constraint propagation algorithm operates by asking each constraint in turn to perform any filtering it can by examining the current domains of the variables that it constrains. Each constraint performs its filtering in a round-robin fashion until either a variable has an empty domain (no possible values) or there is a complete pass where no domains are changed by the constraints. In the first case a *failure* and a backtrack occur. In the second, a fixed point has been reached: if all variables are assigned a value, we have a solution, otherwise branching is required to find a solution or come up with a proof of insolvability. The fixed point reached by constraint propagation does not depend on the order in which constraints filter domain values. The same fixed point results regardless of the ordering.

Figure 8.2 shows how depth-first search with propagation works on our simple problem. As can be seen it is more efficient than the backwards checking version in terms of the number of failures: only one failure now results; all other paths lead to solutions. The search proceeds as follows. At the top of the tree, before any branching, constraint propagation is carried out. In this case, however, no filtering can be done by any of the constraints individually. A branch is then created, and we assign $a = 0$. The constraint $a \neq b$ deduces that $b \neq 0$ and thus $b = 1$. The constraint $a \geq c$ deduces that $c \neq 1$ and thus $c = 0$. The constraint $b \leq c$ deduces that $c \neq 0$ (which causes a failure as $c = 0$) and that $b \neq 1$ (which also causes a failure as $b = 1$). Note that only one failure will occur depending on which deduction is made first, resulting in either b or c having an empty domain. When the failure occurs, the solver backtracks to the root node, and restores the current domains of the variables to that which they were directly before branching. The right branch is then taken by assigning $a = 2$. As at the root node, none of the constraints can reduce domains, and so we branch again, assigning $b = 0$. Again, no filtering occurs, and a final branch to the first solution is made by assigning $c = 0$. Backtracking and assigning $c = 1$ gives the second solution in like manner. We then backtrack to before the assignment of b and take the right branch by assigning $b = 1$. In this case, filtering is performed by the constraint $b \leq c$ which removes the value 0 from the domain of c. This is the only filtering that can be performed at this node, and it results in a solution. Branching on c was not required as it had already been given a unique value by propagation. In this manner the final solution is found.

8.2.3 A Note on Optimization

Up until now, we have described how to solve a decision problem: find an assignment that satisfies all constraints. However, often in real-life situations, we are presented with optimization problems. Typically, these problems are solved as a series of decision problems. When a solution of value v is found, a constraint is added to the problem stating that any solution with value v or greater is illegal. This process is repeated until no better solutions are found, the last one found being the optimal. Each time a new better solution is found, there is no need to restart the search, instead, we can continue to search the tree, taking into account the new bounding constraint. The bounding constraint thus becomes progressively tighter as the search-tree is traversed.

8.2.4 Powerful Modeling and Searching

In the preceding pages, for pedagogical purposes, we have given examples only of simple constraints with simple propagation rules. By contrast, when we are faced with industrial optimization problems, models are more complex, and consequently constraint programming systems provide rich constraints to deal with this complexity. Aside from their richness in modeling, these constraints often have powerful and efficient propagation algorithms, resulting in intensive filtering of domains; see, e.g., Régin (1994), Régin (1996), Regin and Puget (1997). Examples of the more complex types of constraints available in constraint programming libraries are:

- $y = a[x]$: y is constrained to be equal to the xth element of a where a is either an array of constants or variables.
- *all-diff(a)*: all variables in array a take on different values.

- *c=card(a, v)*: variable c is constrained to be the number of occurrences of value v in the array of variables a.

- *min-dist(a, d)*: all pairs of variables in array a must take on values which differ by at least d.

In open constraint programming systems, users can write their own constraints, by giving the appropriate filtering rules to the solving engine. This is often indispensable as the system can never have every constraint built-in which would result in the most natural model.

Constraint programming systems invariably offer the ability to write one's own search goals: that is, describe how variables should be given values in the search tree. This can be very important as the order of variable instantiation can play a large part in how much propagation results. For example, in our three-variable example, no failures result if we instantiate the variables in the order b, a, c instead of a, b, c. As another example, in the more practical setting of routing, it is invariably better to build up a route in one long chain, rather than start on many chains which will eventually join up. Such heuristic orders can often improve performance by more than an order of magnitude. The openness and flexibility of constraint programming systems enables users to write more efficient optimization software than would be possible using a black-box system.

8.3 THE LOCAL SEARCH TOOLKIT

In recent years, many successful optimization applications have been based on constraint programming; see, e.g., Wallace (1996). ILOG Solver is the fastest available constraint programming system (Fernandez and Hill (2000)), which owes its success to the basic techniques of tree-based search and powerful propagation. However, for some problems, even these powerful methods can result in propagation that is too weak for the search to produce high quality solutions within a reasonable time frame. For such problems and others, it is often found that local search can be an effective technique. For this reason, local search extensions was added to ILOG Solver 5.0. We describe these extensions in this section.

8.3.1 Philosophy

ILOG Solver is an open system for constraint programming which is presented as a C++ library. It is open in the sense that users have access to primitives of the library such as constraints and the search process. Importantly, this access is at the programming language (C++) level and not at a parameter-changing level, allowing wide ranging changes or specializations of the library.

The local search extensions to ILOG Solver fit with this philosophy, providing a flexible and open local search system which meshes with the traditional tree-search approach. The tenets of the local search extensions are the following:

- **An object-based view of local search concepts.** ILOG Solver has an object-based view of variables, constraints and goals. Its local search extensions take the same view resulting in objects for the main local search concepts, such as solutions, neighborhoods, and meta-heuristics. The benefits of the object view are numerous, especially when concepts such as neighborhoods are presented as abstract classes which are specialized to give desired behavior for a particular case. An object-based model also allows operations to be carried out on these standard local search concepts; for example, two neighborhoods can be combined to create a third.

- **Control given to the user between moves.** In certain local search systems, for example OpenTS (IBM Open Source Software (2001)), the local search itself is performed as a black box. The number of iterations or stopping condition is given as a parameter to the local search, and the search runs until that condition is met. This is inherently less flexible than a system that gives control back to the user after each movement. In the latter, the user is at liberty to do whatever he wants; for example, he could restart the search from a previously recorded point, change the size of a tabu list, switch to a different neighborhood etc. Without such control between moves, such changes need to be anticipated by the system designer, and done through *call-backs* (pieces of code which are attached to the engine and called after each move). The call-back method, however, is often limited. For instance, if the system designer did not envisage that one could switch neighborhood during search, and did not provide a method for doing this from the call-back, then it is simply impossible.

- **Tree-based search used to find move.** The traditional tree-based search methods of finding solutions to a problem are used to find an acceptable neighbor to move to at each step in the local search process. That is, instead of solving the whole problem using tree-based search, we solve the subproblem of finding an acceptable neighbor given the current solution and a specification of the neighborhood. When a local search is performed, a series of complete searches (one per move made) are launched, each one resulting in the new neighbor to move to from the current solution. This structure is quite appealing because of the fact that the search for a neighbor is an ordinary constraint programming search goal. Thus, it can be mixed with other constraint programming goals to produce methods that perform local and complete search together. An example of how this can be done is presented in Section 8.4. The idea of exploring a neighborhood using constraint programming was first introduced by Pesant and Gendreau (1996) and Pesant and Gendreau (1999).

- **Open local search objects.** Concepts in the local search are not only represented as objects: in the case of neighborhoods and meta-heuristics, these objects are extensible. That is, the user is at liberty to define his own neighborhood and meta-heuristics for a specific problem or problem domain. This means that the toolkit can address problems which were either not thought of or were too rarely occurring for the appropriate support to be put into a general tool. Extensions to local search objects are discussed in detail in Section 8.5.

ILOG Solver's fundamental local search objects are *solutions*, *neighborhoods*, and *meta-heuristics*, whose roles will now be described.

8.3.2 Solutions

Solution objects are simple containers that hold variables in the ILOG Solver model, together with their values corresponding to a solution. Often the value of the objective variable is also stored in the solution so that the quality of the solution can be easily accessed without reference to a solving engine to compute it from the values of the decision variables. Solutions are objects which perform several functions within ILOG Solver.

The first function has little to do with local search: solution objects can be useful for accessing more than one solution to a problem, which can be inconvenient using a single solver which can only be in a single state at a time. Decision and non-decision variables can be present in the solution, meaning that the values of all variables can be read off from the solution, including those whose values are a consequence of the values of the decision variables. In a local search context, a solution object is often used to preserve the best solution seen during a run.

Second, solution objects are used to represent the *current solution* to a problem in a local search context. Here, the solution contains only *decision* variables of the problem, as these are the ones which will be changed by the neighborhoods. Moves will be explored from this current solution and eventually one accepted. The action of accepting a move changes the current solution.

Third, and less obviously, solution objects are used to specify neighborhoods. Each neighbor in the neighborhood is described by a *solution delta*. The solution delta is simply a solution object which contains a subset of the decision variables. This subset is exactly the set of variables which change their values at this neighbor, together with their new values. Solution deltas will be elaborated on in the next section on neighborhoods.

An example of an ILOG Solver code which uses a solution object is shown below:

```
01. IloEnv env;
02. IloModel mdl(env);
03. IloIntVar x(env, 0, 2), y(env, 0, 1);
04. mdl.add(x < y);
05. IloSolution soln(env);
06. soln.add(x);
07. soln.add(y);
08. IloSolver solver(mdl);
09. solver.solve();
10. soln.store(solver);
11. solver.end();
12. cout << "X = " << soln.getValue(x) << endl;
13. cout << "Y = " << soln.getValue(y) << endl;
```

ILOG Solver is based on ILOG Concert Technology, a modeling layer enabling different solving engines to solve the same model. All Concert objects begin with the prefix `Ilo`.

Line 1 of the code creates the Concert *environment* which is an object used for memory allocation and bookkeeping. It is invariably passed when modeling objects are created. Line 2 then creates a Concert model. A model is essentially a bag of modeling objects such as variables and constraints. Line 3 creates two variables which can take only integer values: `x` between 0 and 2, and `y` only the values 0 and 1. Line 4 adds a constraint to the model indicating that x must be strictly less than y. When a constraint is added to the model, any variables mentioned do not need to be added; the solving engine finds them automatically. Lines 5–7 create a solution and add the two variables to it. This means that when the solution is stored, the values of variables `x` and `y` will be preserved within the solution. Line 8 creates the solving engine from the model; this invokes a procedure known as *extraction* which takes the concert model and translates it into structures suited to the solving engine.

Line 9 solves the problem using a *default goal.* (It is also possible to pass a goal to `solver.solve` and we shall see this later.) Solver will deliver the first solution it finds which satisfies the constraints of the problem. As it happens, there is a unique solution (x=0, `y`=1). After the solve, solver is in the *solution state*: we have direct access to variable values directly from the solver, for instance via: `solver.getValue(x)`. However, when we use the solver to solve another problem, the state of the solver will change, and the present solution will be lost unless we store it in a solution object.

Line 10 stores the solution by transferring the values of the variables as set by the solving engine to the solution object. Solution objects are invariant to changes of state in the solver and are thus preserved when the solver changes state or is destroyed. Line 11 destroys the solver. The solution is still preserved in `soln` as lines 12 and 13 demonstrate by printing the stored values of `x` and `y`.

The program produces the following output:

```
X = 0
Y = 1
```

8.3.3 Neighborhoods

A neighborhood in ILOG Solver is the structure which determines the range of solutions one can move to from the current solution. The neighborhood needs not guarantee that each neighboring solution respects the constraints of the problem nor that it meets some acceptance criterion (for example improving the quality of the solution); these are the jobs of the constraint solving engine and the meta-heuristic (see next section), respectively. The requirements of a neighborhood are quite simple and are summarized as follows:

- **The neighborhood receives the current solution.** This solution is the one from which all neighbors are specified. The neighborhood must keep this solution as a reference from which to define neighbors. The current solution is supplied to the neighborhood once for each move.
- **The neighborhood can determine its size s.** Normally, this is determined when the neighborhood receives the current solution.
- **The neighborhood can deliver neighbor i, where $0 \leq i < s$.** Neighborhoods have a random, not sequential, access interface. That is, a neighborhood must

be capable of delivering the ith neighbor at any time. The reason for this is it vastly increases the flexibility of neighborhoods, as will be demonstrated later. A neighbor is defined by delivering a *solution delta*: the fragment of the solution which is changed by the neighbor, together with the new values for the changed part.

For example, suppose we have a neighborhood which changes the value of one of a set of 0-1 variables, a so-called "flip" neighborhood. If there are n variables, then there will be n neighbors in the flip neighborhood. Neighbor i is defined by delivering a solution object containing only variable i. Its value in the solution object should be the value different from that in the current solution. `soln.setValue(var, value)` sets the value of variable `var` to be `value` in solution `soln`.

In addition:

- **The neighborhood is notified that neighbor j was chosen.** This is not a requirement of the neighborhood, but a service to it. When a move is made the neighborhood is told the index of the neighbor which was accepted, and the constrained variables are in a state which represents the new accepted solution. Normally notification does nothing unless the neighborhood desires it. However, receiving the notification that a particular neighbor was taken can be useful in adaptive neighborhoods where some state is incrementally maintained.

We now present a code segment showing the use of neighborhoods. In order to demonstrate how a neighborhood operates, we have to have some mechanism for "driving" the neighborhood. That is, for sending it the current solution, and retrieving the deltas. The pre-defined Solver goal `IloScanNHood` does just this, and is used in the following code.

```
IloEnv env;
IloModel mdl(env);
IloIntVarArray vars(env, 3, 1, 3);
mdl.add(vars);
mdl.add(a[0] < a[2]);
IloSolution soln(env);
soln.add(vars);
soln.setValue(vars[0], 1);
soln.setValue(vars[1], 2);
soln.setValue(vars[2], 3);
IloNHood nhood = IloSwap(env, vars);
IloGoal scan = IloScanNHood(env, nhood, soln);
IloSolver solver(mdl);
soln.startNewSearch(scan);
while (solver.next())
```

```
16.   cout << solver.getValue(a[0])
           << solver.getValue(a[1])
           << solver.getValue(a[2]);
```

In this example, we create a model with three variables in an array, each of which can take on the integer values 1, 2 or 3. The variables are added to the model as well as a constraint stating that the first variable in the array must be strictly less than the last variable. From lines 6–10, we then create a solution object and populate it with all variables, the values of these variables in the solution being set to 1, 2 and 3, respectively. We can see that this is a true solution as it does not violate the only constraint `a[0] < a[2]`. Line 11 creates a pre-defined Solver neighborhood which interchanges the values of any two variables in the array. Line 12 creates the Solver search goal which will exercise this neighborhood, taking each neighbor in turn from the neighborhood and instantiating the variables in `vars` according to the deltas delivered. Line 13 creates the solver from the model. Using `solver.solve(scan)` would produce only a single solution to the neighborhood exploration problem. That is, only the first neighbor would be produced. However, to demonstrate how the neighborhood is scanned, we will produce all solutions to the neighborhood scanning problem. This is achieved by a different set of member functions. `solver.startNewSearch(scan)` primes the solving engine with the goal `scan`. Thereafter, each call to `solver.next()` produces a new solution until a false value is returned, at which point all solutions have been found.

The program produces the following output:

```
213
132
```

Note that the "swap" neighborhood should produce exactly $n(n-1)/2$ neighbors, given an array of size n. In this case three neighbors should be produced. In fact, the constraint `vars[0] < a[2]` means that swapping the values of the first and last elements of the array is forbidden, and such a move is automatically rejected by the constraint solver when an attempt is made to instantiate the variables with the values 321.

Powerful operators and functions are also available to manipulate neighborhoods. The most useful of these is the concatenation operator. Writing `nhood1 + nhood2` produces a neighborhood whose neighbors are made up of those in `nhood1` and `nhood2`, the new neighborhood producing those in `nhood1` first.

Another very useful operator is one which randomizes a neighborhood. Writing `IloRandomize(env, nhood, randGen)` randomly jumbles the neighbors of `nhood` at each move, using random numbers drawn from the generator `randGen`, which is an instance of the Concert class `IloRandom`. Drawing neighbors in a random manner from a neighborhood can be particularly useful in avoiding stagnation and looping in meta-heuristics such as tabu search. The fact that neighborhoods can be randomly accessed means that such a randomization can be efficiently implemented. The newly created neighborhood merely contains a mapping table between the indices of its neighbors and those of `nhood`. Such a mapping is randomly re-created at each move.

To give another example, the neighborhood `IloContinue(env, nhood)` is a neighborhood which behaves as `nhood` except that when a move is made, neighbors for the next move are cyclicly indexed from the index of the neighbor that was taken. So, for instance, for a neighborhood which flips the values of n 0-1 variables, if variable i (indexed from 0) was the last variable flipped, at the next move, flips will be explored from variable $i + 1$ though to variable i, cycling back to variable 0 after variable $n - 1$. This type of neighborhood can be useful as it is often more profitable to explore new neighbors than re-explore old ones. To indicate the uses of neighborhood notification, `IloContinue` is a neighborhood which makes use of notification to change its behavior (to offset the neighbors by the appropriate index) for the next move.

8.3.4 Meta-Heuristics

What is termed a *meta-heuristic* in ILOG Solver can have a meaning which is somewhat different from the commonly used term. The term "meta-heuristic" in the literature is quite a general concept which normally means a method which guides the application of local improvement methods. "Guides" in this context can apply to quite a general set of behaviors including forbidding certain types of moves, encouraging others, restarting the search from some point, perturbing the search in some "intelligent" manner, changing neighborhoods, etc.

In ILOG Solver, a *meta-heuristic object* is an object which performs a much more precise role: it filters neighbors and nothing more. That is, a meta-heuristic can say that certain neighbors are forbidden. This is not as restrictive as one might imagine. As control is given to the user between moves, other types of meta-heuristic behavior can be implemented at that point, including restarting from a previous good solution, switching neighborhoods, changing the tabu tenure in a tabu search method etc. Such is the benefit of having control between moves: it alleviates the problem of trying to "black box" such greatly varying behaviors in the meta-heuristic object itself.

Meta-heuristics nearly always have *state*: for example the tabu list for a tabu search mechanism or the current temperature for a simulated annealing meta-heuristic. It is this state that allows the meta-heuristic to change its filtering from move to move.

Note that Solver's meta-heuristics do not decide which neighbor to choose from among the neighborhood. This decision is left to another object known as a *search selector* which will be introduced in Section 8.3.5.

In order to fulfill the above filtering role, a meta-heuristic must perform the following tasks:

- **The meta-heuristic receives the current solution.** Almost always, the meta-heuristic will make use of the cost of the current solution which it can glean via `soln.getObjectiveValue()`. The current solution is supplied to the meta-heuristic once for each move.

  For meta-heuristics objects, often the most natural way to filter certain moves is to add a constraint to the constraint solver forbidding them. This is usually possible if the forbidden moves have some sort of logical structure to them. For instance, although usually not considered a meta-heuristic, a greedy improvement search is implemented in Solver by the meta-heuristic class `IloImprove`. When an object of this type receives the current solution, it takes the objective from the current solution and adds a constraint to the solver to state that any new solution (i.e. neighbor) found must have a cost which is lower (when minimizing). Then, all neighbors not strictly decreasing the cost will automatically be rejected by the constraint solver.

- **The meta-heuristic rejects or allows a given neighbor.** When the neighbor filtering cannot conveniently be expressed as a constraint, the meta-heuristic can reject the neighbor in an alternative manner. The meta-heuristic has the neighbor presented to it with the decision variables instantiated as they were for the previous code segment demonstrating `IloScanNHood`. In this case, the meta-heuristic has all information regarding the neighbor available, including the values of non-decision variables and the cost of the neighbor. If the filtering was difficult to express as a constraint, then an algorithm can be executed at this point to determine if the neighbor should be filtered or not.

  Note that this "test after instantiation" method of filtering neighbors is not as efficient as using a constraint as all decision variables must be instantiated before the test is executed. When the neighbors to be filtered are expressed as a constraint, then they may be filtered before all variables are instantiated.

In addition:

- **The meta-heuristic is notified when a move is made.** Like the notification of neighborhoods, this is not a requirement of the meta-heuristic, but a service to it. When a move is made the meta-heuristic is told the solution delta of the neighbor which was accepted, and the constrained variables are in a state which represents the new accepted solution. When a *neighborhood* is notified, the norm is that nothing happens, as neighborhood objects do not normally have state (unless they are adaptive as previously discussed). However, as it is normal for meta-heuristics to have state, typically the meta-heuristic changes its state when notified. As an example, in a simulated annealing meta-heuristic, the temperature would be decreased. Or in a tabu search meta-heuristic, some attributes would be dropped from the tabu list (the oldest ones), and others would be added according to the move which had just been made.

  If no neighbors could be legally taken, either because all were illegal or were forbidden by the meta-heuristic, then the meta-heuristic is informed of this fact. The meta-heuristic can then take action at this point to reduce its filtering such that when subsequent moves are attempted, not all will be filtered and progress can be made once more. For example, in the tabu search meta-heuristic supplied with ILOG Solver, when no moves can be taken, the oldest attributes are dropped from the tabu list, without adding any new attributes. This exactly mimics the method suggested in Glover and Laguna (1997) for managing the situation where all legal moves are tabu.

Meta-heuristics, as neighborhoods, can be composed. The expression `mh1 + mh2` results in a meta-heuristic which filters a neighbor if either `mh1` or `mh2` would have

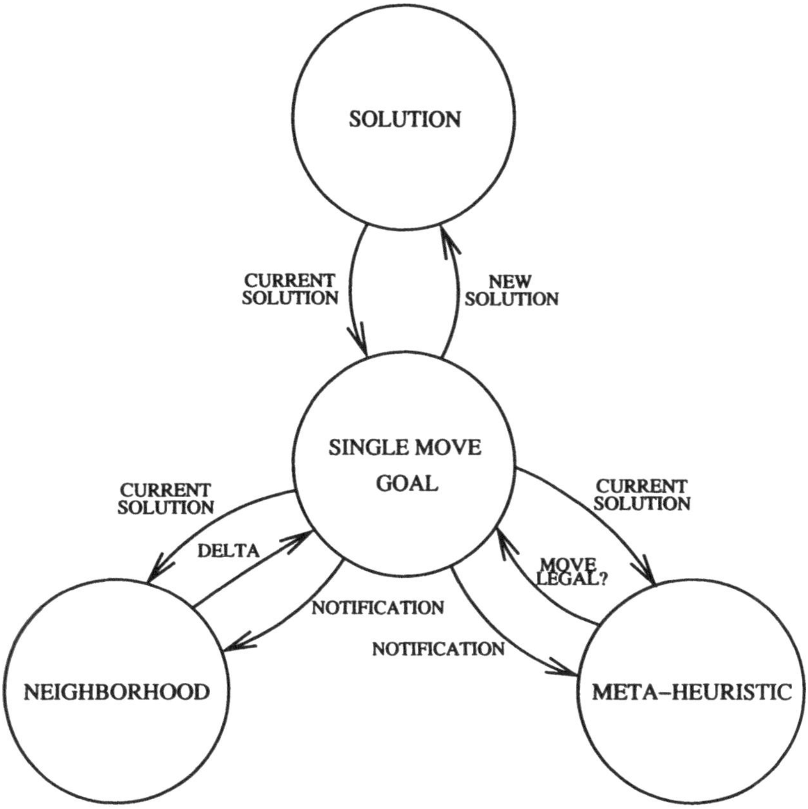

Figure 8.3 Interaction of Local Search Objects

filtered the neighbor alone. Such composition can be useful for creating composite meta-heuristics such as guided tabu search (de Backer et al. (2000)).

The goal `IloScanNHood` was used to explore the neighbors of a neighborhood. In order to combine neighborhoods and meta-heuristics, we use the Solver goal named `IloSingleMove` which combines a neighborhood and a meta-heuristic in order to make one local move in the search space. The operation of this goal is shown in Figure 8.3. We depict the main actors which are a solution, a neighborhood, a meta-heuristic, and the goal itself.

First of all, the current values of decision variables are fetched from the solution after which they are sent to both the neighborhood and the meta-heuristic. The size s of the neighborhood is then fetched (not shown in the figure). The goal then begins asking the neighborhood for solution deltas corresponding to neighbors for indices 0 through $s-1$. For each neighbor, the goal calculates the new values for all constrained variables according to the current solution and the solution delta. The variables are then instantiated with these values and the meta-heuristic is asked if the neighbor should be filtered. (Note that if the meta-heuristic specified its filtering via a constraint, and the neighbor is one which is to be filtered by the meta-heuristic, then the goal will never get to the point where it has to ask the meta-heuristic to check the solution, as the violated constraints would cause a failure before this point.) If the neighbor should be filtered the goal causes a *failure*; otherwise, the goal succeeds in creating a leaf node (legal neighbor).

- In the case where the first legal neighbor of the neighborhood should be taken, the goal move stops at that leaf node: the remainder of the neighbors are not examined.
- In the case where, for instance, the neighbor improving the quality of the solution by the greatest amount is to be chosen, the goal continues examining neighbors but logs the best leaf node it has visited. When the whole search tree corresponding to the neighborhood has been examined, Solver jumps back to the best leaf node. The mechanism by which such selections are made is described in the next section.

In either of the above cases, when the preferred neighbor has been reached, both the neighborhood and meta-heuristic are notified of this. Then the new current assignment of decision variables is stored back in the solution. The performance of multiple moves is effected by repeating the entire process.

A detailed description of the implementation of the underlying mechanisms behind neighborhood exploration goals is given in Shaw et al. (2000).

8.3.5 Selection of a Neighbor

Above, we touched upon different methods of *selecting* a neighbor. The job of selecting which legal neighbor to move to is not performed by any local search object we have yet described. In ILOG Solver, this selection is performed by an object known as a *search selector*. Search selectors are not specific to local search but are objects which can be applied to any search goal.

For instance, consider a standard combinatorial problem with multiple solutions and a goal `g` which searches for those solutions. To find all such solutions, we could use `solver.startNewSearch(g)` and `solver.next()`. Imagine that we wanted to find the solution which minimized the value of some variable `v`. This could be done by using an instance of `IloSolution` to store the solution after each successful call to `solver.next()` when the new solution produced has a value of `v` smaller than the value of `v` in the solution. After the search tree has been fully explored the best solution is preserved.

The above mechanism is quite cumbersome; Solver provides a simpler way to search for the solution we are interested in. A *search selector* transforms a goal into one where only the solutions of interest form leaf nodes. For instance, the line

```
g = IloSelectSearch(g, IloMinimizeVar(env, v));
```

transforms the goal `g` such that it will produce only one leaf node: that which minimizes the value of `v`. This exactly mimics the process described in Section 8.2.3. Solver has other search selectors, notably `IloFirstSolution` which keeps only the first solution produced by the original goal. Finally, the users are at liberty to write their own search selectors to perform the selection they require. For example, "take the first solution of cost less than *c*, but if no such solutions exist, deliver the solution with lowest cost".

One can see that as goals such as `IloScanNHood` are standard Solver goals, one can apply search selectors to them, too. In the context of local search, a selector to choose the best solution chooses the best neighbor.

The implementation of Solver's search selectors is described in ILOG (2000b).

8.3.6 Simple Local Search Example

We now present a complete local search example. We have chosen the well-known *one-max* problem often used for demonstrative purposes. In a one-max problem we are given a set of n 0-1 variables and we are to find a solution with the maximum number of variables set to 1. Of course, this problem is easy, but the local search example which solves it demonstrates all the basic features we are interested in.

To begin, we create an environment, a model, and an integer determining the size of the problem.

```
01. IloEnv env;
02. IloModel mdl(env);
03. IloInt n = 5;
```

We now fill in the details of the model. We first create the set of 0-1 variables `vars` in line 4. Then, a variable which will represent the sum of the number of ones is created. It is constrained to be equal to the sum of the array `vars` by adding a constraint to the model.

```
04. IloIntVarArray vars(env, n, 0, 1);
05. IloIntVar ones(env, 0, n);
06. mdl.add(ones == IloSum(vars));
```

We next create a solution object to use in the local search optimization. We add the decision variables to the solution and an objective, which is to maximize the number of ones.

```
07. IloSolution soln(env);
08. soln.add(vars);
09. soln.add(IloMaximize(env, ones));
```

The model and solution specification is now complete. A solver is then created, and an initial solution generated using the pre-defined Solver goal `IloGenerate`. This goal instantiates variables in the order they appear in the array and chooses low values before higher ones. Thus, the initial solution that results is one where all variables have value zero. The initial solution is stored in the solution object via `soln.store`.

```
10. IloSolver solver(mdl);
11. solver.solve(IloGenerate(env, vars));
12. soln.store(solver);
13. DisplaySolution("Initial Solution: ", solver, vars);
```

`DisplaySolution` is a function which prints a message followed by the values of the decision variables. The values of the decision variables are retrieved from the solver using `solver.getValue`. The function is defined as follows:

```
void DisplaySolution(const char *message,
                     IloSolver solver,
                     IloIntVarArray vars) {
  cout << message;
  for (IloInt i = 0; i < vars.getSize(); i++)
    cout << solver.getValue(vars[i]);
  cout << endl;
}
```

After the initial solution has been generated and stored in the solution, we can progress to the local search. Line 14 creates the neighborhood that we will use: `IloFlip`. This neighborhood can change the value of one of the 0-1 variables in the array handed to it, and is thus capable of moving to the optimal solution which consists purely of ones. Line 15 creates the meta-heuristic, `IloImprove`, which filters any move which does not improve the quality of the solution. In our case, the objective variable, `ones`, must increase at each move. Finally, in line 16, we create the goal which makes one local search move. To create this goal we pass the environment, solution, neighborhood, and meta-heuristic objects.

```
14. IloNHood nhood = IloFlip(env, vars);
15. IloMetaHeuristic greedy = IloImprove(env);
16. IloGoal move = IloSingleMove(env, soln, nhood, greedy);
```

All necessary objects are now prepared, and we enter the loop which improves the solution. The `move` goal is executed repeatedly until it fails, with such a failure indicating that no moves which improve the solution could be found. This means that we have reached a local maximum. In this case, the local maximum coincides with the global maximum.

```
17. while (solver.solve(move))
18.   DisplaySolution("", solver, vars);
```

After the optimization loop, the final solution found is held in the solution object. However, if we wish this solution to be accessible from the solver, the constrained variables need to be instantiated with the values in the solution. This is carried out by the goal `IloRestoreSolution`.

```
19. solver.solve(IloRestoreSolution(env, soln));
20. DisplaySolution("Final Solution: ", solver, vars);
```

The whole program produces the following output. As expected, the number of ones increases monotonically. The moves generated by the "flip" neighborhood which change a 1 to a 0 are filtered out as they decrease the objective.

```
Initial Solution: 00000
10000
11000
11100
11110
11111
Final Solution: 11111
```

8.3.7 *Discussion*

We have presented a toolkit for local search built upon, and now part of, ILOG Solver. The toolkit is based on certain basic design principles: an object model of local search, user control between moves, neighborhood exploration using goals (facilitating meshing between local search and constraint programming), and an open interface.

Each entity is accorded its own specific task, and there is a clean separation between such tasks. *Solution* objects store the values of decision variables and the value of the objective. *Neighborhood* objects are responsible for defining the neighbors of the current solution. *Meta-heuristic* objects filter the neighbors of the current solution. *Search selector* objects choose one of the possible neighbors to move to. Finally, *Local search goals* bring all of these objects together to make moves in neighborhood space.

The adoption of an object model results in a flexible language with which to express local search concepts. For instance, the fact that neighbors should be drawn from a neighborhood in a random fashion is made part of the neighborhood itself, through the use of the neighborhood operator `IloRandomize`. `IloContinue` provides another way of exploring the neighborhood. Neighborhood modifiers are inherently more natural than different possibilities for exploration built into the neighborhood exploration goal (cf. neighborhood iterators in Fink et al. (1999b)).

Likewise, solution objects are useful as they can be cloned, copied, arrays of them kept, sorted etc. The user is not limited to the traditional "current solution" and "best solution". One possibility is to keep a library of elite solution fragments which can then be used to produce new solutions (see, e.g., Rochat and Taillard (1995)). Finally, meta-heuristic objects provide a simple protocol for directing the search via neighborhood filtering.

Operators over meta-heuristics provide methods to combine them: for instance, addition of meta-heuristics form a meta-heuristic which filters a neighbor if any meta-heuristic in the sum would filter the neighbor. Such simple combinations can allow one to produce new meta-heuristics: for instance, guided tabu search can be created from a combination of tabu and guided local search (de Backer et al. (2000)).

The advantages of including local search support in a constraint programming library have not yet been emphasized, and it is perhaps a fitting time to do so. The obvious alternative is to produce a "stand alone" local search toolkit: several academic toolkits are already available; see, e.g., Di Gaspero and Schaerf (2001), Fink et al. (1999b), Michel and van Hentenryck (1997), Michel and van Hentenryck (2000), IBM Open Source Software (2001). The advantages of instead using a constraint programming library as the base are two-fold. The first advantage is the flexibility of modeling provided by constraint programming libraries. Such modeling flexibility is unrivaled by any other technology. The problem variables and constraints can be described in a concise readable form and tools like ILOG Concert Technology (ILOG (2000a)) make it easy to manipulate and decompose models. Secondly, constraint programming comes equipped with another solving method distinct from local search: that of tree-based search. Such methods tend to work well when constraint propagation for a particular problem is strong, while local search methods often fill the gap when propagation is more limited. Thus, between the two a greater solution coverage for real-world problem solving is achieved. A software engineering advantage is that one can run either a complete or local search on the same model, a benefit advocated in Pesant and Gendreau (1999).

Allied to the advantage of the availability of depth-first search mechanisms is the need for an *initial solution* for local search problems. When one is dealing with unconstrained optimization, generating a first solution is not problematic: a random assignment suffices. For problems with constraints, the process of producing a legal first solution is not so clear. Commonly, a specific first solution method is used, but this may then break down if one or two additional side-constraints are added (see, for example, Kilby et al. (2000)). An alternative is to soften the constraint and value its violation. This is often useful, but in many cases results in a loss of search direction, with the local search spending large portions of its time trying to re-establish legality. (By contrast, when using constraint programming constraints can be imposed as hard constraints, or softened and added to the cost function, as desired.) Constraint programming offers a largely constraint-independent method of generating first solutions: tree-search with constraint propagation. Even if for the type of problem considered, constraint propagation is too weak to find optimal or very good solutions, it is often sufficient for finding legal first solutions. These solutions can subsequently be improved by local search.

The benefits of constraint programming's tree-based search approach also aids local search in another important way: we can use complete and local search together. There are various ways in which this can be done. One might simply want to find a good solution using local search which then provides an upper bound on the cost function for a proof of optimality by complete search. More sophisticated is to run local and complete search in parallel. We can communicate the upper bound from the local to the complete search, and communicate solutions found in the other direction. Local search can make use of these as elite solutions; for instance, they may make good points for strategic restart. Finally, we can couple both search methods together more tightly. One method of doing this is via a method known as *shuffling* (Adams et al. (1988), Applegate and Cook (1991)), referent domain optimization (Glover and Laguna (1997)), or large neighborhood search (Shaw (1998)). In this method each local move performs a relaxation of parts of the current solution, followed by a re-optimization of respective parts via complete search. This technique is very natural using a constraint programming library, and results achieved can be highly competitive; see, e.g., Rousseau et al. (2000), Shaw (1998). A demonstration of a quite different hybrid search is given in Section 8.4.5.

An obvious disadvantage of the integration of local search mechanisms into a constraint programming library is that such libraries are typically non-monotonic and perform constraint checking by propagation. Both of these attributes result in local searches which have significant computational overhead compared to approaches which do not have to deal with backtracking and propagation. We have described in Shaw et al. (2000) how to minimize these overheads via specific tree-search mechanisms. It is our belief that for complex real-world problems, the flexibility of the constraint programming library – with its ability to mix search methods – will more than compensate for such inefficiencies.

8.4 INDUSTRIAL EXAMPLE: FACILITY LOCATION

Facility location problems are of significant practical importance in transportation logistics. They concern the allocation of customers to facilities in a cost-effective manner. The facilities perform some service to the customers, be it supply of goods, pickup of goods, maintenance of equipment at the customer site etc. It is often the case that the "customers" and the "facilities" form different parts of the same company.

The particular problem we address here is posed as follows. Given a set of customers to be served, and a set of possible locations where facilities can be constructed, decide which sites to use for the construction of a facility, and which customers to serve from each facility. Each customer has a demand which is the amount of service required, and must be served from only one facility. Each facility has a capacity, dependent upon the construction site, which must be no less than the sum of the demands of the customers served by it. The construction of a facility entails a certain site-dependent cost, while the cost of serving a particular customer is dependent upon which facility serves it. Thus the optimization is a balance between reducing the opening costs and service costs. The former is best done by opening few facilities, whereas the latter is best done by opening many, such that distances from customer to facility are reduced.

Although the basic problem can be enriched with various types of additional constraints (which makes solving it using constraint programming all the more interesting), for simplicity we treat only the basic problem described above.

8.4.1 Problem Description

We are given a set of customers $\mathcal{C}$ and a set of possible facility locations $\mathcal{F}$. Each customer $c \in \mathcal{C}$ demands D_c of service. Each potential facility $f \in \mathcal{F}$ has capacity

C_f. The cost of constructing facility f is given by $cost(f)$ and the cost of serving customer c from facility f is $cost(c, f)$.

We represent a solution by a variable x_c for each customer c, indicating which facility serves that customer. Given such a representation, the constraint that each customer must be served from exactly one facility need not be modeled explicitly. The model can be expressed mathematically as:

Minimize:

$$\sum_{c \in C} cost(c, x_c) + \sum_{\{f \mid \exists c . x_c = f\}} cost(f)$$

Subject to:

$$\forall f \in \mathcal{F} \sum_{\{c \mid x_c = f\}} D_c \leq C_f$$

An ILOG Solver program which implements this abstract model is now given. We assume that the costs of building each facility are present in an array `bc`, while the costs of serving customers from facilities are held in a two dimensional array `sc`. The capacities of the facilities are held in the array `capacities`, while the demands of the customers are held in the array `demands`. With regards to problem size, `nbFac` holds the possible number of facility locations, while `nbCusts` indicates the number of customers to be served.

First of all the model, the basic decision variables, and variables which determine the load on each facility are created. The array `x` models the assignment of customers to facilities, and so each variable in this table is created with a domain from 0 to one less than the number of facilities. The `load` array will later contain variables representing the load on each facility. We assume the environment `env` has already been created.

```
IloModel mdl(env);
IloIntVarArray x(env, nbCusts, 0, nbFac - 1);
IloIntVarArray load(env, nbFac);
```

Now, we create the variables and constraints which maintain and constrain the load on each facility. This is done via a set of intermediate 0-1 variables `here`, which delivers an encoding of the set of customers assigned to a facility. The load on the facility is then formed by the scalar product of this array and the array of demands. Notice that for constraints to take effect, they must be added to the model of the problem `mdl`. Finally, we add a constraint stating that the sum of the loads on the facilities must equal the sum of the demands. Although this constraint is not strictly required, in a constraint programming context, it can help to prune out some possibilities earlier in the search.

```
IloInt i, j;
for (i = 0; i < nbFac; i++) {
  load[i] = IloIntVar(env, 0, capacities[i]);
  IloIntVarArray here(env, nbCusts, 0, 1);
  for (j = 0; j < nbCusts; j++)
    mdl.add(here[j] == (x[j] == i));
```

```
10.   mdl.add(load[i] == IloScalProd(here, demands));
    }
11. mdl.add(IloSum(load) == IloSum(demands));
```

We next add the constraints which maintain the cost of serving each customer from their chosen facility. We use the Solver constraint `IloTableConstraint`, which indexes an array of constants using a constrained variable as the index. On line 12, the array `cc` holds the costs of serving all customers. The variables in `cc` are declared as variables which can have discrete integer values. For instance, `cc[0]` can only take on values which are the costs of serving customer 0 from each facility. In general, the domain of variable `cc[j]` is thus defined to be `sc[j]`, the possible costs of serving customer j (line 14). The table constraint then constrains `cc[j]` to be equal to `sc[j][x[j]]` (line 15).

```
12. IloNumVarArray cc(env, nbCusts);
13. for (j = 0; j < nbCusts; j++) {
14.   cc[j] = IloIntVar(env, sc[j]);
15.   mdl.add(IloTableConstraint(env, cc[j], sc[j], x[j]));
    }
```

Having formed the cost of serving each customer from its chosen facility, we must also form the cost of building the facilities. The 0-1 array `open` indicates which facilities are open: namely those which have a non-zero load. Finally, the total cost of the problem is formed from the sum of the costs of serving the customers and the costs of building the open facilities. The latter term is formed by a scalar product between the `open` array and the build costs `bc`.

```
16. IloIntVarArray open(env, nbFac, 0, 1);
17. for (i = 0; i < nbFac; i++)
18.   mdl.add(open[i] == (load[i] != 0));
19. IloIntVar cost(env, 0, IloIntMax);
20. mdl.add(cost == IloSum(cc) + IloScalProd(open, bc));
```

8.4.2 Solution Methods

We present three alternative approaches to solving the above model using ILOG Solver. The first is to use a standard tree search to find the optimal solution. The second is to use a pure local search method where customers are moved from facility to facility. In this second method we use a short term tabu memory. The third is a mixture of local search and tree-based search methods. We describe each of these in more detail, giving example code of how each is implemented.

Before we begin describing the three methods in more detail, we first introduce some common objects we will need for all methods via the following code:

```
IloObjective obj = IloMinimize(env, cost);
IloSolution best(env);
best.add(obj);
best.add(x);
IloGoal assign = AssignServers(env, x, open, sc, bc);
```

Three objects are created:

- The objective of the problem: minimize the total cost.
- A solution to hold the best solution found to the problem. This solution keeps a note of the facility assigned to each customer and the objective value (via the `add` method).
- A tree-search goal to instantiate the `x` variables. We do not give the code for this goal but describe it.

  The goal makes branching decisions of the form "either customer c is served by facility f, or c is not served by f". c is chosen before f to be the customer with the smallest number of facilities that can serve it. f is then chosen to be the least costly facility where the cost of the facility is taken to be `sc[c][f]`, plus `bc[f]` if no customers are as yet served by f. Ties in the choice of c and f are broken by choosing the lowest indexed customer or facility.

8.4.3 Tree-Based Search

Most of the work in describing a tree-based search is the definition of the appropriate search goal, whose behavior we have just described. Once this is done, search is quite simple. Here, we add the objective to the model to indicate that we are looking for a solution with minimum cost. Then an instance of the class `IloSolver` is created to solve the model, and is primed with goal `assign`. The solutions to the problem are then generated using the `next()` function of the solver. As we are minimizing cost, each new solution produced by `next()` is guaranteed to have better cost than the previous one, such that the last one produced will be the optimal. The solution `best` is stored each time a solution is found, and so the optimal will be preserved therein.

```
mdl.add(obj);
IloSolver solver(mdl);
solver.startNewSearch(assign);
while (solver.next())
   best.store(solver);
```

8.4.4 Pure Local Search

A local search method which can be used on this problem is one where the facility assigned to a customer is changed. One can view this as moving a customer to another facility. The neighborhood is then the movement of each customer to all facilities other than its current one. Such a neighborhood is of size $O(|\mathcal{C}||\mathcal{F}|)$. In order to reduce the probability of becoming trapped in local optima, a short-term tabu memory is used which prohibits a customer moving back to a facility it was moved from for a certain number of iterations.

The first step in any local search, however, is to build a first solution. We achieve this via the following code (line numbers continue from 26 as this code is an alternative code to the tree-based search):

```
26. IloSolver solver(mdl);
27. if (!solver.solve(assign)) {
28.     cout << "No solution" << endl;
29.     env.end();
30.     return 0;
    }
31. best.store(solver);
```

Unlike the tree-based search, we do not add the objective to the model. To do so would mean that we would try to find an optimal solution on line 27. Here, we are interested in any solution satisfying the constraints. If no initial solution could be found, we exit the program.

Next, we move onto the local search mechanism itself. First, we create the neighborhood. Solver's neighborhood `IloChangeValue` has neighbors in which each variable of a specified set can change its value to one within a specified range. When applied to the variables x which decide which facility serves each customer, this neighborhood corresponds to the movement of a customer from one facility to another. We also randomize this neighborhood. The reasons for this will be explained below.

```
32. IloNHood nh = IloChangeValue(env, x, 0, nbFac - 1);
33. IloRandom rnd(env);
34. nh = IloRandomize(env, nh, rnd);
```

We use Solver's built-in tabu search meta-heuristic which is based upon a short term tabu memory of assignments in the current solution. Any assignment which is undone cannot be re-done within the specified tabu period. The tabu period can be adjusted dynamically by the user using a simple interface. The meta-heuristic also includes an aspiration criterion which means that the tabu status of a move is overridden if it would result in a move which is better than any seen so far. Here, we create a tabu search with tenure 15. This means that if a facility f was assigned to a customer, and this assignment is changed, the f cannot be reassigned to the customer for 15 iterations.

```
35. IloMetaHeuristic mh = IloTabuSearch(env, 15);
```

Finally, we construct the goal which performs a local search move. For this, we need a solution object to hold the current solution. This is done on line 36 by cloning the best solution object `best`. We also specify that we would like to move to the minimum cost neighbor (which is non-tabu) via Solver's `IloMinimizeVar` search selector (line 37). This selector specifies that the goal will generate a maximum of one leaf node, and that node will be the one minimizing a specified variable; in this case the total cost. If more than one neighbor has the same cost, the first one encountered will be chosen. It is for this reason that the neighborhood was randomized earlier: it removes any structure or implied order in the neighborhood such that if there is such a tie-break, a neighbor minimizing the cost will be chosen randomly. On line 38, the `IloSingleMove` goal takes the solution, neighborhood, meta-heuristic and best-neighbor selector and produces a goal which will make the move. We attach to this another goal which stores the new solution in `best` if it is better than `best`, thus automatically keeping the best solution found up to date.

```
36. IloSolution soln(best.makeClone(env));
37. IloSearchSelector selMin = IloMinimizeVar(env, cost);
38. IloGoal move = IloSingleMove(env, soln, nh, mh, selMin)
                 && IloStoreBestSolution(env, best);
```

The local search loop can then be entered. In this example, up to thirty moves are made, with the cost of the current solution and best solution displayed at each stage. If at any stage a move could not be taken, this could mean that all moves are tabu. In this case, we make a call to the `complete()` method of the meta-heuristic, which is the notification performed when no move could be made. (This was previously discussed in Section 8.3.4.) `complete` returns true if the meta-heuristic would like to terminate the search (for instance, if it cannot reduce its filtering strength), or false if it wishes to continue. When `complete` is called on Solver's tabu search mechanism, the tabu status of the oldest element in the tabu list is revoked, potentially allowing a move to be made at the next step. A false value is always returned unless the tabu list was empty on entry to `complete`, in which case, true is returned.

```
39. for (i = 1; i <= 30; i++) {
40.   if (solver.solve(move)) {
41.     cout << "Cost = " << solver.getValue(cost)
             << " (Best = " << best.getValue(obj)
             << ")" << endl;
      }
42.   else if (mh.complete()) break;
    }
```

Finally, the best solution is presented for inspection.

```
43. cout << "Final Solution" << endl;
44. for (i = 0; i < nbCust; i++) {
45.   cout << best.getValue(x[i]) " ";
46. cout << endl;
```

8.4.5 *Hybrid Local/Tree-Based search*

Unfortunately, each of the above methods has trouble finding high quality solutions, even when more effort is expended in tuning meta-heuristics or trying different possibilities for the search goals.

When one analyzes the local search procedure, one finds that it can be relatively difficult for a local search based on moves of customers from facility to facility to make bulk transfers of customers from one facility to another (or from two larger ones to three smaller ones, for example.) This is because the benefits of such a bulk transfers lie relatively far away in terms of number of moves, while there are great cost disadvantages on the way there, notably because of the fixed cost of the facilities. On the local search trajectory for such a transfer, more facilities are open than necessary, and this increases the cost. These reasons make the landscape difficult to negotiate for a search procedure which performs simple movements of customers.

There are standard techniques for diversifying the search (for instance, see a range of techniques in Glover and Laguna (1997)), such as the use of frequency memory, searching outside the feasible region, penalties and inducements, strategic oscillation, etc. However, the breadth of available techniques makes even such an investigation costly. We instead examine a different approach for improving the local search – we examine a hybrid method.

Using a constraint programming library, we can exploit the inherent structure in the problem. Although an assignment of facility to each customer is enough to determine a solution, we can look upon the problem as being two-tiered. There is first the problem of determining which facilities should be open, and then there is the secondary problem of determining which open facility should be assigned to which customer. If both of these problems are teased apart somewhat, a two-level search can be effected. The idea is to perform a "coarse" local search over a finer search. We perform a local search over which facilities are open using a move operator which opens or closes a facility. After each such move, the "finer details" are filled in – these being which facility each customer is served from. This could be performed by another local search process, but to illustrate a mixture here, we use a traditional tree search process to assign facilities. (One could also use a mixed integer programming technique to find the optimal assignment, although this may become less possible if the problem is enriched with side-constraints.) This tree search at each step results in large rearrangements of assignments of facilities to customers (bulk transfers) as facilities are opened and closed, largely circumventing the problems described earlier.

For the local search which moved customers from one facility to another we used a short-term tabu memory which disallowed moving a customer back to a facility it had come from for a certain number of iterations. In the hybrid method, we again use a short-term tabu memory, but this time we say that when a facility is opened or closed, it must remain in that state for a certain number of iterations.

We first begin by generating a first solution exactly as for the pure local search method: lines 26–31 in Section 8.4.4, and storing the first solution in the solution `best`. We next must create a new solution with a different structure. `best` contains the assignments of all facilities to customers and as such defines a complete solution. For the hybrid method, we perform local search over a partial solution: the "openness" of facilities, and so we need a solution object to represent this structure. The following code creates a solution containing the appropriate variables and the objective, and stores it.

```
32. IloSolution soln(env);
33. soln.add(obj);
34. soln.add(open);
35. soln.store(solver);
```

We then create a neighborhood which opens or closes a facility. For this, we use the neighborhood `IloFlip` which changes the value of a 0–1 variable. The use of `IloRandomize` ensures that the flips are explored in a random order, the reasons for this being the same as those described in Section 8.4.4.

```
36. IloNHood nh =
        IloRandomize(env, IloFlip(env, open), rnd);
```

As previously described, a short-term tabu memory is used. This is defined below. The parameter 3 ensures that when a facility is opened or closed, it must remain in that state for three iterations.

```
37. IloMetaHeuristic mh = IloTabuSearch(env, 3);
```

Finally, the goal to make a move is constructed. We choose the lowest cost non-tabu move using the `IloMinimizeVar` search selector which ensures that we open/close the facility which leads to a reassignment of facilities to customers at minimum cost. At each move we also store the new solution in `best` if it is the best seen so far.

```
38. IloSearchSelector selMin = IloMinimizeVar(env, cost);
39. IloGoal move =
        IloSingleMove(env, soln, nh, mh, selMin, assign)
     && IloStoreBestSolution(env, best);
```

Figure 8.4 shows graphically how ILOG Solver searches for the move (which facility to open or close) which minimizes the overall cost. The search-tree follows a two level structure. The top part corresponds to the local search action of opening or closing a facility, and works on the `open` variables, whereas the bottom part of the tree deals with assigning open facilities to customers. Although these are drawn differently in the figure, both are standard Solver tree-based searches which marry together in the standard way; the triangles merely represent larger search trees which cannot be depicted in detail. At the top level of the tree $|\mathcal{F}|$ open/close operations are explored – one for each facility. In this example we assume there are six facilities. Imagine that the current state of facilities is CCOOCO (C for closed and O for open), and that the previous three moves closed facility 2, and opened facilities 3 and 6 (counting from the left). It is thus tabu to open facility two, or close either of facilities 3 and 6 at this iteration. Such moves will immediately be rejected and will be dead-ends in the search tree. The $\otimes$ symbols with TABU alongside indicate these illegal moves. Some "closure" moves may also not be possible if closing a facility would mean that the sum of demands of the customers exceeded the sums of the capacities of the open facilities. $\otimes$ symbols with CAP alongside indicate such failed closures for facilities 3 and 4 in our example. For all the remaining open/close moves, the additional `assign` goal will be

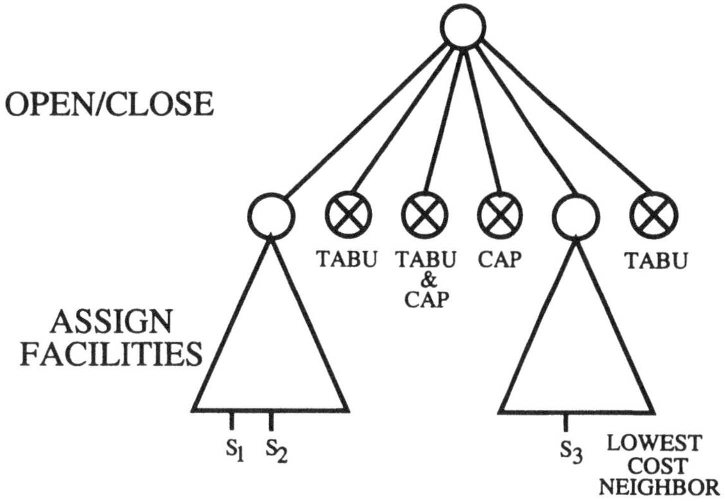

Figure 8.4 Finding the Lowest Cost Move

executed which assigns all customers to the open warehouses. The execution of these search trees is indicated by the triangles in the lower part of the figure. A number of complete assignments meeting all constraints may be found in each of these sub-trees. Each time one is found, the `IloMinimizeVar` search selector ensures that a constraint is added automatically by Solver stating that any solutions found in the entire tree after that point must have a cost which is strictly lower than the one just found. When the entire search tree has been traversed, `IloMinimizeVar` makes the search jump back to the best solution found. In this case it is S_3 indicating that the best legal non-tabu move to make is one which opens facility 5.

8.4.6 Experiments Using the Three Methods

We investigated the behavior of the three methods described in Sections 8.4.3, 8.4.4 and 8.4.5. We tested on all the capacitated facility location problems in the OR-Library (Beasley (1988), Beasley (1990)) of up to 50 customers and 50 facilities. The test code for each of the three methods corresponds almost exactly to the code presented. The main difference is the use of limited discrepancy search (LDS) (Harvey and Ginsberg (1995)) over depth-first search for the tree-based method and the hybrid method. LDS was chosen over depth-first search as the results are significantly better for the tree-based search. A small but non-negligible improvement in the hybrid search also results. ILOG Solver's search mechanism makes this improvement trivial to program. It is sufficient to add the line

```
assign = IloApply(env, assign, IloLDSEvaluator(env));
```

between the declaration of the `assign` goal and its use.

We set a time limit of two minutes on our experiments. For the pure and hybrid local search approaches, the initial solution is the first one generated by the complete

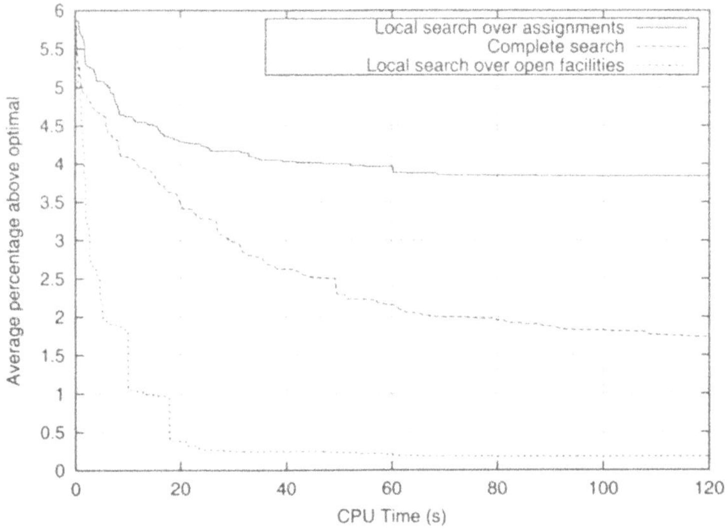

Figure 8.5 Comparison of Different Search Methods on Facility Location Problems

search technique, and so all methods begin at the same solution at the start of search. Figure 8.5 shows run time against the solution quality averaged over each of the location problems. (The average was formed as the geometric mean of ratios of cost to optimal cost. This average was then represented as a percentage by subtracting 1 and multiplying by 100.)

We immediately see the benefit of using the hybrid approach. It produces solutions around 0.2% above optimal after around one minute, whereas the other two methods do not attain the 2% mark in that time. It is clear that the tabu search over the assignment of facilities to customers quickly stagnates, despite our efforts to improve this state of affairs by using a varying tabu tenure. Complete search using limited discrepancy search works better, but its improvement tails off at near 1.5% from optimal. What we believe is important in this result is that the hybrid method required minimal tuning. For instance, it was still significantly better than the other two methods regardless of whether we used limited discrepancy search or depth-first search to perform the final assignments. We spent significantly more time attempting to improve the performance of the other two methods. Also significant is that such a hybrid search method would probably never have come about without the ability to simply integrate local and complete search using a constraint programming toolkit.

8.4.7 Note on Hybrid Approaches

The facility location problem as we have described here is also well suited to solution by mixed integer programming approaches. Such a method is easily implemented using ILOG tools (see, for example, van Hentenryck (1999)). We have concentrated on a constraint programming/local search approach here for simplicity. The reader should be aware however, that constraint programming, local search and linear programming can all be used to create a more complex and powerful hybrid.

8.5 EXTENDING THE TOOLKIT

ILOG Solver's local search mechanisms can be extended in two ways. Users are free to write their own neighborhoods or their own meta-heuristics. This considerably widens the scope of applicability of the local search mechanism as users are not limited to the neighborhoods or meta-heuristics supplied with the library. In this section, we demonstrate how new neighborhoods can be written; the mechanism for writing a new meta-heuristic is similar in style.

We have already seen that the neighborhood object is a class called `IloNHood`. In fact, `IloNHood` is a *handle class*. Handle classes are used throughout Solver and are classes which simply contain a pointer to a so-called *implementation class* which contains the object data. The use of handles means that pointers can generally be avoided, resulting in simpler code. The implementation class associated with `IloNHood` is `IloNHoodI` and it is with this second class that a new neighborhood is created: the relevant behavior is defined by sub-classing the `IloNHoodI` class.

In this section, we demonstrate how one could write the `IloFlip` neighborhood. For this example, the essentials of the `IloNHoodI` class are the following:

```
class IloNHoodI {
public:
  IloNHoodI(IloEnv env);
  virtual void start(IloSolver s, IloSolution currSoln);
  virtual IloInt getSize(IloSolver s) = 0;
  virtual IloSolution define(IloSolver s, IloInt idx) = 0;
};
```

The constructor of the class takes an environment. Any subclass of `IloNHoodI` must call this constructor. `start` is called before the neighborhood exploration begins. The current solution is passed as a parameter. For `IloFlip`, we simply need to keep the current solution. `getSize` is called after start and must deliver the number of neighbors in the neighborhood. For `IloFlip`, the number of neighbors is constant and equal to the number of variables. `define` is called numerous times with different indices in order to generate *solution deltas* representing the neighbors. For `IloFlip`, each delta will be a solution object containing one variable; the one which changes its value. In the above three methods, the active solver is passed as the first parameter.

The sub-class of `IloNHoodI`, `IloFlipI`, is shown below. The class has only two local variables (lines 2 and 3): the array `_vars` holds the array of variables to flip and `_currSoln` holds the current solution. The constructor (line 4) builds the subclass and initializes the array `_vars`. The `start` method (line 5) simply keeps the current solution in `_currSoln`. `getSize` (line 7) returns the size of the neighborhood, which is equal to the number of variables.

```
01. class IloFlipI : public IloNHoodI {
    private:
02.   IloNumVarArray _vars;
03.   IloSolution _currSoln;
    public:
04.   IloFlipI(IloEnv env, IloNumVarArray vars) :
        IloNHoodI(env), _vars(vars) { }
05.   void start(IloSolver, IloSolution currSoln) {
06.     _currSoln = currSoln;
      }
07.   IloInt getSize(IloSolver) { return _vars.getSize(); }
08.   IloSolution define(IloSolver, IloInt index);
    };
```

The define method is the only non-trivial method of `IloFlipI`. This method must return a solution delta with the variable being flipped set to its new value. Line 10 creates the delta, and line 11 adds a variable to it. For `IloFlip`, neighbor index i corresponds exactly to variable index i, and so the correct variable can be fetched using the neighbor index. Line 12 retrieves the value of this variable in the current solution, and line 13 sets the value in the delta to be "flipped". Finally, line 14 returns the solution delta.

```
09. IloSolution IloFlipI::define(IloSolver s, IloInt idx) {
10.   IloSolution delta(s.getEnv());
11.   delta.add(_vars[idx]);
12.   IloInt oldVal = _currSoln.getValue(_vars[idx]);
13.   delta.setValue(_vars[idx], 1 - oldVal);
14.   return delta;
    }
```

This concludes the definition of the neighborhood. Once created, user-defined neighborhoods behave exactly as Solver's built-in neighborhoods. For instance, they can be added, made the subject of randomization, etc. The library extensions become, for the user, part of the library itself.

8.6 SPECIALIZING THE TOOLKIT: ILOG DISPATCHER

Vehicle routing is one of the fields where local search is most widely used. This is mainly due to the fact that the problems encountered are usually large and difficult, and that most of the time, a "good solution" (in opposition with the best solution) is enough.

This section concentrates on how the Local Search Toolkit has been specialized to solve vehicle routing problems within the framework of ILOG Dispatcher. ILOG Dispatcher is a commercial product based on ILOG Solver. It is a C++ component library that users can tightly integrate with their own applications. It both extends Solver's modeling layer, by providing various high-level classes related to vehicle routing (such as vehicles, visits, nodes, or dimensions), and Solver's search engine by offering first solution heuristics, specialized local search neighborhoods and meta-heuristics.

8.6.1 Basic Concepts

In the same way that we have variables and constraints in ILOG Solver, we have modeling classes in ILOG Dispatcher that help us model generic vehicle routing problems. Such problems suppose that there are a certain number of orders from customers located at nodes which have to be performed by a fleet of vehicles. In Dispatcher, the action of performing an order, whether it is a delivery or pickup of goods or service (maintenance, for example), is called a visit. The basic variables of the problem are the ones describing the tour of each vehicle. For each visit, there are three variables, respectively representing:

- the visit immediately after the current visit, which we refer to as the next variable,
- the visit immediately before the current visit, which we refer to as the prev variable,
- the vehicle performing the current visit, which we refer to as the vehicle variable.

Here is an example of code which creates node, visit, and vehicle objects and which builds a route using constraints:

```
01. IloEnv env;
02. IloModel mdl(env);
03. IloNode node(env);
04. IloVisit visit(node);
05. mdl.add(visit);
06. IloNode depot(env);
07. IloVisit first(depot);
08. IloVisit last(depot);
09. IloVehicle vehicle(first, last);
10. mdl.add(vehicle);
11. mdl.add(visit.getPrevVar() == first);
12. mdl.add(visit.getNextVar() == last);
13. mdl.add(visit.getVehicleVar() == vehicle);
```

Lines 3 and 4 create a node and a visit `visit` located at that node. Line 4 adds the visit to the model. Line 6 creates a depot node, where a vehicle (line 9) starts and ends its tour (represented by visits at lines 7 and 8). Line 10 adds the vehicle to the model. Lines 11, 12 and 13 add constraints to the model specifying that the vehicle leaves the depot to perform `visit` and comes back to the depot. Note that lines 11 and 12 are equivalent to writing:

```
mdl.add(first.getNextVar() == visit);
mdl.add(last.getPrevVar() == visit);
```

This is due to the fact that next and prev variables are implicitly linked by the following constraint:

```
visit.getPrevVar().getNextVar() == visit;
```
and
```
visit.getNextVar().getPrevVar() == visit;
```

Line 13 is not mandatory because another constraint states that for two visits `v1` and `v2`:

if `v1.getNextVar() == v2`
then `v1.getVehicleVar() == v2.getVehicleVar()`

and that first and last visits of a vehicle have their vehicle variable bound to this vehicle.

The most innovative class in Dispatcher is probably dimensions. Dimensions represent quantities which are accumulated along routes. There are two types of dimensions: extrinsic and intrinsic. Extrinsic dimensions, instances of `IloDimension2`, represent quantities which depend on two visits to be evaluated. These dimensions usually represent the duration or the distance travelled on a route. Intrinsic dimensions, instances of `IloDimension1`, represent quantities which only depend on one visit. This can be quantities of good picked up or delivered at a visit expressed in weight, volume or number of pallets, for instance. Both classes are subclasses of `IloDimension` which share common features. For each visit there exist two dimension variables per dimension:

a) the cumul variable representing the amount of dimension accumulated when arriving at the visit; it can be accessed using:

`IloVisit::getCumulVar(IloDimension)`,

b) the transit variable representing the amount of dimension added to the cumul variable when going from one visit to another; it can be accessed using:

`IloVisit::getTransitVar(IloDimension)`.

These variables are linked together by the following constraint:

```
if v1.getNextVar() == v2
then v2.getCumulVar(dim)
  == v1.getCumulVar(dim) + v1.getTransitVar(dim)
```

Extrinsic dimensions add three extra dimension variables:

a) the delay variable representing the amount of dimension needed to perform a visit; it can represent service time, for example,

b) the travel variable representing the amount of dimension needed to travel from one visit to another; this usually represents distance,

c) the wait variable representing the slack on dimension between two visits that can be used to perform breaks.

These variables are constrained by the following:

```
v.getTransitVar(dim) == v.getDelayVar(dim)
                      + v.getTravelVar(dim)
                      + v.getWaitVar(dim)
```

and

```
if v1.getNextVar() == v2 and v1.getVehicleVar() == w
then v1.getTravelVar(dim) == dim.getDistance(v1, v2, w)
```

where `IloDimension2::getDistance()` returns the distance between visit `v1` and visit `v2` using vehicle w for dimension `dim`.

8.6.2 Specialized Neighborhoods and Meta-Heuristics

The recommended approach for solving routing problems using ILOG Dispatcher is the standard two phase approach: find a first solution and then improve it using local search. Therefore, Dispatcher extends Solver's local search classes by providing neighborhoods and meta-heuristic specific to vehicle routing.

8.6.2.1 Neighborhoods. The neighborhoods predefined in Dispatcher are the usual move operators used in vehicle routing. Here are the five most generic ones (see Figure 8.6):

a) 2-opt which inverts sub-parts of a tour.

b) Or-opt which moves sub-parts with a maximal length of three visits inside a tour.

c) Relocate which moves a visit from a tour to another.

d) Exchange which exchanges the positions of two visits from two different tours.

e) Cross which exchanges the end of a tour with the end of another tour.

In Dispatcher, Relocate and Exchange are generalized to moving pairs of visits linked by same-vehicle constraints. These constraints ensure that two visits are performed by the same vehicle. They are used to model pickup and delivery problems, for instance.

Dispatcher also provides more specific move operators which are used when performing some visits is not mandatory. Such cases can arise when there is more than one possible destination for a delivery but only one visit must be performed or when the problem is over-constrained and not all visits can be performed. The three move operators provided are:

a) MakePerformed which makes an unperformed visit performed by placing it after a performed visit.

b) MakeUnperformed which makes a performed visit unperformed.

c) SwapPerform which makes a performed visit unperformed and replaces it by an unperformed visit.

8.6.2.2 Meta-Heuristics. ILOG Dispatcher provides two specialized meta-heuristics: Guided local search and a specific tabu search.

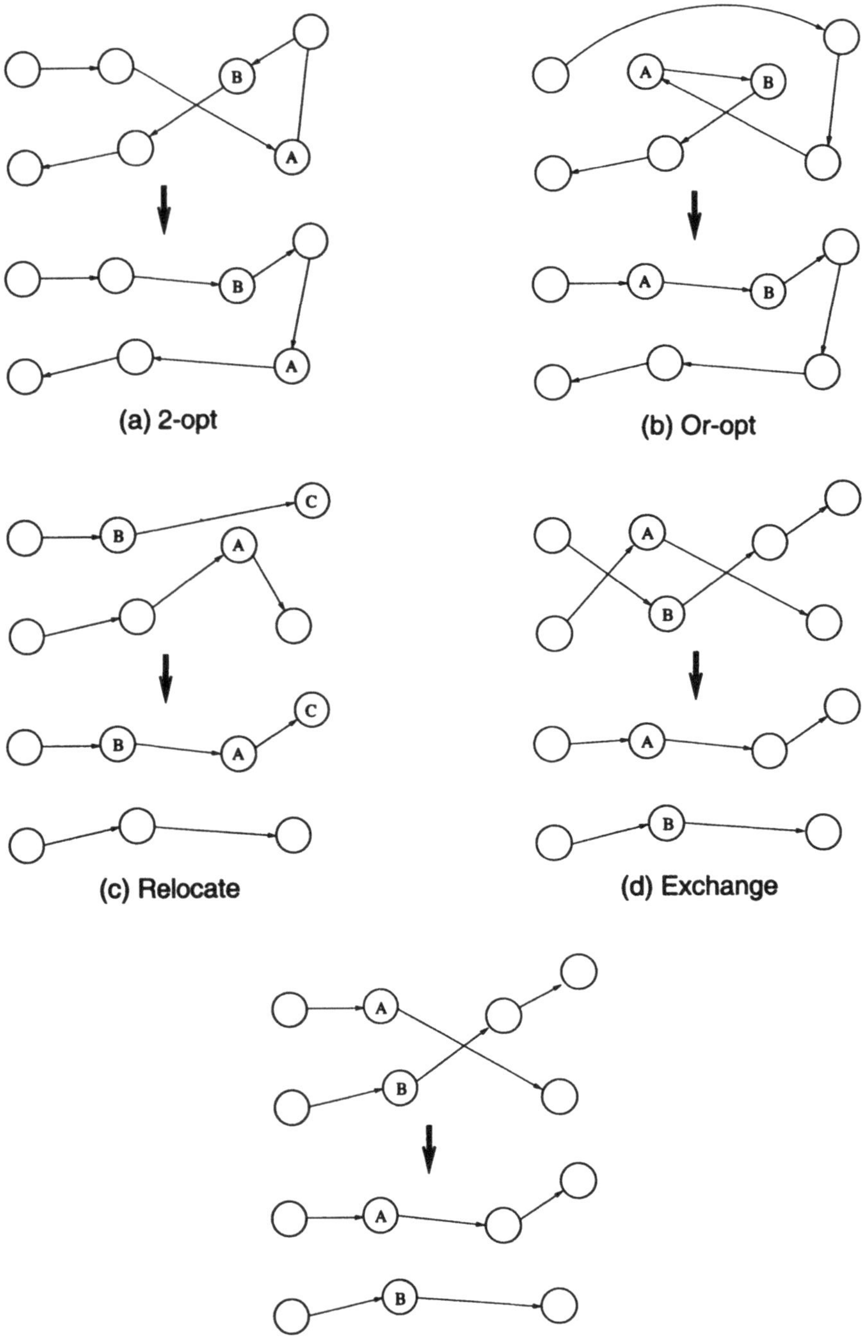

(a) 2-opt

(b) Or-opt

(c) Relocate

(d) Exchange

(e) Cross

Figure 8.6 Move Operators

Guided local search has shown to work very well on vehicle routing problems. At each iteration of this search, a greedy search is performed to a local minimum taking into account a penalized cost function. The penalized cost function is created by adding a penalty term to the true cost function. This term is the sum of all penalties for possible arcs in the routing problem. Initially the penalty for each arc is equal to zero. Each time a local minimum is reached, the most costly and less penalized arc is penalized and a new greedy search is started.

The tabu search procedure used in Dispatcher is rather simple. Each time a move is performed, some arcs (pairs of connected visits) are added to the solution and some are removed. The added arcs are kept in a `keep` list and the removed ones in a `forbid` list. Arcs remain in a list for a certain number of moves known as their tenure. When a new move is considered, the number of arcs it adds which are in the `forbid` list is counted and as well as the number of arcs it removes which are in the `keep` list. If this number is above a certain tabu number, specific to each move operator, then the move is rejected: the move is tabu.

8.6.3 Example: Solving a Vehicle Routing Problem

The problem we address here is the popular capacitated vehicle routing problem with time windows. Given a set of deliveries of goods to customers which have to be performed within a time window, and a set of vehicles which can carry up to a certain quantity of goods (capacity), find the routes of minimum length. Each customer is positioned according to its coordinates and the distance between two customers is the Euclidean distance. This problem can be enriched with various types of side constraints, such as precedence constraints, same vehicle constraints, lateness and earliness costs, but for the sake of simplicity we treat only the problem described above.

8.6.3.1 Problem Description. We assume that all vehicles are identical and have a capacity of `capacity` and that their depot opens at `openTime` and closes at `closeTime` and is located at coordinates `(depotX, depotY)`. We also assume that the demands of the visits are held in an array `quantity`, that their coordinates are held in the arrays `x` and `y`, that time spent at each visit to unload (drop time) is held in the array `dropTime` and that their time windows are held in the arrays `minTime` and `maxTime`.

First an environment (line 1), a model (line 2) and the dimensions of the problem are created and added to the model (lines 3 to 8). There is one intrinsic dimension `weight`, which represents the weight carried by the vehicles, and two extrinsic dimensions, `time` and `length`, respectively, representing the time and the length of the routes. The extrinsic dimensions use the distance object `IloEuclidean` to compute distances.

```
01. IloEnv env;
02. IloModel mdl(env);
03. IloDimension1 weight(env, "Weight");
04. mdl.add(weight);
05. IloDimension2 time(env, IloEuclidean, "Time");
06. mdl.add(time);
07. IloDimension2 length(env, IloEuclidean, "Length");
08. mdl.add(length);
```

Now we create the vehicles and add them to the model. Each vehicle starts and ends at visits located at a depot node `depot`. The cumul variables of these start and end visits are constrained to take into account the opening and closing hours of the depot (lines 12 and 14). The capacity of the vehicles is set to `capacity` (line 17) and the cost of the route is set to be equal to its length (line 16).

```
09. IloNode depot(env, depotX, depotY);
10. for (IloInt j = 0; j < nbOfTrucks + 10; j++) {
11.   IloVisit start(depot, "Depot");
12.   mdl.add(start.getCumulVar(time) >= openTime);
13.   IloVisit end(depot, "Depot");
14.   mdl.add(end.getCumulVar(time) <= closeTime);
15.   IloVehicle vehicle(start, end);
16.   vehicle.setCost(length, 1.0);
17.   vehicle.setCapacity(weight, capacity);
18.   mdl.add(vehicle);
19. }
```

After the vehicles, the visits are created and added to the model. Each of them is located at a node `customer` and the dimension variables are constrained to take into account the drop time (by constraining the delay variable for `time`, line 23), the quantity of goods the customer demands (by constraining the transit variable for `weight`, line 25) and its time window (by constraining the cumul variable for `time`, line 27).

```
20. for (IloInt i = 0; i < nbOfVisits; i++) {
21.   IloNode customer(env, x[i], y[i]);
22.   IloVisit visit(customer);
23.   mdl.add(visit.getDelayVar(time) == dropTime[i]);
24.   IloNumVar transit = visit.getTransitVar(weight);
25.   mdl.add(visit.transit == quantity[i]);
26.   IloNumVar cumul = visit.getCumulVar(time);
27.   mdl.add(minTime[i] <= cumul <= maxTime[i]);
28.   mdl.add(visit);
29. }
```

These lines finish stating the model and we can move on to solving the problem.

8.6.3.2 Finding a Solution. We present the standard two phase approach, first finding a solution using a first solution heuristic, then improving it with local search. We propose to decompose the local search phase into two sub-phases: a first phase of first accept greedy search, and a second phase based on tabu search.

The following lines show how to create a first solution using the savings heuristic based on Clarke and Wright (1964).

```
30. IloSolver solver(mdl);
31. IloDispatcher dispatcher(solver);
32. IloRoutingSolution solution(mdl);
33. IloNumVar cost = dispatcher.getCostVar();
34. IloGoal instCost =
35.         IloDichotomize(env, cost, IloFalse);
36. IloGoal goal = IloSavingsGenerate(env) && instCost;
37. if (!solver.solve(goal)) {
38.   cout << "No solution" << endl;
39.   env.end();
40.   return 0;
41. }
42. solution.store(solver);
```

We create a solver, which extracts the model (line 30), and a dispatcher (line 31), which is an object whose purpose is to let the user access the state and values of Dispatcher objects during the search. A specific type of solution, containing the visits of the model, is created in line 32. This solution is nothing more than a standard Solver solution with a routing-oriented programming interface. In line 36 we create the Solver goal which uses the savings heuristic to find a solution. The goal created at lines 34 and 35 is used to instantiate the cost variable. The search starts in line 37 and if a solution is found it is stored (line 42). Now the local search phase can begin.

We start by a simple greedy search.

```
43. IloGoal improve = IloSingleMove(env,
44.                                 solution,
45.                                 IloTwoOpt(env)
46.                                 + IloOrOpt(env)
47.                                 + IloRelocate(env)
48.                                 + IloExchange(env)
49.                                 + IloCross(env),
50.                                 IloImprove(env),
51.                                 instCost);
52. while (solver.solve(improve)) {
53.   cout << "Cost = "
54.        << dispatcher.getTotalCost() << endl;
55. }
```

The `IloSingleMove` goal is used once again, taking as parameter the Dispatcher move operators described in the preceding section (lines 45 to 49). The meta-heuristic `IloImprove` is used meaning a downhill greedy search is going to be performed.

`IloSolver::solve()` is called (line 52) and the current value of the objective is displayed (lines 53, 54) until no improving move can be found and that a local minimum is reached.

To climb out of the local minimum, we continue with a tabu search phase very similar to the one described in Section 8.4.4.

```
56. cout << "Starting tabu" << endl;
57. IloRoutingSolution best = solution.makeClone(env);
58. IloDispatcherTabuSearch dts(env, 12);
59. IloSearchSelector sel = IloMinimizeVar(env, cost);
60. IloGoal tabuMove = IloSingleMove(env,
61.                                  solution,
62.                                  IloTwoOpt(env)
63.                                  + IloOrOpt(env)
64.                                  + IloRelocate(env)
65.                                  + IloExchange(env),
66.                                  dts,
67.                                  sel,
68.                                  instCost);
69. tabuMove = tabuMove
70.         && IloStoreBestSolution(env, best);
71. for (IloInt i = 0; i < 150; i++) {
72.   if (i == 70) dts.setTenure(20);
73.   if (i == 85) dts.setTenure(25);
74.   if (i == 105) dts.setTenure(12);
75.   if (solver.solve(tabuMove))
76.      cout << "Cost = "
77.           << dispatcher.getTotalCost() << endl;
78.   else
79.      if (dts.complete()) break;
80. }
```

The current solution is cloned (line 57) in order to keep track of the best solution during the search (using the goal at line 70). We use the Dispatcher-specific version of tabu search, described above, to move out of local mimima (line 58). The move operators considered are the same as during the greedy search phase, except for `IloCross` which can lead to too many symmetric neighbors (lines 62 to 65). Note that it is possible to modify the tenure during the search, as shown in lines 72 to 74. The resulting effect is an intensification or a diversification of the search, depending if the tenure is decreased or increased. The tabu search is run for 150 iterations (line 71) or until no non-tabu moves are left (line 79).

Finally, the best solution is restored and presented for inspection.

```
solver.solve(IloRestoreSolution(env, best)
             && instCost);
cout << "Final Solution" << endl;
cout << "Cost = "
     << dispatcher.getTotalCost() << endl;
cout << dispatcher << endl;
```

8.7 RELATED WORK

The work presented in this article can be related to several different areas.

Numerous constraint programming systems have been developed since the mid eighties, starting with CHIP (Dincbas et al. (1988)). These were based on the Prolog programming language and, therefore, implemented only depth-first search. The first implementations mixing local search and constraint programming can be traced back to the mid nineties, with works such as Pesant and Gendreau (1996), and the GreenTrip project (de Backer and Furnon (1999)) with an application to vehicle routing.

Our work was highly influenced by that of Pesant and Gendreau (1996) and later Pesant and Gendreau (1999). However, our approach differs from theirs in that they use *variables and constraints* to represent the neighborhood. *Interface constraints* then connect these new variables with the original model variables. In this way, propagation can proceed between the original model variables and the new "neighborhood" variables. These interface constraints depend on the current solution to the problem and are added anew to the constraint solver each time a local move is to be made. Thereafter, the neighborhood exploration proceeds by generating all possible assignments to the neighborhood variables using a standard tree search. The interface constraints interpret the assignments to the neighborhood variables and perform the required propagation into the model variables. One difference with our approach lies in the fact that some variables and constraints have to be added, thus increasing the memory consumption. The most notable difference, however, is the lower algorithmic complexity guarantee of our approach (Shaw et al. (2000)).

Caseau and Laburthe (1998) gives a description of Salsa, which is essentially a set of enhanced search structures that enable to mix local and complete search within a constraint programming framework. Our approach shares this feature by extending the notion of goals.

LOCALIZER (Michel and van Hentenryck (2000)) consists in the combination of a modeling language for combinatorial problems and a local search solver for these problems. It retains some nice features of constraint programming, for example the ability to describe a problem through constraints. The solution process is somewhat different, since propagation over the domains of variables is not used. When the value of a decision variable is changed, data structures known as invariants update the values of all dependent variables, including the objective. Such maintenance of only complete assignments makes it difficult to combine complete and local search methods.

Although not toolkits or frameworks as such, work has also been performed on solving constraint satisfaction problems using local search techniques. Probably the most well-known is the GSAT/WalkSat family of algorithms (Selman et al. (1994)) which is applied to propositional satisfiability problems. More recently, more complex strategies have been applied (McAllester et al. (1997)). Also notable is the seminal work of Minton et al. (1992) on min-conflicts methods. An interesting approach has also be advocated by Prestwich (2000) who performs local search not on the decision variables themselves, but on the decisions made by a constructive algorithm. The approach thus fits well into a constraint programming framework.

EasyLocal++ (Di Gaspero and Schaerf (2001)), HOTFRAME (Fink et al. (1999b)) and OpenTS (IBM Open Source Software (2001)) represent other attempts to provide reusable components for local search. EasyLocal++ and HOTFRAME are a set of C++ classes for implementing various meta-heuristics, while OpenTS is written in Java and provides the basic elements for implementing a tabu search. The main difference between these approaches and ours is that some coding is needed to implement the enforcement of the constraints representing the problem at hand, since there is no way to model the problem using constraints.

Finally, Mautor and Michelon (1997) present a combination of local and complete search which can be easily implemented using our approach.

8.8 CONCLUSION

We have described the local search facilities of the ILOG Solver constraint programming library. This local search toolkit is flexible and open, giving users the ability to experiment with different types of search strategies, neighborhoods, and mixes of complete and local search. The toolkit is also extensible, providing the basis for creating new neighborhoods and meta-heuristics. The toolkit was described both in terms of its primary goals, oriented towards usability, and in terms of its orthogonal object model; the latter description being made explicit by the inclusion of ILOG Solver code samples. Of particular interest is the way in which neighborhoods and meta-heuristics can be composed. We have also described the benefits that traditional constraint programming can bring to local search, and thus why it is advantageous to embed local search support in a constraint programming library.

A real-life example of facility location has been described with code demonstrating the ease by which complete, local, and hybrid search methods can be applied. Changing search methods results in only a small change to the ILOG Solver program, which means that alternative strategies can be investigated quickly. Importantly, this can be done without changing the model as the search and model are separate entities; an uncommon occurrence for local search algorithms. We have shown that a hybrid method – which uses local search to decide which facilities to open and tree search to assign facilities to customers – can be an effective technique.

The ease with which the toolkit can be extended was highlighted by an example which describes a neighborhood. Neighborhoods are defined explicitly by stating, for each neighbor, what parts of the solution change and how. This form means that very general and even unstructured neighborhoods can be created, for example, which use candidate lists and learning (Glover and Laguna (1997)). Such flexibility is not so readily available when neighborhoods are implicitly defined.

Finally, ILOG Dispatcher was described. Dispatcher is a *vertical* library built upon ILOG Solver and dedicated to solving vehicle routing problems. The verticalization theme is carried consistently through four different aspects of Dispatcher: modeling objects for routing, constraints for routing, tree-search algorithms and heuristics for routing, and local search neighborhoods and meta-heuristics for routing. Each one of these verticalizations is made possible by the openness of ILOG Solver in all of these aspects. The result is a product that is both cleanly designed and easy to use. The coded ILOG Dispatcher example clearly demonstrates that industrial problems can be modeled and solved in an intuitive manner.

9 THE MODELING LANGUAGE OPL – A SHORT OVERVIEW

Pascal Van Hentenryck and Laurent Michel

Brown University, Box 1910
Providence, RI 02912, USA
{pvh,ldm}@cs.brown.edu

Abstract:
OPL is a modeling language for mathematical programming and combinatorial optimization problems. It is the first modeling language to combine high-level algebraic and set notations from modeling languages with a rich constraint language and the ability to specify search procedures and strategies that is the essence of constraint programming. In addition, OPL models can be controlled and composed using OPLSCRIPT, a script language that simplifies the development of applications that solve sequences of models, several instances of the same model, or a combination of both as in column-generation applications. This paper illustrates some of the functionalities of OPL for combinatorial optimization using frequency allocation, sport scheduling, and project scheduling applications. It also gives a brief overview of OPLSCRIPT and of the code generation facilities of the development environment of OPL.

9.1 INTRODUCTION

Combinatorial optimization problems are ubiquitous in many practical applications, including scheduling, resource allocation, planning, and configuration problems. These problems are computationally difficult (i.e., they are $\mathcal{NP}$-hard) and require considerable expertise in optimization, software engineering, and the application domain.

The last two decades have witnessed substantial development in tools to simplify the design and implementation of combinatorial optimization problems. Their goal is to decrease development time substantially while preserving most of the efficiency of specialized programs. Most tools can be classified in two categories: mathematical modeling languages and constraint programming languages. Mathematical modeling languages such as AMPL (Fourer et al. (1993)) and GAMS (Bisschop and Meeraus (1982)) provide very high-level algebraic and set notations to express concisely mathematical problems that can then be solved using state-of-the-art solvers. These modeling languages do not require specific programming skills and can be used by a wide audience. Constraint programming languages such as CHIP (Dincbas et al. (1988)), PROLOG III and its successors (Colmerauer (1990)) and OZ (Smolka (1995)) have orthogonal strengths. Their constraint languages, and their underlying solvers, go beyond traditional linear and nonlinear constraints and support logical, high-order, and global constraints. They also make it possible to program search procedures to specify how to explore the search space. However, these languages are mostly aimed at computer scientists and often have weaker abstractions for algebraic and set manipulation.

The work described in this paper originated as an attempt to unify modeling and constraint programming languages and their underlying implementation technologies. It led to the development of the optimization programming language OPL (see van Hentenryck (1999)), its associated script language OPLSCRIPT (see van Hentenryck and Michel (2000)), and its development environment.

OPL is a modeling language sharing high-level algebraic and set notations with traditional modeling languages. It also contains some novel functionalities to exploit sparsity in large-scale applications, such as the ability to index arrays with arbitrary data structures. OPL shares with constraint programming languages their rich constraint languages, their support for scheduling and resource allocation problems, and the ability to specify search procedures and strategies. OPL also makes it easy to combine different solver technologies for the same application.

OPLSCRIPT is a script language for composing and controlling OPL models. Its motivation comes from the many applications that require solving several instances of the same problem (e.g., sensibility analysis), sequences of models, or a combination of both as in column-generation applications. OPLSCRIPT supports a variety of abstractions to simplify these applications, such as OPL models as first-class objects, extensible data structures, and linear programming bases to name only a few.

The development environment of OPL and OPLSCRIPT provides, beyond traditional support for the traditional "edit, execute, and debug" cycle, automatic visualizations of the results (e.g., Gantt charts for scheduling applications), visual tools for debugging and monitoring OPL models (e.g., visualizations of the search space), and C++ code generation to integrate an OPL model in a larger application. The code generation produces a class for each model object and makes it possible to add/remove constraints dynamically and to overwrite the search procedure.

The purpose of this paper is to review some of the novel features of OPL. As a consequence, it does not describe the use of OPL for mathematical programming but it focuses on illustrating constraint programming features through a number of applications. The hope is that these applications convey the spirit and the potential of the novel features of OPL for some classes of combinatorial optimization problems and motivate readers to have a closer look at the OPL system.

The rest of the paper is organized as follows. Section 9.2 describes a model for a frequency allocation application that illustrates how to use high-level algebraic and set manipulation, how to exploit sparsity, and how to implement search procedures in OPL. Section 9.3 describes a model for a sport-scheduling application that illustrates the use of global constraints in OPL. Section 9.4 describes an application that illustrates the support for job-shop scheduling applications and for search strategies in OPL. Section 9.5 describes a more substantial scheduling application. It illustrates some advanced features of OPL, some uses of OPLSCRIPT, and the code generation facilities. Section 9.6 describes new research directions. In particular, it shows how many advanced modeling features can be elegantly supported in object-oriented constraint programming libraries. The material in this chapter is based on van Hentenryck et al. (1999a), van Hentenryck et al. (1999b) and Michel and van Hentenryck (2001b).

9.2 FREQUENCY ALLOCATION

The frequency-allocation problem illustrates a number of interesting features of OPL: the use of complex quantifiers, and the use of a multi-criterion ordering to choose which variable to assign next. It also features an interesting data representation that is useful in large-scale linear models.

The frequency-allocation problem consists of allocating frequencies to a number of transmitters so that there is no interference between transmitters and the number of allocated frequencies is minimized. The problem described here is an actual cellular phone problem where the network is divided into cells, each cell containing a number of transmitters whose locations are specified. The interference constraints are specified as follows:

- The distance between two transmitter frequencies within a cell must not be smaller than 16.
- The distances between two transmitter frequencies from different cells vary according to their geographical situation and are described in a matrix.

The problem of course consists of assigning frequencies to transmitters to avoid interference and, if possible, to minimize the number of frequencies. The rest of this section focuses on finding a solution using a heuristic to reduce the number of allocated frequencies.

Figure 9.1 shows an OPL statement for the frequency-allocation problem and Figure 9.2 describes the instance data. Note the separation between models and data which is an interesting feature of OPL. The model data first specifies the number of cells (25 in the instance), the number of available frequencies (256 in the instance), and their associated ranges. The next declarations specify the number of transmitters needed for each cell and the distance between cells. For example, in the instance, cell 1 requires eight transmitters while cell 3 requires six transmitters. The distance between cell 1 and cell 2 is 1. The first interesting feature of the model is how variables are declared:

```
int nbCells = ...;
int nbFreqs = ...;
range Cells 1..nbCells;
range Freqs 1..nbFreqs;
int nbTrans[Cells] = ...;
int distance[Cells,Cells] = ...;

struct TransmitterType { Cells c; int t; };
{TransmitterType} Trans = { <c,t> |     c in Cells
                                      &  t in 1..nbTrans[c] };
var Freqs freq[Trans];

solve {
   forall(c in Cells & ordered t1, t2 in 1..nbTrans[c])
      abs(freq[<c,t1>] - freq[<c,t2>]) >= 16;

   forall(ordered c1, c2 in Cells :  distance[c1,c2] > 0)
      forall(t1 in 1..nbTrans[c1] & t2 in 1..nbTrans[c2])
         abs(freq[<c1,t1>] - freq[<c2,t2>]) >= distance[c1,c2];
};

search {
   forall(t in Trans ordered by
            increasing <dsize(freq[t]),nbTrans[t.c]>)
      tryall(f in Freqs ordered by decreasing nbOccur(f,freq))
         freq[t] = f;
};
```

Figure 9.1 The Frequency-Allocation Problem (`alloc.mod`)

```
struct TransmitterType { Cells c; int t; };
{TransmitterType} Trans = { <c,t> |  c in Cells
                                   & t in 1..nbTrans[c] };
var Freqs freq[Trans];
```

As is clear from the problem statement, transmitters are contained within cells. The above declarations preserve this structure, which will be useful when stating constraints. A transmitter is simply described as a record containing a cell number and a transmitter number inside the cell. The set of transmitters is computed automatically from the data using

```
{TransmitterType} Trans = { <c,t> |  c in Cells
                                   & t in 1..nbTrans[c] };
```

which considers each cell and each transmitter in the cell. OPL supports a rich language to compute with sets of data structures and this instruction illustrates some of this functionality. The model then declares an array of variables

```
var Freqs freq[Trans];
```

indexed by the set of transmitters; the values of these variables are of course the frequencies associated with the transmitters. This declaration illustrates a fundamental aspect of OPL: arrays can be indexed by arbitrary data. In this application, the arrays of variables `freq` is indexed by the elements of `transmitters` that are records. This functionality is of primary importance to exploit sparsity in large-scale models and to simplify the statement of many combinatorial optimization problems.

There are two main groups of constraints in this model. The first set of constraints handles the distance constraints between transmitters inside a cell. The instruction

```
forall(c in Cells & ordered t1, t2 in 1..nbTrans[c])
    abs(freq[<c,t1>] - freq[<c,t2>]) >= 16;
```

enforces the constraint that the distance between two transmitters inside a cell is at least 16. The instruction is compact mainly because we can quantify several variables in `forall` statements and because of the keyword `ordered` that makes sure that the statement considers triples `<c,t1,t2>` where `t1 < t2`. Of particular interest are the expressions `freq[<c,t1>]` and `freq[<c,t2>]` illustrating that the indices of array `freq` are records of the form `<c,t>`, where `c` is a cell and `t` is a transmitter. Note also that the distance is computed using the function `abs`, which computes the absolute value of its argument (which may be an arbitrary integer expression).

The second set of constraints handles the distance constraints between transmitters from different cells. The instruction

```
forall(ordered c1, c2 in Cells :  distance[c1,c2] > 0)
    forall(t1 in 1..nbTrans[c1] & t2 in 1..nbTrans[c2])
       abs(freq[<c1,t1>] - freq[<c2,t2>]) >= distance[c1,c2];
```

considers each pair of distinct cells whose distance must be greater than zero and each two transmitters in these cells, and states that the distance between the frequencies of these transmitters must be at least the distance specified in the matrix `distance`.

```
nbCells = 25;
nbFreqs = 256;
nbTrans = [8 6 6 1 4 4 8 8 8 8 4 9 8 4 4 10 8 9 8 4 5 4 8 1 1];
distance = [
   [16 1 1 0 0 0 0 0 1 1 1 1 1 2 2 1 1 0 0 0 2 2 1 1 1]
   [1 16 2 0 0 0 0 0 2 2 1 1 1 2 2 1 1 0 0 0 0 0 0 0 0]
   [1 2 16 0 0 0 0 0 2 2 1 1 1 2 2 1 1 0 0 0 0 0 0 0 0]
   [0 0 0 16 2 2 0 0 0 0 0 0 0 0 0 0 0 1 1 1 0 0 0 1 1]
   [0 0 0 2 16 2 0 0 0 0 0 0 0 0 0 0 0 1 1 1 0 0 0 1 1]
   [0 0 0 2 2 16 0 0 0 0 0 0 0 0 0 0 0 1 1 1 0 0 0 1 1]
   [0 0 0 0 0 0 16 2 0 0 1 1 1 0 0 1 1 1 1 2 0 0 0 1 1]
   [0 0 0 0 0 0 2 16 0 0 1 1 1 0 0 1 1 1 1 2 0 0 0 1 1]
   [1 2 2 0 0 0 0 0 16 2 2 2 2 2 2 1 1 1 1 1 1 1 0 1 1]
   [1 2 2 0 0 0 0 0 2 16 2 2 2 2 2 1 1 1 1 1 1 1 0 1 1]
   [1 1 1 0 0 0 1 1 2 2 16 2 2 2 2 2 2 1 1 2 1 1 0 1 1]
   [1 1 1 0 0 0 1 1 2 2 2 16 2 2 2 2 2 1 1 2 1 1 0 1 1]
   [1 1 1 0 0 0 1 1 2 2 2 2 16 2 2 2 2 1 1 2 1 1 0 1 1]
   [2 2 2 0 0 0 0 0 2 2 2 2 2 16 2 1 1 1 1 1 1 1 1 1 1]
   [2 2 2 0 0 0 0 0 2 2 2 2 2 2 16 1 1 1 1 1 1 1 1 1 1]
   [1 1 1 0 0 0 1 1 1 1 2 2 2 1 1 16 2 2 2 1 2 2 1 2 2]
   [1 1 1 0 0 0 1 1 1 1 2 2 2 1 1 2 16 2 2 1 2 2 1 2 2]
   [0 0 0 1 1 1 1 1 1 1 1 1 1 1 1 2 2 16 2 2 1 1 0 2 2]
   [0 0 0 1 1 1 1 1 1 1 1 1 1 1 1 2 2 2 16 2 1 1 0 2 2]
   [0 0 0 1 1 1 2 2 1 1 2 2 2 1 1 1 1 2 2 16 1 1 0 1 1]
   [2 0 0 0 0 0 0 0 1 1 1 1 1 1 1 2 2 1 1 1 16 2 1 2 2]
   [2 0 0 0 0 0 0 0 1 1 1 1 1 1 1 2 2 1 1 1 2 16 1 2 2]
   [1 0 0 0 0 0 0 0 0 0 0 0 0 1 1 1 1 0 0 0 1 1 16 1 1]
   [1 0 0 1 1 1 1 1 1 1 1 1 1 1 1 2 2 2 2 1 2 2 1 16 2]
   [1 0 0 1 1 1 1 1 1 1 1 1 1 1 1 2 2 2 2 1 2 2 1 2 16]];
};
```

Figure 9.2 Instance Data for the Frequency-Allocation Problem (`alloc.dat`)

Another interesting part of this model is the search strategy. The basic structure is not surprising: OPL considers each transmitter and chooses a frequency nondeterministically. The interesting feature of the model is the heuristic. OPL chooses to generate a value for the transmitter with the smallest domain and, in case of ties, for the transmitter whose cell size is as small as possible. This multi-criterion heuristic is expressed using a tuple `<dsize(freq[t]),nbTrans[t.c]>` to obtain

```
forall(t in Trans ordered by
         increasing <dsize(freq[t]),nbTrans[t.c]>)
```

Each transmitter is associated with a tuple $< s, c >$, where s is the number of possible frequencies and c is the number of transmitters in the cell to which the transmitter belongs. A transmitter with tuple $< s_1, c_1 >$ is preferred over a transmitter with tuple $< s_2, c_2 >$ if $s_1 < s_2$ or if $s_1 = s_2$ and $c_1 < c_2$.

Once a transmitter has been selected, OPL generates a frequency for it in a nondeterministic manner. Once again, the model specifies a heuristic for the ordering in which the frequencies must be tried. To reduce the number of frequencies, the model says to try first those values that were used most often in previous assignments. This heuristic is implemented using a nondeterministic `tryall` instruction with the order specified using the `nbOccur` function (`nbOccur(i,a)` denotes the number of occurrences of `i` in array `a` at a given step of the execution):

```
forall(t in Trans ordered by
         increasing <dsize(freq[t]),nbTrans[t.c]>)
    tryall(f in Freqs ordered by decreasing nbOccur(f,freq))
       freq[t] = f;
```

This search procedure is typical of many constraint satisfaction problems and consists of using a first heuristic to dynamically choose which variable to instantiate next (variable choice) and a second heuristic to choose which value to assign nondeterministically to the selected variable (value choice). The `forall` instruction is of course deterministic, while the `tryall` instruction is nondeterministic: potentially all possible values are chosen for the selected variable. Note that, on the instance depicted in Figure 9.2, OPL returns a solution with 95 frequencies in about three seconds.

9.3 SPORT SCHEDULING

This section considers the sport-scheduling problem from McAloon et al. (1997) and Régin (1998). The problem consists of scheduling games between n teams over $n - 1$ weeks. In addition, each week is divided into $n/2$ periods. The goal is to schedule a game for each period of every week so that the following constraints are satisfied:

1. Every team plays against every other team.
2. A team plays exactly once a week.
3. A team plays at most twice in the same period over the course of the season.

A solution to this problem for eight teams is shown in Figure 9.3. In fact, the problem can be made more uniform by adding a "dummy" final week and requesting that

Week	1	2	3	4	5	6	7
period 1	0 vs 1	0 vs 2	4 vs 7	3 vs 6	3 vs 7	1 vs 5	2 vs 4
period 2	2 vs 3	1 vs 7	0 vs 3	5 vs 7	1 vs 4	0 vs 6	5 vs 6
period 3	4 vs 5	3 vs 5	1 vs 6	0 vs 4	2 vs 6	2 vs 7	0 vs 7
period 4	6 vs 7	4 vs 6	2 vs 5	1 vs 2	0 vs 5	3 vs 4	1 vs 3

Figure 9.3 A Solution to the Sport-Scheduling Application with Eight Teams

all teams play exactly twice in each period. The rest of this section considers this equivalent problem for simplicity.

The sport-scheduling problem is an interesting application for constraint programming. On the one hand, it is a standard benchmark (submitted by Bob Daniel) to the well-known MIP library and it is claimed in McAloon et al. (1997) that state-of-the-art MIP solvers cannot find a solution for 14 teams. The `OPL` models presented in this section are computationally much more efficient. On the other hand, the sport-scheduling application demonstrates fundamental features of constraint programming including global and symbolic constraints. In particular, the model makes heavy use of arc consistency (see Mackworth (1977)), a fundamental constraint satisfaction technique from artificial intelligence.

The rest of this section is organized as follows. Section 9.3.1 presents an `OPL` model that solves the 14-teams problem in about 44 seconds. Section 9.3.2 shows how to specialize it further to find a solution for 14 to 30 teams quickly. Both models are based on the constraint programs presented in Régin (1998).

9.3.1 A Simple OPL Model

The simple model is depicted in Figure 9.4. Its input is the number of teams `nbTeams`. Several ranges are defined from the input: the teams `Teams`, the weeks `Weeks`, and the extended weeks `EWeeks`, i.e., the weeks plus the dummy week. The model also declares an enumerated type `slot` to specify the team position in a game (`home` or `away`). The declarations

```
int occur[t in Teams] = 2;
int values[t in Teams] = t;
```

specify two arrays that are initialized generically and are used to state constraints later on. The array `occur` can be viewed as a constant function always returning 2, while the array `values` can be thought of as the identify function over teams.

The main modeling idea in this model is to use two classes of variables: team variables that specify the team playing on a given week, period, and slot and the game variables specifying which game is played on a given week and period. The use of game variables makes it simple to state the constraint that every team must play against each other team. Games are uniquely identified by their two teams. More precisely, a game consisting of home team `h` and away team `a` is uniquely identified by the integer `(h-1)*nbTeams + a`. The instruction

```
var Teams team[Periods,EWeeks,Slots];
```

```
int nbTeams = ...;
range Teams 1..nbTeams;
range Weeks 1..nbTeams-1;
range EWeeks 1..nbTeams;
range Periods 1..nbTeams/2;
range Games 1..nbTeams*nbTeams;
enum Slots = { home, away };

int occur[t in Teams] = 2;
int values[t in Teams] = t;

var Teams team[Periods,EWeeks,Slots];
var Games game[Periods,Weeks];

struct Play { int f; int s; int g; };
{Play} Plays = { <i,j,(i-1)*nbTeams+j> | ordered i,
                                        j in Teams };
predicate link(int f,int s,int g) in Plays;

solve {
   forall(w in EWeeks)
      alldifferent(all(p in Periods & s in Slots)
                   team[p,w,s]) onDomain;
   alldifferent(game) onDomain;
   forall(p in Periods)
      distribute(occur,values,all(w in EWeeks & s in Slots)
                 team[p,w,s])
         extendedPropagation;
   forall(p in Periods & w in Weeks)
      link(team[p,w,home],team[p,w,away],game[p,w]);
};

search {
   generate(game);
};
```

Figure 9.4 A Simple Model for the Sport-Scheduling Model

```
var Games game[Periods,Weeks];
```

declares the variables. These two sets of variables must be linked together to make sure that the game and team variables for a given period and a given week are consistent. The instructions

```
struct Play { int f; int s; int g; };
{Play} Plays = { <i,j,(i-1)*nbTeams+j> | ordered i,
                                         j in Teams };
```

specify the set of legal games `Plays` for this application. For eight teams, this set consists of tuples of the form

```
<1,2,2>
<1,3,3>
...
<7,8,56>
```

Note that this definition eliminates some symmetries in the problem statement since the home team is always smaller than the away team. The instruction

```
predicate link(int f,int s,int g) in Plays;
```

defines a symbolic constraint by specifying its set of tuples. In other words, `link(h,a,g)` holds if the tuple `<h,a,g>` is in the set `Plays` of legal games. This symbolic constraint is used in the constraint statement to enforce the relation between the game and the team variables.

The constraint declarations in the model follow almost directly the problem description. The constraint

```
alldifferent(all(p in Periods & s in Slots)
             team[p,w,s]) onDomain;
```

specifies that all the teams scheduled to play on week `w` must be different. It uses an aggregate operator `all` to collect the appropriate team variables by iterating over the periods and the slots and an annotation `onDomain` to enforce arc consistency. See Régin (1994) for a description on how to enforce arc consistency on this global constraint. The constraint

```
distribute(occur,values,all(w in EWeeks & s in Slots)
                          team[p,w,s])
    extendedPropagation
```

specifies that a team plays exactly twice over the course of the "extended" season. Its first argument specifies the number of occurrences of the values specified by the second argument in the set of variables specified by the third argument that collects all variables playing in period `p`. The annotation `extendedPropagation` specifies to enforce arc consistency on this constraint. See Régin (1996) for a description on how to enforce arc consistency on this global constraint.

The constraint

```
alldifferent(game) onDomain;
```

specifies that all games are different, i.e., that all teams play against each other team. These constraints illustrate some of the global constraints of OPL. Other global constraints in the current version include a sequencing constraint, a circuit constraint, and a variety of scheduling constraints. Finally, the constraint

```
link(team[p,w,home],team[p,w,away],game[p,w]);
```

is most interesting. It specifies that the game `game[p,w]` consists of the teams `team[p,w,home]` and `team[p,w,away]`. OPL enforces arc consistency on this symbolic constraint.

The search procedure in this statement is extremely simple and consists of generating values for the games using the first-fail principle. Note also that generating values for the games automatically assigns values to the team by constraint propagation. As mentioned, this model finds a solution for 14 teams in about 44 seconds on a PC (400mhz).

9.3.2 *A Round-Robin Model*

The simple model has many symmetries that enlarge the search space considerably. In this section, we describe a model that uses a round-robin schedule to determine which games are played in a given week. As a consequence, once the round-robin schedule is selected, it is only necessary to determine the period of each game, not its schedule week. In addition, it turns out that a simple round-robin schedule makes it possible to find solutions for large numbers of teams. The model is depicted in Figures 9.5 and 9.6.

The main novelty in the statement is the array `rr` that specifies the games for every week. Assuming that n denotes the number of teams, the basic idea is to fix the set of games of the first week as

$$< 1,2 > \cup \{< p+1, n-p+2 > \mid p > 1\}$$

where p is a period identifier. Games of the subsequent weeks are computed by transforming a tuple $< f, s >$ into a tuple $< f', s' >$ where

$$f' = \begin{cases} 1 & \text{if } f = 1 \\ 2 & \text{if } f = n \\ f+1 & \text{otherwise} \end{cases}$$

and

$$s' = \begin{cases} 2 & \text{if } s = n \\ s+1 & \text{otherwise} \end{cases}$$

This round-robin schedule is computed in the `initialize` instruction and the last instruction computes the game associated with the teams. The instruction

```
{int} domain[w in Weeks] = { rr[w,p].g | p in Periods };
```

```
int nbTeams = ...;
range Teams 1..nbTeams;
range Weeks 1..nbTeams-1; range EWeeks 1..nbTeams;
range Periods 1..nbTeams/2;
range Games 1..nbTeams*nbTeams;
enum Slots = { home, away };
int occur[t in Teams] = 2;
int values[t in Teams] = t;
var Teams team[Periods,EWeeks,Slots];
var Games game[Periods,Weeks];
struct Play { int f; int s; int g; };
{Play} Plays = { <i,j,(i-1)*nbTeams+j> | ordered i,
                                          j in Teams };
predicate link(int f,int s,int g) in Plays;

Play rr[Weeks,Periods];
initialize {
   rr[1,1].f = 1;
   rr[1,1].s = 2;
   forall(p in Periods :  p > 1) {
      rr[1,p].f = p+1;
      rr[1,p].s = nbTeams - (p-2);
   };
   forall(w in Weeks:  w > 1) {
      forall(p in Periods) {
         if rr[w-1,p].f <> 1 then
            if rr[w-1,p].f = nbTeams then rr[w,p].f = 2
            else rr[w,p].f = rr[w-1,p].f + 1 endif
         else
            rr[w,p].f = rr[w-1,p].f;
         endif;
         if rr[w-1,p].s = nbTeams then rr[w,p].s = 2
         else rr[w,p].s = rr[w-1,p].s + 1 endif;
      }
   };
   forall(w in Weeks, p in Periods)
      if rr[w,p].f < rr[w,p].s then
         rr[w,p].g = nbTeams*(rr[w,p].f-1) + rr[w,p].s
      else
         rr[w,p].g = nbTeams*(rr[w,p].s-1) + rr[w,p].f
      endif;
};
{int} domain[w in Weeks] = { rr[w,p].g | p in Periods };
```

Figure 9.5 A Round-Robin Model for the Sport-Scheduling Model (Part I)

```
solve {
   forall(p in Periods & w in Weeks)
      game[p,w] in domain[w];
   forall(w in EWeeks)
      alldifferent(all(p in Periods & s in Slots)
                  team[p,w,s]) onDomain;
   alldifferent(game) onDomain;
   forall(p in Periods)
      distribute(occur,values,all(w in EWeeks & s in Slots)
                              team[p,w,s])
          extendedPropagation;
   forall(p in Periods & w in Weeks)
      link(team[p,w,home],team[p,w,away],game[p,w]);
};

search {
   forall(p in Periods) {
      generateSeq(game[p]);
      forall(po in Periods :  po > 1)
          generate(game[po,p]);
   };
};
```

Figure 9.6 A Round-Robin Model for the Sport-Scheduling Model (Part II)

nb. of teams	14	16	18	20	22	24	26	28	30
CPU time	3.91	4.97	1.00	6.41	10.36	11.81	45.66	36.2	42.38

Figure 9.7 Experimental Results for the Sport-Scheduling Model

defines the games played in a given week. This array is used in the constraint

```
game[p,w] in domain[w];
```

which forces the game variables of period p and of week w to take a game allocated to that week.

The model also contains a novel search procedure that consists of generating values for the games in the first period and in the first week, then in the second period and the second week, etc. Figure 9.7 depicts the experimental results for various numbers of teams (CPU times are given in seconds). It is possible to improve the model further by exploiting even more symmetries; see Régin (1998) for complete details.

9.4 JOB-SHOP SCHEDULING

One of the other significant features of OPL is its support for scheduling applications. OPL has a variety of domain-specific concepts for these applications that are translated into state-of-the-art algorithms. To name only a few, they include the concepts of activities, unary, discrete, and state resources, reservoirs, and breaks as well as the global constraints linking them.

Figure 9.8 describes a simple job-shop scheduling model. The problem is to schedule a number of jobs on a set of machines to minimize completion time, often called the *makespan*. Each job is a sequence of tasks and each task requires a machine. Figure 9.8 first declares the number of machines, the number of jobs, and the number of tasks in the jobs. The main data of the problem, i.e., the duration of all the tasks and the resources they require, are then given. The next set of instructions

```
ScheduleHorizon = totalDuration;
Activity task[j in Jobs, t in Tasks](duration[j,t]);
Activity makespan(0);
UnaryResource tool[Machines];
```

is most interesting. The first instruction describes the schedule horizon, i.e., the date by which the schedule should be completed at the latest. In this application, the schedule horizon is given as the summation of all durations, which is clearly an upper bound on the duration of the schedule. The next instruction declares the activities of the problem. Activities are first-class objects in OPL and can be viewed (in a first approximation) as consisting of variables representing the starting date, the duration, and the end date of a task, as well as the constraints linking them. The variables of an activity are accessed as fields of records. In our application, there is an activity associated with each task of each job. The instruction

```
int nbMachines = ...;
range Machines 1..nbMachines;
int nbJobs = ...;
range Jobs 1..nbJobs;
int nbTasks = ...;
range Tasks 1..nbTasks;
Machines resource[Jobs,Tasks] = ...;
int+ duration[Jobs,Tasks] = ...;
int totalDuration = sum(j in Jobs, t in Tasks) duration[j,t];

ScheduleHorizon = totalDuration;
Activity task[j in Jobs, t in Tasks](duration[j,t]);
Activity makespan(0);
UnaryResource tool[Machines];

minimize
   makespan.end
subject to {
   forall(j in Jobs)
      task[j,nbTasks] precedes makespan;
   forall(j in Jobs & t in 1..nbTasks-1)
      task[j,t] precedes task[j,t+1];
   forall(j in Jobs & t in Tasks)
      task[j,t] requires tool[resource[j,t]];
};

search {
   LDSearch() {
      forall(r in Machines ordered by
               increasing localSlack(tool[r]))
         rank(tool[r]);
   }
}
```

Figure 9.8 A Job-Shop Scheduling Model (`jobshop.mod`)

```
UnaryResource tool[Machines];
```

declares an array of unary resources. Unary resources are, once again, first-class objects of OPL; they represent resources that can be used by at most one activity at anyone time. In other words, two activities using the same unary resource cannot overlap in time. Note that the makespan is modeled for simplicity as an activity of duration zero.

Consider now the problem constraints. The first set of constraints specifies that the activities associated with the problem tasks precede the makespan activity. The next two sets specify the precedence and resource constraints. The resource constraints specify which activities require which resource. Finally, the search procedure

```
search {
    LDSearch() {
       forall(r in Machines ordered by
                increasing localSlack(tool[r]))
          rank(tool[r]);
    }
}
```

illustrates a typical search procedure for job-shop scheduling and the use of limited discrepancy search (LDS) (Harvey and Ginsberg (1995)) as a search strategy. The search procedure

```
forall(r in Machines ordered by increasing localSlack(tool[r]))
    rank(u[r]);
```

consists of ranking the unary resources, i.e., choosing in which order the activities execute on the resources. Once the resources are ranked, it is easy to find a solution. The procedure ranks first the resource with the smallest local slack (i.e., the machine that seems to be the most difficult to schedule) and then considers the remaining resource using a similar heuristic. The instruction `LDSearch()` specifies that the search space specified by the search procedure defined above must be explored using limited discrepancy search. This strategy, which is effective for many scheduling problems, assumes the existence of a good heuristic. Its basic intuition is that the heuristic, when it fails, probably would have found a solution if it had made a small number of different decisions during the search. The choices where the search procedure does not follow the heuristic are called *discrepancies*. As a consequence, LDS systematically explores the search tree by increasing the number of allowed discrepancies. Initially, a small number of discrepancies is allowed. If the search is not successful or if an optimal solution is desired, the number of discrepancies is increased and the process is iterated until a solution is found or the whole search space has been explored. Note that, besides the default depth-first search and LDS, OPL also supports best-first search, interleaved depth-first search, and depth-bounded limited discrepancy search. It is interesting to mention that this simple model solves MT10 in about 40 seconds and MT20 in about 0.4 seconds (see Muth and Thompson (1963) for an early description of these problem instances).

9.5 THE TROLLEY APPLICATION

This section illustrates the use of OPL for a more advanced scheduling application. It discusses abstractions such as discrete and state resources and transition times. It also shows how to use OPLSCRIPT to find good solutions quickly and demonstrates the use of code generation. To ease understanding, the application is presented in stepwise refinements starting with a simplified version of the problem and adding more sophisticated concepts incrementally.

9.5.1 *The Basic Model*

Consider an application where a number of jobs must be performed in a shop equipped with a number of machines. Each job corresponds to the processing of an item that needs to be sequentially processed on a number of machines for some known duration. Each item is initially available in area A of the shop. It must be brought to the specified machines with a trolley. After it has been processed on all machines, it must be stored in area S of the shop. Moving an item from area x to area y consists of (1) loading the item on the trolley at area x; (2) moving the trolley from area x to area y and (3) unloading the item from the trolley at area y. The goal is to find a schedule minimizing the makespan. In this version of the problem, we ignore the time to move the trolley and we assume that the trolley has unlimited capacity. Subsequent sections will remove these limitations.

The specific instance considered here consists of six jobs, each of which requires processing on two specified machines. As a consequence, a job consists of eight tasks:

1. Load the item on the trolley at area A.
2. Unload the item from the trolley at the area of the first machine required by the job.
3. Process the item on the first machine.
4. Load the item on the trolley at the area of this machine.
5. Unload the item from the trolley at the area of the second machine required by the job.
6. Process the item on the second machine.
7. Load the item on the trolley at the area of this machine.
8. Unload the item from the trolley at area S.

Figures 9.9 and 9.10 depict an OPL model for this problem, while Figure 9.11 describes the instance data.

The statement starts by defining the set of jobs, the set of tasks to be performed by the jobs, and the possible locations of the trolley. As can be seen in the instance data (see Figure 9.11), the tasks correspond to the description given previously. The trolley has five possible locations, one for each available machine, one for the arrival area, and one for the storage area. The statement then defines the machines and the data for

```
enum Jobs ...;
enum Tasks ...;
enum Location ...;
{Location} Machines = ...;
struct jobRecord {
   Location machine1;
   int durations1;
   Location machine2;
   int durations2;
};
jobRecord job[Jobs] = ...;
int loadDuration = ...;

Location location[Jobs,Tasks];
initialize {
   forall(j in Jobs) {
      location[j,loadA] = areaA;
      location[j,unload1] = job[j].machine1;
      location[j,process1] = job[j].machine1;
      location[j,load1] = job[j].machine1;
      location[j,unload2] = job[j].machine2;
      location[j,process2] = job[j].machine2;
      location[j,load2] = job[j].machine2;
      location[j,unloadS] = areaS;
   };
};

int duration[Jobs,Tasks];
initialize {
   forall(j in Jobs) {
      duration[j,loadA] = loadDuration;
      duration[j,unload1] = loadDuration;
      duration[j,process1] = job[j].durations1;
      duration[j,load1] = loadDuration;
      duration[j,unload2] = loadDuration;
      duration[j,process2] = job[j].durations2;
      duration[j,load2] = loadDuration;
      duration[j,unloadS] = loadDuration;
   }
};
```

Figure 9.9 The Trolley Problem: Part I

```
scheduleHorizon = 2000;
UnaryResource machine[Machines];
StateResource trolley(Location);
Activity act[i in Jobs,j in Tasks](duration[i,j]);
Activity makespan(0);

minimize
   makespan.end
subject to {
   forall(j in Jobs & ordered t1, t2 in Tasks)
      act[j,t1] precedes act[j,t2];
   forall(j in Jobs) {
      act[j,process1] requires machine[job[j].machine1];
      act[j,process2] requires machine[job[j].machine2];
   };
   forall(j in Jobs,
          t in Tasks :  t <> process1 & t <> process2)
      act[j,t] requiresState(location[j,t]) trolley;
   forall(j in Jobs)
      act[j,unloadS] precedes makespan;
};

search {
   setTimes(act);
};
```

Figure 9.10 The Trolley Problem: Part II

```
Jobs = {j1,j2,j3,j4,j5,j6};
Tasks = {loadA,unload1,process1,load1,
         unload2,process2,load2,unloadS};
Location = {m1,m2,m3,areaA,areaS};
Machines = { m1, m2, m3 };
job = [
   <m1,80,m2,60>, <m2,120,m3,80>, <m2,80,m1,60>,
   <m1,160,m3,100>, <m3,180,m2,80>,<m2,140,m3,60> ];
loadDuration = 20;
```

Figure 9.11 The Trolley Problem: The Instance Data

the jobs, i.e., it specifies the two machines required for each job and the duration of the activities to be performed on these machines. The machines are identified by their locations for simplicity. The statement also specifies the duration of a loading task, which concludes the description of the input data.

The remaining instructions in Figure 9.9 specify derived data that are useful in stating the constraints. The instruction

```
Location location[Jobs,Tasks];
initialize {
    forall(j in Jobs) {
       location[j,loadA] = areaA;
       location[j,unload1] = job[j].machine1;
       location[j,process1] = job[j].machine1;
       location[j,load1] = job[j].machine1;
       location[j,unload2] = job[j].machine2;
       location[j,process2] = job[j].machine2;
       location[j,load2] = job[j].machine2;
       location[j,unloadS] = areaS;
    };
};
```

specifies the locations where each task of the application must take place, while the next two instructions specify the durations of all tasks. The subsequent instructions, shown in Figure 9.10, are particularly interesting. The instruction

```
UnaryResource machine[Machines];
```

declares the machines of this application. Note that the array `machine` is indexed by a set of values. The instruction

```
StateResource trolley(Location);
```

defines the trolley as a state resource whose states are the five possible locations of the trolley. A state resource is a resource that can only be in one state at any given time: Hence any two tasks requiring a different state cannot overlap in time. The instructions

```
Activity act[i in Jobs,j in Tasks](duration[i,j]);
Activity makespan(0);
```

define the decision variables for this problem. They associate an activity with each task of the application and an activity to model the makespan. Note how the subscripts `i` and `j` are used in the declaration to associate the proper duration with every task. These generic declarations are often useful to simplify problem description. The rest of the statement specifies the objective function and the constraints. The objective function consists of minimizing the end date of the makespan activity. The instruction

```
forall(j in Jobs & ordered t1, t2 in Tasks)
    act[j,t1] precedes act[j,t2];
```

specifies the precedence constraints inside a job. It also illustrates the rich aggregate operators in OPL. The instruction

```
forall(j in Jobs) {
    act[j,process1] requires machine[job[j].machine1];
    act[j,process2] requires machine[job[j].machine2];
};
```

specifies the unary resource constraints, i.e., it specifies which task uses which machine. The instruction

```
forall(j in Jobs, t in Tasks :  t <> process1 & t <> process2)
    act[j,t] requiresState(location[j,t]) trolley;
```

specifies the state resource constraints for the trolley, i.e., it specifies which tasks require the trolley to be at a specified location. The instruction

```
forall(j in Jobs)
    act[j,unloadS] precedes makespan;
```

makes sure that the makespan activity starts only when all the other tasks are completed. The search procedure

```
search {
    setTimes(act);
};
```

is rather simple in this model and uses a procedure `setTimes(act)` that assigns a starting date to every task in the array `act` by exploiting dominance relationships. The solution produced by OPL for this application is of the following form:

```
act[j1,loadA] = [0 -- 20 --> 20]
act[j1,unload1] = [40 -- 20 --> 60]
act[j1,process1] = [60 -- 80 --> 140]
act[j1,load1] = [140 -- 20 --> 160]
act[j1,unload2] = [160 -- 20 --> 180]
act[j1,process2] = [380 -- 60 --> 440]
act[j1,load2] = [440 -- 20 --> 460]
act[j1,unloadS] = [460 -- 20 --> 480]
...
act[j6,unloadS] = [540 -- 20 --> 560]
makespan = [560 -- 0 --> 560]
```

It displays the starting date, the duration, and the completion time of each activity in the model.

9.5.2 Transition Times

Assume now that the time to move the trolley from an area to another must be taken into account. This new requirement imposes transition times between successive activities. In OPL, transition times can be specified between any two activities requiring the same unary or state resource. Given two activities a and b, the transition time between a and b is the amount of time that must elapse between the end of a and the beginning of b when a precedes b. Transition times are modelled in two steps in OPL. First, a transition type is associated with each activity. Second, a transition matrix is associated with the appropriate state or unary resource. To determine the transition time between two successive activities a and b on a resource r, the transition matrix is indexed by the transition types of a and b.

In the trolley application, since the transition times depend on the trolley location, the key idea is that each activity may be associated with a transition type that represents the location where the activity is taking place. For instance, task `unload1` of job `j1` is associated with state `m1` if the first machine of `j1` is machine 1. The state resource can be associated with a transition matrix that, given two locations, returns the time to move from one to the other. The model shown in the previous section can thus be enhanced easily by adding a declaration

```
int transition[Location,Location] = ...;
```

and by modifying the state resource and activity declarations to become

```
StateResource trolley(Location,transition);
UnaryResource machine[Machines];
Activity act[i in Jobs,j in Tasks](duration[i,j])
    transitionType location[i,j];
```

Using a transition matrix of the form

```
[
    [ 0, 50, 60, 50, 90 ],
    [ 50, 0, 60, 90, 50 ],
    [ 60, 60, 0, 80, 80 ],
    [ 50, 90, 80, 0,120 ],
    [ 90, 50, 80,120, 0 ]
];
```

would lead to an optimal solution of the following form:

```
act[j1,loadA] = [0 -- 20 --> 20]
act[j1,unload1] = [70 -- 20 --> 90]
act[j1,process1] = [90 -- 80 --> 170]
act[j1,load1] = [370 -- 20 --> 390]
act[j1,unload2] = [530 -- 20 --> 550]
act[j1,process2] = [550 -- 60 --> 610]
act[j1,load2] = [850 -- 20 --> 870]
act[j1,unloadS] = [920 -- 20 --> 940]
...
act[j6,unloadS] = [920 -- 20 --> 940]

makespan = [940 -- 0 --> 940]
```

9.5.3 *Capacity Constraints*

Consider now adding the requirement that the trolley has a limited capacity, i.e., it can only carry so many items. To add this requirement in OPL, it is necessary to model the trolley by two resources: a state resource as before and a discrete resource that represents its capacity. Several activities can require the same discrete resource at a given time provided that their total demand does not exceed the capacity. In addition, it is necessary to model the tasks of moving from a location to another. As a consequence, each job is enhanced by three activities that represents the move from area A to the first machine, from the first machine to the second machine, and from the second machine to area S. Each of these trolley activities uses one capacity unit of the trolley. The declarations

```
int trolleyMaxCapacity = 3;
DiscreteResource trolleyCapacity(trolleyMaxCapacity);
enum TrolleyTasks {onTrolleyA1,onTrolley12,onTrolley2S};
Activity tact[Jobs,TrolleyTasks];
```

serve that purpose. It is now important to state that these activities require the trolley capacity and when these tasks must be scheduled. The constraints

```
forall(j in Jobs, t in TrolleyTasks)
    tact[j,t] requires trolleyCapacity;
```

specify the resource consumption, while the constraints

```
forall(j in Jobs) {
    tact[j,onTrolleyA1].start = act[j,loadA].start;
    tact[j,onTrolleyA1].end = act[j,unload1].end;
    tact[j,onTrolley12].start = act[j,load1].start;
    tact[j,onTrolley12].end = act[j,unload2].end;
    tact[j,onTrolley2S].start = act[j,load2].start;
    tact[j,onTrolley2S].end = act[j,unloadS].end;
};
```

specify the temporal relationships, e.g., that the activity of moving from area A to the first machine in a job should start when the item is being loaded on the trolley and is completed when the item is unloaded. The trolley application is now completed and the final model is depicted in Figures 9.12 and 9.13. This last model in fact is rather difficult to solve optimally despite its reasonable size.

9.5.4 *Controlling Execution Time*

We now discuss how to control the execution of this model using OPLSCRIPT, a language for composing and controlling OPL models. OPLSCRIPT is particularly appropriate for applications that require solving several instances of the same model, a sequence of models, or a combination of both as in column-generation applications. See van Hentenryck and Michel (2000) for an overview of these functionalities. OPLSCRIPT can also be used for controlling OPL models in order to find good solutions quickly or

```
enum Jobs ...;
enum Tasks ...;
enum Location ...;
{Location} Machines = ...;
struct jobRecord {
   Location machine1;
   int durations1;
   Location machine2;
   int durations2;
};
jobRecord job[Jobs] = ...;
int loadDuration = ...;
int transition[Location,Location] = ...;
int trolleyMaxCapacity = ...;
Location location[Jobs,Tasks];
initialize {
   forall(j in Jobs) {
      location[j,loadA] = areaA;
      location[j,unload1] = job[j].machine1;
      location[j,process1] = job[j].machine1;
      location[j,load1] = job[j].machine1;
      location[j,unload2] = job[j].machine2;
      location[j,process2] = job[j].machine2;
      location[j,load2] = job[j].machine2;
      location[j,unloadS] = areaS;
   };
};
int duration[Jobs,Tasks];
initialize {
   forall(j in Jobs) {
      duration[j,loadA] = loadDuration;
      duration[j,unload1] = loadDuration;
      duration[j,process1] = job[j].durations1;
      duration[j,load1] = loadDuration;
      duration[j,unload2] = loadDuration;
      duration[j,process2] = job[j].durations2;
      duration[j,load2] = loadDuration;
      duration[j,unloadS] = loadDuration;
   }
};
```

Figure 9.12 The Final Trolley Model: Part I

```
scheduleHorizon = 2000;
UnaryResource machine[Machines];
StateResource trolley(Location);
DiscreteResource trolleyCapacity(trolleyMaxCapacity);
Activity act[i in Jobs,j in Tasks](duration[i,j])
   transitionType location[i,j];
Activity tact[Jobs,TrolleyTasks];
Activity makespan(0);

minimize
   makespan.end
subject to {
   forall(j in Jobs & ordered t1, t2 in Tasks)
      act[j,t1] precedes act[j,t2];
   forall(j in Jobs) {
      act[j,process1] requires machine[job[j].machine1];
      act[j,process2] requires machine[job[j].machine2];
   };
   forall(j in Jobs,
          t in Tasks :  t <> process1 & t <> process2)
      act[j,t] requiresState(location[j,t]) trolley;
   forall(j in Jobs, t in TrolleyTasks)
      tact[j,t] requires trolleyCapacity;
   forall(j in Jobs) {
      tact[j,onTrolleyA1].start = act[j,loadA].start;
      tact[j,onTrolleyA1].end = act[j,unload1].end;
      tact[j,onTrolley12].start = act[j,load1].start;
      tact[j,onTrolley12].end = act[j,unload2].end;
      tact[j,onTrolley2S].start = act[j,load2].start;
      tact[j,onTrolley2S].end = act[j,unloadS].end;
   };
   forall(j in Jobs)
      act[j,unloadS] precedes makespan;
};

search {
   setTimes(act);
};
```

Figure 9.13 The Final Trolley Model: Part II

```
Model m("trolley.mod","trolley.dat");
int fails := 1000;
m.setFailLimit(fails);
int solved := 0;
while m.nextSolution() do {
   solved := 1;
   cout << "solution with makespan:  " << m.objectiveValue() << endl;
   m.setFailLimit(m.getNumberOfFails() + fails);
}
m.setFailLimit(m.getNumberOfFails() + fails);
if solved then {
   m.restore();
   cout << "final solution with makespan:  "
         << m.objectiveValue() << endl;
}

cout << "Time:  " << m.getTime() << endl;
cout << "Fails:  " << m.getNumberOfFails() << endl;
cout << "Choices:  " << m.getNumberOfChoicePoints() << endl;
cout << "Variables:  " << m.getNumberOfVariables() << endl;
```

Figure 9.14 A Script for the Trolley Problem (`trolley.osc`)

to improve efficiency by exploiting more advanced techniques (e.g., shuffling in job-shop scheduling). This section illustrates how OPLSCRIPT can be used to find a good solution quickly on the final trolley application.

The motivation here is that it is sometimes appropriate to limit the time devoted to the search of an optimal solution by restricting the number of failures, the number of choice points, or the execution time. Figure 9.14 depicts a script for the trolley problem that limits the number of failures when searching for a better solution. The basic idea of the script is to allow for an initial credit of failures (say, i) and to search for a solution within these limits. When a solution is found with, say, f failures, the search is continued with a limit of $i + f$ failures, i.e., the number of failures needed to reach the last solution is increased by the initial credit. Consider now the script in detail. The instruction

```
Model m("trolley.mod","trolley.dat");
```

defines an OPLSCRIPT model in terms of its model and data files. Models are first-class objects in OPLSCRIPT: They can be passed as parameters of procedures and stored in data structures and they also support a variety of methods. For instance, the method `nextSolution` on a model can be used to obtain the successive solutions of a model or, in optimization problems, to produce a sequence of solutions, each of which improves the best value of the objective function found so far. The instruction

```
m.setFailLimit(fails);
```

specifies that the next call to `nextSolution` can perform at most `fails` failures, i.e., after `fails` failures, the execution aborts and `nextSolution()` returns 0. The instructions

```
while m.nextSolution() do {
    solved := 1;
    cout << "solution with makespan:  "
          << m.objectiveValue() << endl;
    m.setFailLimit(m.getNumberOfFails() + fails);
}
```

make up the main loop of the script and produce a sequence of solutions, each of which having a smaller makespan. Note the instruction

```
m.setFailLimit(m.getNumberOfFails() + fails);
```

that retrieves the number of failures needed since the creation of model `m` and sets a new limit by adding `fails` to this number. The next call to `nextSolution` takes into account this new limit when searching for a better solution. Note also the instruction `m.restore()` to restore the last solution found by `nextSolution()`. This script displays a result of the form

```
solution with makespan:  2310
solution with makespan:  2170
solution with makespan:  2080
...
solution with makespan:  1690
solution with makespan:  1650
solution with makespan:  1620
final solution with makespan:  1620
Time:  7.0200
Fails:  3578
Choices:  3615
Variables:  204
```

9.5.5 Code Generation

Once a reasonable model has been successfully designed in OPL, it can be integrated in a larger application through C++ code generation. This is one of the novel aspects of the development environment of OPL. It makes it possible to develop models at a very high level of abstraction, while allowing a smooth integration of the model in a larger context. In addition, the generated code directly refers to the modeling objects of the OPL model, thus simplifying the integration work. We briefly illustrate these features on the trolley application.

The basic idea behind code generation consists of producing a C++ class associated with each object in the model and a top-level class for the model. In other words, the generated code is specialized to the model and is strongly typed. Theses classes can then be used to access and modify the data and, of course, to solve the model

```
int main(int argc,char* argv[])
{
   IloSolver_trolleyComplete solver;
   if (solver.nextSolution()) {
      IloArray_act act = solver.get_act();
      IloEnum_Jobs Jobs = solver.get_Jobs();
      IloEnum_Tasks Tasks = solver.get_Tasks();
      IloEnumIterator_Jobs iteJobs(Jobs);
      while (iteJobs.ok()) {
         IloEnumIterator_Tasks iteTasks(Tasks);
         while (iteTasks.ok()) {
            cout << "act[" << *iteJobs << "," << *iteTask << "]=" ;
            cout << act[*iteJobs][*iteTask] << endl;
            ++iteTasks;
         }
         ++iteJobs;
      }
      cout << endl;
   }
   solver.end();
   return 0;
}
```

Figure 9.15 C++ Code for the Trolley Problem

and collect the results. Figure 9.15 depicts C++ code to obtain the first solution to the trolley application and to display the main activities. Instruction

```
IloSolver_trolleyComplete solver;
```

defines an instance `solver` that encapsulates the functionality of the trolley model. The class definition is available in the `.h` that is not shown here. The instruction

```
IloArray_act act = solver.get_act();
```

is used to obtain the array of activities, while the instructions

```
IloEnum_Jobs Jobs = solver.get_Jobs();
IloEnum_Tasks Tasks = solver.get_Tasks();
IloEnumIterator_Jobs iteJobs(Jobs);
```

obtain the enumerated types and define an iterator for the jobs. The remaining instructions iterate over the enumerated sets and display the activities.

9.6 RESEARCH DIRECTIONS

One of the main contributions of OPL is to merge, inside a modeling language, the high-level functionalities of mathematical programming modeling languages and the rich constraint and search languages of constraint programming.

```
void magic(int n)
{
   ModCPManager mgr;
   ModIntRange Dom(mgr,0,n-1);
   ModIntVarArray s(mgr,Dom,Dom,"s");
   ModIntParameter i(mgr), j(mgr);

   mgr.post(   ModForall(i,Dom,s[i] == ModSum(j,Dom,s[j]==i)));
   mgr.post(ModSum(j,Dom,s[j]*j)==n);

   if (mgr.newSearch().search(ModGenerate(s)))
      cout << s << endl;
}
```

Figure 9.16 Magic Series in `Modeler++`

Our current work on `Modeler++` shows how to integrate these orthogonal features inside object-oriented constraint libraries. `Modeler++` is a C++ modeling layer for constraint programming which combines the salient characteristics of both modeling approaches and which allows C++ statements that are almost as concise as `OPL` models. The rest of this section illustrates `Modeler++` on two applications. See Michel and van Hentenryck (2001b) for a more comprehensive coverage of `Modeler++`.

9.6.1 The Magic Series Problem

The magic-series problem is a traditional constraint programming benchmark. The problem consists of finding a magic series, i.e., a sequence of numbers $S = (s_0, s_1, \dots, s_n)$ such that s_i represents the number of occurrences of i in S. For instance, $(1, 2, 1, 0)$ is a magic series of size 4, since there is one occurrence of 0, two occurrences of 1, one occurrence of 1 and zero occurrences of 3. Figure 9.16 depicts a solution to the magic series problem. It introduces the use of parameters and aggregate operators in `Modeler++`. The instruction

```
ModIntParameter i(mgr), j(mgr);
```

declares two integer parameters `i` and `j` to be used in aggregate operators. The instruction

```
mgr.post(ModForall(i,Dom,s[i] == ModSum(j,Dom,s[j]==i));
```

states all cardinality constraints using a `ModForall` operator. It states constraints of the form

```
s[i] == ModSum(j,Dom,s[j]==i)
```

where `ModSum(j,Dom,s[j]==i)` counts the occurrences of `i` in `s`. The instruction

```
mgr.post(ModSum(j,Dom,s[j]*j)==n);
```

reuses parameter `j` to state the redundant constraint typically found in this problem. Finally, the instruction

```
mgr.newSearch().search(ModGenerate(s))
```

explores the search tree specified by `ModGenerate(s)` which assigns values to variables using the first-fail principle. Observe the similarity between the `Modeler++` and `OPL` statements (van Hentenryck (1999)): the main difference is the declaration of the parameters needed in C++.

9.6.2 Sport Scheduling

We now turn to the sport-scheduling problem presented earlier in this paper. Figure 9.17 presents a function to express the sport-scheduling model of Figure 9.4. Once again, there is almost a one-to-one correspondence between the `OPL` and `Modeler++` statements. The instruction

```
ModTupleSet setGames(
    ModSetCollect(ModOrdered(i,j),Teams,
                  ModTuple(i,j,(i-1)*n+j)));
```

computes a set of tuples of the form `<i,j,nb>` where `i` and `j` are, respectively, the home and away teams of game `nb`. The instructions

```
ModIntVarMatrix<3> teams(mgr,Periods,EWeeks,Slots,
                         Teams,"teams");
ModIntVarMatrix<2> games(mgr,Periods,Weeks,Games,"games");
```

declare the two arrays of variables. Observe that `teams` is a three-dimensional array. The constraint

```
ModAlldifferent(ModCollect(p,Periods,s,Slots,teams[p][w][s]))
```

specifies that all the teams scheduled to play on week `w` must be different. It uses an aggregate operator `ModCollect` to collect the appropriate team variables by iterating over the periods and the slots. The constraint

```
ModExactly(2,Teams,ModCollect(w,EWeeks,s,Slots,teams[p][w][s]))
```

specifies that a team plays exactly twice over the course of the "extended" season. It specifies that the values in `Teams` (second argument) must occur twice (first argument) in teams playing in period `p` (third argument). The constraint

```
ModAlldifferent(games)
```

specifies that all games are different, i.e., that all teams play against every other team. Finally, the constraint

```
void sport(int n)
{
   int np = n/2;
   ModCPManager mgr;
   ModIntParameter i(mgr), j(mgr), p(mgr), w(mgr), s(mgr);
   ModIntRange Teams(mgr,1,n);
   ModIntRange Weeks(mgr,1,n-1);
   ModIntRange EWeeks(mgr,1,n);
   ModIntRange Periods(mgr,1,np);
   ModIntRange Slots(mgr,0,1);
   ModIntRange Games(mgr,1,(n-1)*n);

   ModTupleSet setGames(
      ModSetCollect(ModOrdered(i,j),Teams,ModTuple(i,j,(i-1)*n+j-1)));
   ModIntVarMatrix<3> teams(mgr,Periods,EWeeks,Slots,Teams,"teams");
   ModIntVarMatrix<2> games(mgr,Periods,Weeks,Games,"games");

   mgr.post(
      ModForall(w,EWeeks,
         ModAlldifferent(ModCollect(p,Periods,
                     s,Slots,teams[p][w][s]))));
   mgr.post(
      ModForall(p,Periods,
         ModExactly(2,Teams,ModCollect(w,EWeeks,
                     s,Slots,teams[p][w][s]))));
   mgr.post(ModAlldifferent(games));
   mgr.post(
      ModForall(p,Periods,ModForall(w,Weeks,
         ModElementOf(teams[p][w][0],teams[p][w][1],
                     games[p][w],setGames));
   mgr.post(
      ModForall(p,Periods,
         ModForall(w,EWeeks,
                    teams[p][w][0] < teams[p][w][1]);

   if (mgr.newSearch().search(ModAnd(ModGenerate(games),
                                     ModGenerate(teams))))
      cout << teams << endl;
}
```

Figure 9.17 A Simple Model for the Sport-Scheduling Model in `Modeler++`

```
ModElementOf(teams[p][w][0],teams[p][w][1],games[p][w],setGames)
```

is most interesting. It specifies that the game `game[p][w]` consists of the teams `team[p][w][0]` and `team[p][w][1]`. More precisely, it ensures that the tuple

```
<teams[p][w][0],teams[p][w][1],games[p][w]>
```

belongs to the set `setGames`, effectively linking the team and the game variables. The search procedure in this statement is extremely simple and consists of generating first values for the games. Note also that generating values for the games automatically assigns values to the teams by constraint propagation, except for the teams in the dummy week.

It is worth emphasizing the conciseness of the statement. As a consequence, it seems that many high-level modeling features, which were found significant in modeling languages, will find their way into constraint programming languages and libraries in the future.

9.7 CONCLUSION

The purpose of this paper was to review, through five applications, a number of constraint programming features of OPL to give a basic understanding of the expressiveness of the language. These features include very high-level algebraic notations and data structures, a rich constraint programming language supporting logical, higher-level, and global constraints, support for scheduling and resource allocation problems, and search procedures and strategies. The paper also introduced briefly OPLSCRIPT, a script language to control and compose OPL models, and the code generation features. These five applications should give a preliminary, although very incomplete, understanding of how OPL may decrease development time significantly and complement existing modeling tools for combinatorial optimization.

10 GENETIC ALGORITHM OPTIMIZATION SOFTWARE CLASS LIBRARIES

Andrew R. Pain and Colin R. Reeves

School of Mathematical and Information Sciences,
Coventry University,
Priory Street, Coventry CV1 5FB,
United Kingdom
andrew.pain@virgin.net, C.Reeves@coventry.ac.uk

Abstract: Software libraries provide more-or-less adaptable building blocks for application-specific software systems. Optimization libraries in particular provide useful abstractions for manipulating algorithm and problem concepts. Such libraries are usually built and applied using object-oriented software technology. To enable portability and assist in adaptation, these libraries are often provided at source code level. These class libraries support the development of application-specific software systems by providing a collection of adaptable classes that are intended to be reused.

In this chapter, the definition and typical requirements for a *genetic algorithm* class library are established, and a number of interesting software products for this task are evaluated. A brief survey of a variety of genetic algorithm optimization software products is presented, including some that are not actual class libraries, although they are constructed with similar objectives.

10.1 INTRODUCTION

10.1.1 Genetic Algorithms

The *genetic algorithm* (GA) was developed by John Holland at the University of Michigan in the 1960s to mimic some of the features of natural evolution. Although optimization was not initially the main focus of attention for a GA, as is clear from Holland's book (Holland (1975)), the study of its effectiveness for such tasks by DeJong (1975), and subsequent research in the 1980s showed that GAs were very useful tools for optimization. This early work is comprehensively surveyed by Goldberg (1989). For a modern statement of the position of GA research, see Reeves and Rowe (2002).

For the purposes of this book, it is sufficient to observe that genetic algorithms are techniques that constantly adapt a population of solutions in order to obtain increasingly better solutions to complex problems. Like other heuristic search algorithms (Reeves (1993)), GAs use an evaluation function to link the problem with the search technique; however, it differs from most other techniques in that it searches a population of strings, which it does using genetic *operators* such as crossover and mutation. In GA terminology, the encoded solution to a problem is commonly known as a *chromosome*, while the variables within the chromosome are known as *genes*. The possible values of the genes are called *alleles* and the position of a gene within a chromosome is referred to as its *locus*.

The basic technique shown in Figure 10.1 is really just a high-level overview, since there are many sub-algorithms that must also be specified: the selection process, the genetic operators (crossover, for example, can take many forms), the acceptance of the children into the population, the way in which the population evolves and the termination conditions for the search. Further, this really describes a *steady-state* GA, where only some parents are replaced at each iteration. Although some workers recommend using the steady-state version for optimization, *generational* GAs, where a new population of children is created to supplant the parents, are also popular.

```
Initialize a population of chromosomes
Evaluate each chromosome
Repeat
   Select two parent chromosomes
   Create children using crossover and mutation operators
   Delete members of the population and add new children
   Evaluate the new child chromosomes
Until the termination condition is true
```

Figure 10.1 The Steady-State Genetic Algorithm

Some of the more common mechanisms and operators for producing more robust algorithms, or for specific performance enhancements, as described by Davis (1991), include:

- Conversion of natural objective function values into fitness values

- Population control using elitism and generational replacement
- Steady-state reproduction without duplicates
- Modification of parameters such as population size and operator probabilities
- Competition amongst multiple operators
- Modifications of the crossover operator such as uniform two-point crossover and multi-point crossover
- The use of interpolation to modify the algorithm's parameters dynamically

Similarly, Goldberg (1989) discusses some alternative selection mechanisms:

- Deterministic sampling
- Remainder stochastic sampling without replacement
- Stochastic sampling without replacement
- Remainder stochastic sampling with replacement
- Stochastic sampling with replacement
- Stochastic tournament selection (Wetzel ranking)

These brief lists illustrate the fact that an apparently simple algorithm has been, and continues to be, the subject of much research. Goldberg (1989) has observed that the early period of GA research consisted of many attempts to find the ultimate robust search algorithm by using a variety of operators and complex mechanisms. Since the effects of tailoring a GA to specific applications have become understood, more recent research has focused on the applications of GAs to a diverse range of problems. A recent survey (Reeves (1997)) provides an extensive list of published applications in optimization. As more uses are found, it becomes ever more important for a readily available implementation of a GA, such as a class library.

10.1.2 Languages, Classes, Libraries and Frameworks

10.1.2.1 Programming Languages and Libraries. A programming language can make some tasks easy, others difficult and yet others impossible. In this way, the language imposes restrictions, discourages the programming of certain tasks and encourages the programmer to take the "easy route". These programming practices can be relaxed by using languages in which every task is equally easy to program. Most successful languages are somewhere in the middle, balancing the encouragement (or enforcement) of good programming practices with the flexibility to program almost any task. The languages that are suitable for the tasks here are those that support object-oriented programming, with the most common language of choice being C++ (cf. Stroustrup (1997)). There is, however, a growing acceptance of the benefits of more stringent languages (since they restrict the programmer into following good programming practices), highlighted by the popular transition from C++ to Java (Horstmann and Cornell (1998)).

Early programming languages (such as PASCAL in 1971) provided all of the functions that the designers considered as being required for all uses of the language, and this eventually led to an explosion of built-in functions. Since then, there has been a move to take specific functionality out of the language (such as with C in 1978). Modular languages thus began to appear (including Modula-2 in 1982) in which function libraries were used – such libraries could be modified without requiring any changes to the programming language, and additional libraries could be added. More recent implementations and developments have seen the rise of object-oriented languages (Smalltalk in 1980, C++ in 1986, Eiffel in 1990 and Java in 1995), often providing general purpose and operating system specific classes in place of (or in addition to) the function libraries of old. The driving force behind the advances in programming languages has been due to many things, including the reuse of code, the ease of use, the speed of development, the implementation of the latest programming methodologies, advances in hardware and operating system design and (as always) experience.

The current situation and immediate future point to the use of component oriented systems that build on existing object-oriented approaches to provide reusable software components that are easily incorporated into large applications (Szyperski (1998)).

10.1.2.2 Classes and Objects. Classes form the basis of object-oriented software technology – an approach that is considered to be the leading methodology for programming (Rumbaugh et al. (1991)). Many concepts are incorporated into this technology, including abstraction, encapsulation, inheritance, and polymorphism, leading to better understanding of requirements, cleaner designs, reusable code and more easily maintainable systems. An object is an application-specific element that has some unique identity and is thus distinguishable from other objects by its existence, not by the properties that it may have. An object has a state, possibly persistent, and it encapsulates its state and behaviour. A class describes a group of related objects, such as objects with common behaviour, properties and/or logical relationships. For example, a chromosome class will contain methods for initialization and manipulation of a chromosome. These methods are applicable to all chromosome objects. A particular chromosome object will represent a specific chromosome – i.e., it will contain the data that describes one particular chromosome exactly.

A recent addition to the class concept is the use of template classes. A template class provides a set of functionalities as do standard classes, but operates on multiple data types. Thus instead of requiring a separate standard class for each data type, just one template class with multiple type parameters is all that is required. The data type is simply passed as a parameter when the class is instantiated – thus all of the benefits of polymorphism are provided without the need for a multi-class hierarchy. In this way, template classes can be reused more efficiently than standard classes. Consider a chromosome template class. Unlike a standard class that will restrict the data type of a chromosome to one form (such as a bit string or an integer array), a single template class can operate on a variety of different data types, replacing many standard classes and so reducing the amount of code (which in turn reduces errors and maintenance).

10.1.2.3 Class Libraries. Class libraries are collections of related classes that are intended to be used and reused by any and all applications that require their functionality. Many implementations of class-based programming languages will contain built-in class libraries for common functionality such as abstract data types (strings, lists, queues, trees, hash-tables, etc.), graphical components (dialogs, frames, panels, buttons, labels, etc.) and file processing (input/output streams, buffering, character/binary specific file processing, etc.). These class libraries contain functionally related building blocks that can be used as presented, extended or (implementation allowing) replaced. These building blocks must be sufficiently general to make them useful as reusable elements in many different applications. Broadly speaking, the more adaptable that a class is, the higher its chances of being used, yet a simple class may be chosen over a more comprehensive but complex class if the simple class is sufficiently functional for the desired task. Additionally, the more complex a class is, the higher the learning curve to use it effectively – although the usual case is that simpler sub-classes will be available, providing a reduced interface that is tuned to a specific aspect of the original class. For example, an all-encompassing file reading class may have many functions for reading all types of primitive data types and more complex objects, functions for buffering, for describing the modes in which the file is opened in (binary, text or object), functions to describe how the file is shared by other processes, etc. A far simpler sub-class could handle the reduced functionality required for reading simple text files, with one mode of opening, and a few simple functions for reading bytes, strings, lines or arrays of characters from the file.

10.1.2.4 Class Frameworks. Frameworks are a further development of class libraries. They must additionally provide support for extending the classes, and they must guide the design of the resulting applications. A framework consists of a set of cooperating classes that comprise a reusable design that is specific to a particular concept. Frameworks keep a number of their classes open for implementation inheritance (i.e., they can be sub-classed). Some classes may be abstract, requiring sub-classing to make the framework complete. It is common, though, to provide default implementations of all classes to reduce the burden on lightweight usage of the framework. Thus a user simply replaces the defaults that are inappropriate for the application.

Sub-classing tends to require knowledge of the super-class implementations, which is often called "white box" reuse. Conversely, a framework based on forwarding and relying on the interfaces of the objects within it is called "black box" reuse.

The main role of a framework is its regulation of the interaction that parts of the framework engage in – inter-operation aspects are fixed, so that the creation of specific solutions out of the partially complete framework becomes significantly faster. The main disadvantages of frameworks over class libraries are the need to understand how they are used and the possible performance implications that may be imposed.

10.1.2.5 Summary. The goals of libraries, class libraries, template libraries and frameworks are very similar, with each evolution being better than the previous one for their particular use. For instance, there is a significant reduction in programming effort, the application will be far more robust, there will be a greatly reduced code size and they are easier to support and maintain.

Class libraries provide adaptable building blocks for application-specific software systems, using object-oriented software technology. Some software is provided in source code form to enable adaptation by directly modifying the source code and rebuilding the application. A class library specifically provides a collection of adaptable classes for modification, expansion and general reuse. The main disadvantages are that they can be complex to construct correctly, they are more difficult to understand than procedural programming and can lead to loss of performance rather than the expected gain if badly implemented (such as using a large class hierarchy of calls), and they can be difficult to debug (owing to polymorphism – which method is actually being called?).

10.1.3 Genetic Algorithms Class Libraries

So, what do we expect to see in a genetic algorithm class library? Clearly, there must be classes to encapsulate the problem definition, to define the string encoding and to handle a user-defined evaluation function. There must also be a class to allow the GA to search the space until some specified termination condition has been reached. This is the bare minimum – such a class library hides the actual GA code, allowing flexibility in the problem specification only. The next step up from this would be to provide the source code for the GA class, so that at least the GA can be altered, albeit in an inflexible way. The other extreme case would be to have classes that cover a large number of functions, breaking down the GA into small elements and encapsulating each small element in its individual class. Mechanisms for incorporating the user's own classes to add to or override the standard classes will also be required. Additional classes for providing a GUI, search analysis and for defining complex evaluation and string encoding would also be required. A class could encapsulate a complete genetic algorithm – useful in applications where a "black box" search technique is all that is required, but somewhat inadequate for any experimentation, expansion or enhancement. It is better to follow good object-oriented design guidelines (Rumbaugh et al. (1991)) and encapsulate small elements into separate classes.

A very simple GA class library could contain classes that encapsulate the main data items (chromosomes and populations) and the search algorithm itself (see Figure 10.2). The chromosome class could be a base class from which specific types of chromosome classes are derived (such as bit strings or permutations). Alternatively, the chromosome class could remain as it is but the data it contains could be designed as an array of gene classes. A gene base class would then be used to derive specific types of gene. A template class could also be used to achieve the same effects. Other classes could be created: for example, a crossover class could allow the type of crossover to be selected (n-point, uniform etc.), and the class could have methods for querying where these crossover points occurred.

There are few guidelines to suggest the best way of representing a particular task using objects and classes. Generally, an object encapsulates a set of data and a set of methods that relate to the data, based on one particular entity (such as a person, or a binary string). However, some entities do not have clear boundaries and not all entities

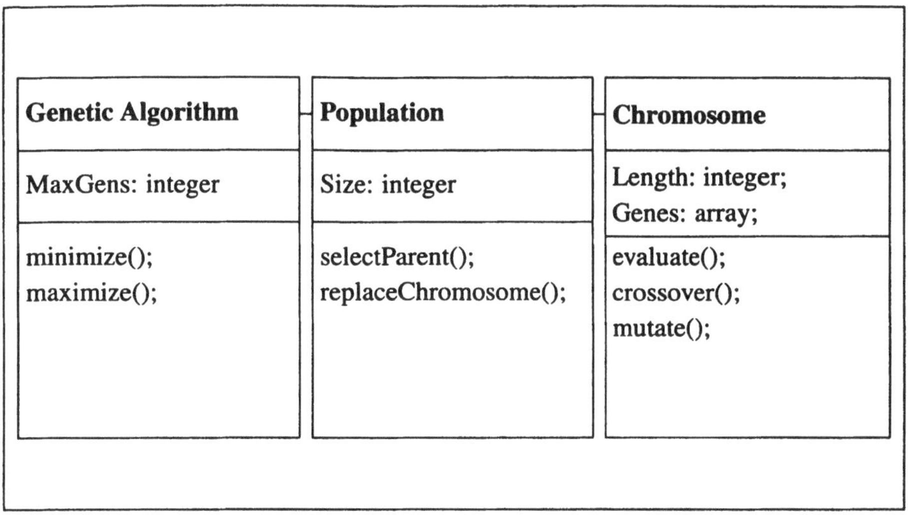

Figure 10.2 Simple Genetic Algorithm Classes

have a set of data and methods, so where do they fit? Consider the `crossover` method. It performs an operation on a string, but it could be an optional function of a particular algorithm (in this case the genetic algorithm). The same string could be used for local search which does not have crossover functions, so the crossover method could belong to the genetic algorithm object; but perhaps it should be part of a genetic operators object. There are many alternative options and so nearly every class library will be different. As a consequence, some will perform better at certain tasks than others, although this may not even be apparent to the creator until the library has been completed!

The provision of source code is not required and it is debatable as to whether or not it is beneficial, although certain schools of thought do see it as extremely useful (Sing (1997)). Many class libraries in existence only expose an interface to the underlying functionality, and in most cases the code is proprietary anyway. The main reason that many class libraries are available as source code is for portability, so that the source can be compiled on whatever platform the user requires, although some platform-specific changes to the source may be necessary.

Genetic algorithms are computational in nature and must therefore be coded to be used. They are used in many subject areas for many problem solving purposes and for algorithm research. It seems then rather a waste of time if everyone who wants to use a GA has to write one from scratch (and since the areas of use include groups of people who typically are not expert computer programmers, this could be some task!). The basic GA is well described, robust and a good starting point for any problem solving or algorithm experimentation. Access to pre-existing code is very beneficial in that it was probably written by someone knowledgeable in GAs and computer programming, it has had extensive testing and it is being used by many people thus making it easier to share extensions. Reusable code in the form of classes has additional benefits: classes form the basis of object-oriented software technology, which is generally thought of as a very good programming model, and they can be adapted or expanded using inheritance and polymorphism.

10.1.4 Using Genetic Algorithm Class Libraries

Typically, a class library would be made available to the user either with or without source code, and usually accompanied by some documentation and some examples. If source code is not supplied, there will certainly be limits on the systems that can be used with the library. Libraries with source code almost always need to be compiled on the user's own system, so that additional problems may arise if there are any differences between this system and those with which the class library has been tested.

To use any of the class libraries, the user must supply an objective function to compute the values of the solutions to their specific problem. Almost all of the class libraries will require the user to define the form of the chromosome (such as a bit string or permutation), while some systems will limit the chromosome to a specific type. The choice of chromosome will usually involve selecting an appropriate class from those supplied within the class library. Finally, an application will need to be built to create, initialize and configure the classes, usually following a specific sequence as guided by the class library implementation, documentation and/or examples (see Figure 10.3).

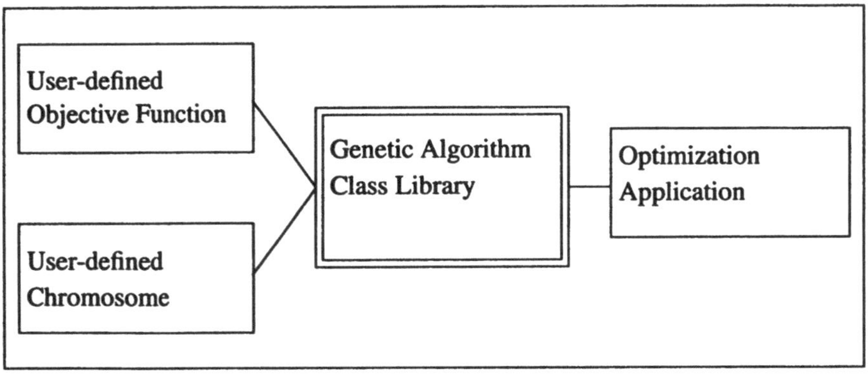

Figure 10.3 Basic Usage of a Typical Genetic Algorithm Class Library

If the default GA does not fit the requirements of the user, there are two methods that can be followed to make all necessary alterations. The easiest changes are made by substituting classes for alternatives that are supplied within the class library, and by using the configuration, parameters and methods of the classes to alter their behaviour. If the required modifications are not available, users must create their own methods and/or classes to achieve their objectives. Most class libraries will have base classes that are provided for users to derive their own versions of that type of class. For example, a base class for parent selection may be supplied, along with a number of classes that provide standard implementations (such as roulette wheel selection, or stochastic universal selection; Hancock (1994)). The base class provides a standard interface through which other classes use the selection technique, so that alternatives can be substituted without requiring any changes elsewhere. Thus the user's own class can be derived from the appropriate base class, and the class library can be instructed to use this new class.

A high-quality class library will provide a good variety of alternative techniques for the various parts of a GA (such as parent selection, genetic operators and recombination), and most of these will also contain configuration parameters to permit alteration of various aspects (such as the mutation rate or the number of points in a multi-point crossover operator). Thus, it is possible to produce an efficient GA application using object-oriented methodologies with a fairly small amount of work being required from the user, achieving a significant amount of code reuse through the utilization of the class library.

10.1.5 Genetic Algorithm Class Library Software

The field of genetic algorithm optimization software is still in its infancy: real-world applications only began emerging in the early 1990s, and since then advances in GA research and in programming languages and programming techniques have meant that the research, development and implementation of reusable software has been in constant evolution.

For these reasons, the software available is generally not as advanced as that in other more established areas such as operating systems, word processors, graphics and audio applications, which are driven by industry demands and a ready-made market. The mainstream is still new to the idea of GAs (perhaps the concept of a computer only being able to get a "good" solution is hard to comprehend for those with little knowledge of optimization problems), but major companies are starting to use them. For example, the UK based television company Channel 4 is using software based on GAs for advert scheduling (Liston (1993)), while Phillips used a GA for configuring their Fast Component Mounter robot for printed circuit board assembly, saving large sums of money (Schaffer and Eshelman (1996)).

Many people have been involved in programming GAs, using a wide range of programming languages, styles and techniques. The Internet contains a significant amount of this work, the vast majority of which has been written by students and individuals who have an interest in this subject. The more complex and more comprehensive software usually comes from two sources: academic groups or commercial enterprises. While the academics provide a significant amount of the work, the quality of the software is variable: in nearly all cases the software proves a point and successfully demonstrates the aspect that it was originally designed for, but the overall quality and usefulness of the finished software is not always apparent. Commercial efforts usually aim their particular software at specific tasks, although some more general applications can be found. The benefit of using commercial software is that a level of support is received that is not available elsewhere, while the software tends to be more "polished" and the documentation is often of a higher standard. There are, of course, exceptions to these generalizations, the details of which can be found over the following sections of this chapter.

The software packages that are detailed below illustrate some of the current class libraries that are readily available now, almost all of which are easily accessible via the Internet. The next section details several applications that have been written in C++, described in an increasing order of complexity. This is followed by a look at a class library written in Java, including a comparison of the significant differences that these two languages make when implementing a GA class library. A brief survey of other class library software follows, along with examples of software that does not use a class library. The final section provides a summary and an attempt is made to describe the developments that may be made in the future.

10.2 C++ CLASS LIBRARY SOFTWARE

10.2.1 GAGS: Genetic Algorithm from Granada, Spain

The GAGS application (Merelo (1994), Merelo (1995), GAGS (2001)) offers a C++ class library of genetic algorithm functions. It has been designed specifically for Unix and requires several Unix utilities to operate. GAGS has primarily been written as a GA application generator, and uses Unix scripting utilities to build source code from the user's requirements. The software is supplied in source code form and has been successfully used on Sun Solaris and Silicon Graphics machines. GAGS has been designed to operate at three levels:

- User Level: GAGS is run as an application generator, simply requiring an objective function and configuration details
- Programmer Level: the software is used as a standard class library
- Wizard Level: the software is used as a basis for user-defined extensions

10.2.1.1 User Level. GAGS can be used as a GA application generator: the user must write a fitness function to evaluate the chromosomes, and then run one of the supplied scripts (providing answers to questions about the genetic algorithm), which creates the required source files, compiles the source and runs the application.

The fitness function must be written using C, following some specified rules (see Merelo (1994)). The application generator builds a source file that includes the standard `main()` function and the user-defined fitness function and calls to the class library that are required for the GA. It compiles, and optionally runs, the newly created GA, so that the user can use GAGS as a black-box engine to solve any suitable optimization problem. Configuration scripts can be created for setting the parameters of the GA. There are two options for this: a command-line based PERL script (Wall et al. (1991), PERL (2001)) can be used, or a TCL/TK script (Welch (1997), Raines and Tranter (1999), TCL/TK (2001)) will provide the same level of parameter editing, but in a graphical form. The parameters that the user can adjust include the following:

- The number of bytes per gene (chromosomes are defined as bit-strings divided into a number of genes, each of which having the same length)
- The gene parameter ranges, either fixed for all genes or set for each individual gene
- The mutation rate (crossover always occurs)
- The population size
- The maximum number of generations for termination
- The selection method (roulette-wheel, elitist or tournament selection)
- The output interval (for statistics), and whether to save the best chromosome or its fitness
- The input file (for training) and the output file (for saving recorded data)
- Options for using the Gnuplot utility (Gnuplot (2001))

10.2.1.2 Programmer Level. The libGAGS class library has been written to match closely the way in which GAs work (Merelo (1995)). The class library is divided into three sets of classes (see Figure 10.4):

- The chromosome hierarchy
- The population hierarchy
- Auxiliary classes (for system and user interaction)

The Chromosomes Hierarchy. To follow the natural philosophy of GAs, a chromosome has been defined as a set of genes created at birth, and which thereafter cannot be altered. The classes reflect this in that the GA operators can only be applied within the class constructor; all other methods of the class only allow inspection of the genes. Each chromosome contains a number of genes, which are bit-strings, with all genes having the same length. Each gene can have its own numeric type and range. The genetic operators that can be applied to the chromosomes when created are:

- Binary operators (requiring two parents), using n-point crossover or uniform crossover with or without mutation
- Unary operators (requiring one parent), using either bit-flip mutation or transposition
- Variable length operators, using random increment, kill/eliminate a gene, duplicate and mutate a gene

These classes can be summarized as follows:

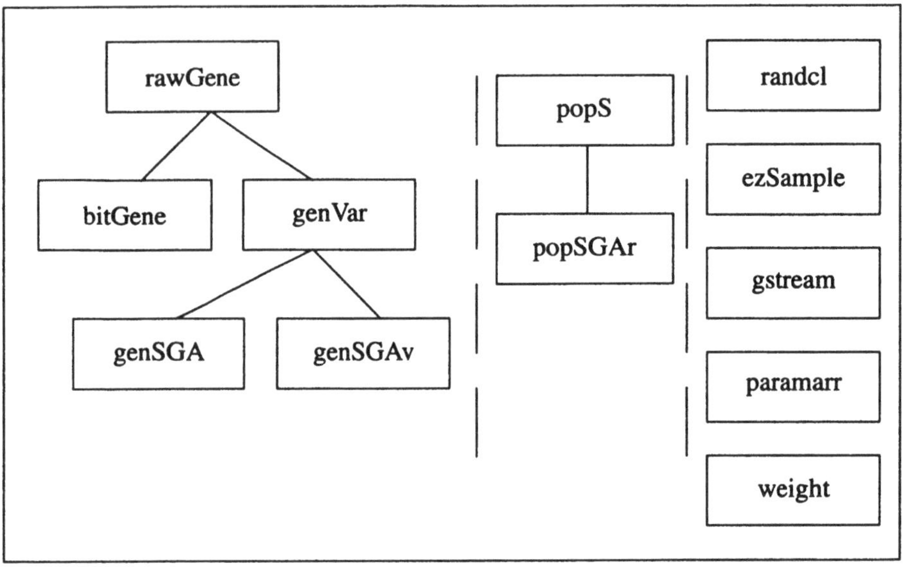

Figure 10.4 GAGS Chromosome (Gene), Population (pop) and Auxiliary Class Hierarchy

rawGene This is a base class for variable length genotypes based on raw bits. It contains default gene methods (for mutation and crossover), plus standard methods for data querying and bit manipulation.

bitGene This class is derived from rawGene, handling multiple genes within the same chromosome. Each gene is represented using bits of a fixed length, and decoded into unsigned integer values, thus having a range from 0 to $2^n - 1$. Methods are provided for initialization constructors (for random initialization), "binary" constructors (where a bitGene is created from two parents) and "unary" constructors (where one parent is used).

genVar This class is also derived from rawGene; it is used for deriving classes containing variable length genotypes.

genSGA Derived from genVar, this class is used for representing genes that can take a range of real values. Each gene within the chromosome has the same data range, and uses the same number of bits.

genSGAv Using this derived class, each gene within the chromosome can represent a different range of values, although they still contain the same number of bits.

The Population Hierarchy. The population class is used as a container for the chromosome classes. It contains methods for selection and replacement schemes, and methods for building a generation. The population class thus provides all of the methods necessary for a GA search.

popS This abstract template class handles a population of fixed-length genomes, with the template allowing the fitness data-type to be specified. It contains selection methods (such as ranking and roulette wheel), replacement schemes (such as elitism), genetic operators (mutation and crossover) and data querying methods (for obtaining such data items as the population size and the best fitness).

popSGAr Derived from popS, this class contains two templates, one for defining the fitness data-type, and the other for defining the type of the gene. Additional constructors for this class allow the number of genes within each chromosome to be specified in addition to the data range that each gene represents. A new-Generation method is provided to build the next complete generation from the current data within the class.

Auxiliary classes. Additional classes have been provided in order to supply supplementary functionality to the search algorithm classes to enable a complete application to be developed.

randcl This class contains methods for a variety of random number generators.

ezSample This is a template class for loading the data within a file into a matrix, using the specified template data-type.

gstream This class turns Gnuplot into a stream, to enable other functions to print to it as if it were a standard output stream.

paramarr Parameter handling is encapsulated within this class. It parses a file and places the values of the parameters into an array indexed by the first letter of the parameter; thus a total of 26 parameters can be used.

weight This is an implementation of a vector template class (a very simple basis for neural net programming).

scanFile This source file module does not actually contain a class, but provides a number of standard functions for file parsing.

10.2.1.3 Wizard Level. This level of usage for GAGS is currently undocumented. The suggestion is that wizard users would derive their own classes from those available or add completely new classes to the class library. Doing so would mean that the application generation scripts would also need to be modified so that they could use these new classes; alternatively, and more probably, the user would program the class library directly, as discussed below.

```
float fitness (const genSGA<float> & myGen)
{
  // Calculate the fitness from the genes in myGen
}

main (int argc, char ** argv)
{
  popSGAr<float, genSGA<float>> myPop (popSize, mutRate, genSize,
                                  locusSize, rangeStart, rangeEnd);

  for (g= 0; g < generations ; g++)
  {
    // Evaluate population fitness
    for (p = 0; p < myPop.getPopSize(); p ++ )
      myPop.setFitness (p, fitness(*(myPop [p])));
    myPop.newGeneration (ROULWHEEL, 0);
  }

  cout << myPop.bestIndiv() << endl;
  cout << myPop.bestFitness() << endl;
}
```

Figure 10.5 A Simple Genetic Algorithm Using GAGS

10.2.1.4 Using the GAGS Class Library. A genetic algorithm can be created from the GAGS class library by using the classes directly. A simple GA requires a fitness function, a population of genes, and a loop to create the desired number of generations. The example of Figure 10.5 uses a floating-point data type for the genes using the template classes genSGA and popSGAr. The population class constructor is used to supply the parameters for the population including genetic operator configuration (the mutation rate) and the configuration of the genes (the number of genes in each chromosome, their size and their data ranges). The generation loop simply calls the fitness function for each member of the population, followed by the newGeneration function (supplying a selection method type) to create the next population.

10.2.2 GAlib: A C++ Library of Genetic Algorithm Components

GAlib is described by Wall as "... a C++ library of genetic algorithm objects. ... it can be used in any C++ program using any representation and any genetic operators" (Wall (1996), p.1). See also:

`http://lancet.mit.edu/ga/`

This class library (some of which are template classes) is supplied in source code form, allowing it to be built on a variety of systems (Windows, MacOS and Unix). It consists of a set of base classes (the primary ones being Genome and GeneticAlgo-

GAGeneticAlgorithm	GASelectionScheme	GAScalingScheme
GASimpleGA	GARankSelector	GANoScaling
GASteadyStateGA	GARouletteWheelSelector	GALinearScaling
GAIncrementalGA	GATournamentSelector	GAPowerLawScaling
GADemeGA	GAUniformSelector	GASharing
	GASRSSelector	GASigmaTruncationScaling
	GADSSelector	
GAStatistics		
GAParameterList		
GAPopulation		
	GAGenome Derived Classes:	
GAList<T>	GABinaryString	GAArray<T>
GAListGenome<T>	GA1DbinaryStringGenome	GA1DArrayGenome<T>
	GABin2DecGenome	GA1DarrayAlleleGenome<T>
GATree<T>	GA2DbinaryStringGenome	GA2DArrayGenome<T>
GATreeGenome<T>	GA3DbinaryStringGenome	GA2DarrayAlleleGenome<T>
		GA3DarrayGenome<T>

Figure 10.6 The GAlib Class Library Hierarchy

rithm) covering all of the major elements of traditional GAs, and a number of more specific classes that have been derived from these base classes (see Figure 10.6).

GAlib has several distinguishing features:

- It can be used to evolve populations and/or individuals in parallel, employing multiple CPUs by using the "Parallel Virtual Machine" (PVM (2001)).
- Many inbuilt variations of GAs are supplied including population types, termination, replacement strategies, selection mechanisms, chromosome initialization and genetic operators, all of which can be customized or replaced.
- The recording of various statistics is available and includes off-line and on-line measures and their minima and maxima.

The GA classes contain statistics, the replacement strategy and operational parameters. A population object is used to contain the genome, the statistics (specifically relating to the population), the selection strategy and the scaling strategy classes. The user must at least undertake to provide an objective function and construct a simple function to initialize the classes and instruct them to run the search.

10.2.2.1 The Main Classes.

GAGenome. The *genome class* is used to encapsulate the genome and the functions that operate on that genome. Static functions are used to define the basic default set of methods for the class and include initialization, comparison, mutation, crossover and evaluation methods. These are default methods since they have been designed to be overridden by supplying alternative functions at run-time. Additional methods exist for copying and cloning the genomes, for loading and saving data to a stream, and, optionally, a comparator function can be defined to calculate the degree of difference between individual genomes. The genome class should typically use multiple-inheritance to sub-class the base genome class (GAGenome) and a data type class (such as GABinaryString or a user-defined problem-specific class).

GAPopulation. This class acts as a container for genomes. It provides methods for initialization, evaluation, statistical recording, selection and scaling.

GAGeneticAlgorithm. Alternative GA classes can be created by sub-classing any of the existing classes and modifying them by overriding any of their methods. Derivatives of this class are used to determine how population generations are constructed, such as by steady state or incremental means. This class also contains static termination functions, which have a GA class passed to them so that they can query any relevant data, and they must return either true or false to indicate whether or not the GA should terminate.

GAScalingScheme. This class transforms raw objective function return values to scaled fitness values. Scaling scheme classes can be created by sub-classing the abstract GAScalingScheme class. An evaluate function is used to perform the scaling calculations on the original fitness scores. For non-trivial data types, a clone and copy function must be provided.

GASelectionScheme. This class is used to perform the function of picking genomes from a population. Additional selection scheme classes can be created by deriving from one of the existing selection classes. There are two methods to define: update and select. The update method is called before selection so that any pre-selection data transformations can be performed. The select method picks an individual from the population, returning a reference to the genome.

In this way the class library establishes the means for using and modifying a GA, and includes many of the traditional GA variations in addition to many different data types for the genome. It appears to be a very useful starting point for any GA-specific work.

10.2.2.2 Using GAlib. A typical genetic algorithm can be constructed from the class library by using just the default classes and a user defined objective function (see Figure 10.7).

Modifications can be applied by using any of the variety of class methods supplied for this purpose, or by using one of the alternative classes. For example, methods exist for configuration modifications such as changing the population size and altering the mutation rate, and alternative classes can be used, such as a steady-state GA, to alter the behaviour of the search technique (see Figure 10.8).

Two primary mechanisms exist for extending the built-in objects: sub-classing and function substitution (see Figure 10.9). Many of the classes provide methods that accept a user-defined function as a parameter, which is then used instead of the original function, so that simple modifications can be made without the need for sub-classing. This is achieved by using static functions within the classes. Static functions are not

```
float Objective (GAGenome & g)
{
// compute the value of the genome, g
}
main()
{
GA1DBinaryStringGenome    genome (length, Objective);
GASimpleGA    ga (genome);
ga.evolve ();
cout << ga.statistics() << endl;
}
```

Figure 10.7 A Minimal Genetic Algorithm in GAlib

```
ga.populationSize (newSize);
ga.pMutation
(newMutationProbability);
GASteadyStateGA    ga (genome);
```

Figure 10.8 Making Simple Modifications in GAlib

like standard class methods; they exist without requiring an instance of the class to be created, but as a consequence they cannot access any non-static methods or data items within the class. This is not much of a disadvantage since they can be passed a pointer to the class once it has been instantiated, thus gaining access to any required data or class methods. The main advantage of using static functions, though, is that they can be set at run-time to modify the behaviour of the GA at any desired point before or during its execution.

10.2.3 Templar

Templar is a framework using an object-oriented design, consisting of a set of components made up of C++ classes (cf. Jones et al. (1998), Templar (2001) as well as Chapter 2). It has been built on the premise that object-oriented application frameworks allow a developer to rapidly produce an application – the gain in speed is mostly due to code reuse. Templar is not restricted to genetic algorithms as it supports general combinatorial optimization techniques (Reeves (1993), Rayward-Smith et al. (1996)), and can operate in a distributed environment. Among the objectives of the Templar framework are the following:

- Embedable components: its components should be embedable in other systems as well as in the Templar system itself
- Component consistency: each component must accept the same control mechanism, thus allowing alternatives to be substituted easily

```
// Derive a new class to use an alternative mutate operator
class newGenome: public GAGenome
{
public:
   static int newMutator (GAGenome &, float);
   newGenome() {    // Class constructor
      mutator (newMutator)
   }
}

// Use a function substitution method to provide
// an alternative crossover operator
genome.crossover (newCrossover);
```

Figure 10.9 User-Defined Modifications in GAlib

- Component flexibility: components must be extensible and able to form aggregate systems; hybrids must be able to be constructed and components must be able to communicate with each other
- Abstraction: abstract data-types must be used wherever possible to avoid specific code rewriting
- GUI support: although not essential, GUI support is provided to allow graphics to be attached to objects for parameter setting and visual feedback
- System portability: the basic framework is written to be system independent (the system must have a compatible C++ compiler); system specific code, such as GUIs, are layered on top
- Component distribution: components can use multiple processes (threads), processors or machines

Search techniques are implemented as "engine" objects, while problem-specific information is handled by "problem" objects. The objects that implement a search technique do so independently of the problem they are searching. The framework allows multiple engines to run concurrently.

10.2.3.1 The Framework. One of the major benefits of using a framework is for fast prototyping. Once the fundamentals of the framework are known, the user should be able to code a problem quickly and easily. The basic framework consists of three main components:

- Representations: define objects that store and manipulate solutions
- Problems: define specific problem classes (e.g. the famous travelling salesman problem; Lin (1965)) and include methods for loading instances and evaluating solutions

- Engines: define specific engines (e.g. GA) and include methods for generating and manipulating solutions to a problem

The components are structured so that Representations are independent of the other components, while Problem components refer to specific Representations, and Engine components refer to both Representation and Problem components.

Distribution. The framework is intended to execute each Engine and Problem in its own thread. Thus Engines can operate concurrently. On a multi-processor machine, this arrangement should have a noticeable performance increase over a single-threaded approach. If multiple machines are available (either single or multi-processor machines), then the framework can execute the search Engines and Problems on the remote machines. The components that are to be executed on the remote machines are copied, with one copy remaining on the local machine and the other being placed on its designated remote machine. The controlling (local) machine then uses the local copies as normal, and these delegate the work to the corresponding remote copy. The details of the multi-threaded and object duplication mechanisms are hidden from the developer by encapsulating the functionality within the base class, *TrComponent.*

Cooperation. Cooperation between individual search Engines has been implemented using messages. The *TrComponent* class contains methods for a message passing system. A message queue is used for other objects to send messages, which have been designed to work in the multi-threaded and distributed environments. A tag is used to identify the type of message being passed. Several standard tags exist, such as best solution and current solution. The receiving object acts on the tag to provide a sensible response to the message sender. A message filter can be applied to prevent unwanted messages from entering an object's message queue.

10.2.3.2 The Classes.

TrParamObject. *TrParamObject* provides methods for manipulating parameters. It provides the basis for meeting the consistency objective, and is sub-classed by all of the Problem and Engine classes. There are four data types for parameters: textual, double, integer and Boolean. Parameters are identifiable by a name and a data type. A parameter with a data type of textual can be restricted to a value from a list of values, while integer and double types can be limited by specifying a numeric range. Parameter objects contain methods for testing the existence of a parameter, setting and getting values, testing for valid values, saving and loading their data. When a new Problem or Engine class is created it adds any required parameters to the list.

TrComponent. *TrComponent* contains methods that are common to the Problem and Engine classes. Amongst its methods are functions for multi-threading, message passing and running in a distributed environment.

TrProblem. *TrProblem* is the base class for deriving all Problem classes, providing a consistent interface and common methods. A Problem class may be connected to many Engine classes. Methods exist for loading a problem instance, preparing a solution, checking that a solution has been prepared, checking that a solution is valid and generating the fitness value for the current solution. During construction of a problem object, its abilities are made available to the engine that attaches to it. Problem instances are typically loaded from files.

TrEngine. *TrEngine* is another base class and is used for deriving Engine classes, which are responsible for generating and manipulating solutions. Derived classes must be problem-independent, enabling any Engine class to be reusable for any problem. Classes of this type are expected to run concurrently. An engine may be connected to one Problem class and any number of other Engine classes. Methods exist for resetting the internal state of the engine, performing work on the current problem, starting and stopping, testing to see if the engine is currently working, saving results and statistics to a file, obtaining the best solution, getting, setting and seeding a solution and estimating how close the engine is to completion.

TrSystem. *TrSystem* is a container class for *TrComponent* derived classes (i.e., it contains all of the problems and search techniques). Multiple *TrSystem* classes may exist within an application.

TrRepresentation. *TrRepresentation* is a base class used to derive classes for problem solution representations. Four derived classes are supplied with Templar, to handle bit strings (*TrBitString*), permutations (*TrPermutation*), integer arrays (*TrIntArray*) and floating point arrays (*TrDoubleArray*). Representations are independent of all other components. Methods for representation classes include manipulation of the size of the solution, manipulation of the values of parts of the solution and solution initialization. These classes are intentionally vague so that Engine classes can manipulate solutions blindly, thus achieving the desired independence. Problem classes contain the ability to add or remove features contained within the representation classes.

***TrGUIComponent, TrGUIDisplays* and *TrGUIControls*.** These are abstract base classes providing the support for implementing GUI classes. The use of these classes is optional.

Fitness. Supplied data types for the computation of fitness values include integer, long and double. Fitness values must convert to the standard data type (a double), have a textual representation, must be comparable and must indicate if a change to the solution is an improvement. User-defined fitness types are supported, provided that the class provides the necessary functions as determined by the framework.

Ability Groups. Abilities manipulate solutions or data structures that will be used to manipulate solutions. "Operator Groups" group abilities of the same type in lists of name-function pairs. Initializers manipulate solutions to make them valid in the context of the current problem.

Driver. A driver is used to control the Engine and Problem classes. This could be a GUI or possibly another engine.

Avatar. In addition to embedding engines and problems in an application, the control language "Avatar" can be used. The avatar program can be used by supplying it with an engine and problem, or an avatar object can be embedded within an application that links to an engine and problem. A control script is used to control the avatar. Methods exist in the avatar object for loading the control script, determining if it is in a consistent state, performing a dry run of the script to check for errors, starting and stopping.

Additional Classes. Supporting classes exist for random number generation, for a timer designed to time sections of code execution, and for a statistics box for storing and retrieving data and string classes.

10.2.3.3 Using Templar. A Problem class must be created to represent the problem being solved. This involves the creation of several functions that relate directly to the problem. A file format for problem instances is required to enable the object to load the data that is associated with the problem (i.e. the parameters of the problem). A suitable representation is required for the solution to the problem (the solution in genetic algorithm terms being the chromosome). Methods are required for creating and destroying instances of the Problem class, for loading a problem instance from a file and for evaluating the solution. Additional methods are required for preparing a solution (initializing any data such as the length of a bit string) and for testing that the solution has been correctly prepared (such as checking that the bit string is of the correct length).

Once the main methods of the Problem class have been defined, any required GUI components would then be added. It can be very useful to add GUI components at an early stage since visual feedback will make testing and debugging of the Problem class a much easier task. Once the Problem class has been tested, it is usual to add any problem specific abilities. This would include adding such things as a "delta" function to provide quick evaluations of small changes in the solution.

For example, consider the travelling salesman problem. A file format would consist of a list of the towns and their relative distances. The solution would naturally be represented by the built-in permutation object, *TrPermutation*. An evaluation function would compute the length of the tour described by the solution. The prepare function would simply set the length of the permutation to the number of towns. The validation function will check if a solution is prepared by checking that the permutation is of the correct length. This completes all required methods for the new Problem class (see Figure 10.10).

Once the Problem class has been created, the application can easily be created by the framework. Scripts are supplied that build the Templar application; providing that Problem class files are placed in the correct directories, the build scripts will include them in the application.

```
void TSP::Load(istream& instream)
{
   // Load the TSP data from the input stream
}
void TSP::PrepareSolution(TSPRepn& p)
{
   p.SetLength(numTowns);
}
bool TSP::SolutionIsPrepared(const TSPRepn& p) const
{
   return (p.GetLength() == numTowns) && consistent;
}
TSPFit TSP::Evaluate(const TSPRepn& p)
{
   TSPFit objvalue = Dij(p[numTowns-1], p[0]);
   for (int i = 0; i < numTowns-1; ++i)
   {
      objvalue += Dij(p[i], p[i+1]);
   }
   return objvalue;
}
```

Figure 10.10 Some Methods from a TSP Problem-Class in the Templar Framework

10.3 JAVA CLASS LIBRARY SOFTWARE

10.3.1 JDEAL: Java Distributed Evolutionary Algorithms Library

JDEAL is a Java class library for genetic algorithms and generic evolution strategies, with the more specialized feature of being able to run in a distributed and parallel manner, making use of several machines for performing function evaluation (Costa et al. (1999), JDEAL (2001)). The class library is supplied in compiled form (Java byte-code class files) in addition to the full set of source code and a complete set of class documentation. For the full class hierarchy, see Figure 10.11.

- **EvolutionaryAlgorithm**
  - GeneticAlgorithm
    - SimpleGA
    - GenerationalGA
  - **EvolutionStrategy**
- **Chromosome**
  - BitStringChromosome
  - ESChromosome
  - IntegerChromosome
  - RealChromosome
- **Mutation**
  - FlipMutator
  - SwapMutator
  - GaussianMutator
- ESMutator

- **Crossover**
  - GACrossover
    - OnePointCrossover
    - TwoPointCrossover
    - NpointCrossover
    - UniformCrossover
    - EvenOddCrossover
    - PartialMatchCrossover
    - ArithmeticCrossover
    - BlendCrossover
  - ESCrossover
    - DiscreteCrossover
    - PanmicticDiscreteCrossover
    - IntermediateCrossover
    - GeneralizedIntermediateCrossover
    - PanmicticGeneralizedIntermediateCrossover
- **ChromosomeEvaluator**
- **PopulationEvaluator**
  - DefaultPopulationEvaluator
  - DefaultDistributedPopulationEvaluator
- **Selector**
  - RankSelector
  - UniformSelector
  - RouletteWheelSelector
  - TournamentSelector
- **Scaling**
  - PowerLawScaling
  - LinearScaling
  - SigmaTruncationScaling
  - WindowScaling
- **Termination**
  - GenerationTerminator
  - EvaluationTerminator
  - ConvergenceTerminator
  - TimeTerminator
- **ScoresWriter**
- **ChromosomesWriter**
- **JDSWriter**

Figure 10.11 The JDEAL Class Library Hierarchy

10.3.1.1 The Class Library. Any evolutionary algorithm that can be created is derived from the *EvolutionaryAlgorithm* class. This base class implements methods for initialization, evolution, termination and finalization. Methods also exist for chromosome and population initialization and evaluation, selection, statistics recording and information. The Java language has facilities for generating events and setting up event-listeners that will be executed when the event they are listening to has occurred. Events are generated at the start of evolution, at initialization, when a new generation is started, when a generation is completed and at the end of the evolution.

All GA classes are derived from the *GeneticAlgorithm* class, which is itself a subclass of the *EvolutionaryAlgorithm* class. The derived class adds methods for the two genetic operators, crossover and mutation. Two standard GA classes are available in the class library: *SimpleGA* (in which each new generation entirely replaces the previous one) and *GenerationalGA* (in which each generation only partially replaces the previous one). For generic evolution strategies, the *EvolutionStrategy* class is provided. This technique is optimized for solving problems with real functions. This class can be configured to act like a (μ, λ) or a $(\mu + \lambda)$ strategy (Michalewicz (1999)).

To define a new Chromosome class, the base *Chromosome* class is sub-classed. Usually, it will also be necessary to define new mutation and crossover genetic operators to operate correctly on the new type of chromosome. Chromosome derived classes that are supplied with JDEAL include bit-string, integer and real chromosomes.

The abstract class, *Crossover*, is used to derive new genetic crossover operators. Due to two types of search algorithm being implemented in JDEAL, there are two corresponding classes for deriving appropriate crossover classes: *GACrossover* and *ESCrossover*. A number of crossover classes are supplied in the class library, including classes for one point, two point, multi-point, uniform and even/odd crossover operators.

Mutation operators are derived from the *Mutator* abstract base class. Supplied derivations include classes that implement flip, swap and Gaussian mutation operators. The abstract class, *Selector*, is used to derive selection scheme classes. This class may also include a scaling object to modify the fitness values of the chromosomes. Supplied classes include rank, uniform, roulette wheel and tournament selection schemes. Scaling methods are implemented by creating a class using the scaling interface. JDEAL provides scaling classes that implement power law, linear, sigma truncation and window scaling methods.

Termination rules are defined from the *Terminator* interface. Supplied classes that provide termination conditions based on this interface do so when a certain number of generations have been made, when a number of evaluations have been computed, when the fitness values converge and/or when the specified execution time is reached.

The most frequent change that the user will make is to define a new evaluation function for a specific problem. To do this, a new class must be created from the *ChromosomeEvaluator* base class, and the *doEvaluation* method must be overridden to perform the required evaluation. The *PopulationEvaluator* abstract class should be sub-classed to create a population evaluator class. This class will use a *ChromosomeEvaluator* to perform the evaluations. The standard class provided in the JDEAL class library is *DefaultPopulationEvaluator*, which simply calls the *ChromosomeEvaluator* for each member of the population.

The distributed aspect of the class library can be obtained by using the alternative *PopulationEvaluator* class, *DefaultDistributedPopulationEvaluator*. A cluster of machines can then be used to share the processing when evaluating the chromosomes in the population. This is achieved using the supplied Java Distribution Service; the evaluator sends the chromosomes to the distribution service and waits until it receives the results. This is a highly desirable feature when the evaluation task is complex, and thus requires much computation.

```
RealChromosome    chromosome = new RealChromosome (length, min_value, max_value);
GenerationalGA    ga = new GenerationalGA (chromosome, pop_size);

ga.setCrossover (new OnePointRealCrossover() );
ga.setMutator (new SwapRealMutator() );
ga.setSelector (new RouletteWheelSelector() );
ga.setTerminator (new GenerationalTerminator(max_gens) );

ga.setChromosomeEvaluator (new ProblemModelFunction() );

ga.minimize();
```

Figure 10.12 Using the JDEAL Class Library

Finally, statistics *Writer* classes provide the means for recording a variety of information including: best and worst chromosome, deviation scores, averages, online and offline performances, and computation time.

10.3.1.2 Creating Applications. There are three main elements that are required for creating a simple application from the JDEAL class library:

- Representation: define the chromosome – the data type and its encoding
- Genetic/Evolutionary Algorithm: define the functions that comprise the desired algorithm
- Objective Function: model the problem that is to be optimized

Users would first create a chromosome object using an appropriate representation for their particular application. The chromosome object is passed as a parameter to a GA object. The GA object can be altered to the user's requirements by modifying various aspects through the class interface of the object. Finally, the search algorithm is executed by calling the member function in the GA object that is most suitable for the problem, such as *minimize* (see Figure 10.12).

As an alternative to using the classes in a Java application, a properties file can be created to instruct the JDEAL application how to operate. The file name of this properties file must be passed as a parameter to the *EALauncher* application class. Thus, providing that all required classes exist, users can create their evolutionary algorithms without the need to use Java. Although users with few or no skills in Java programming can experiment with JDEAL in this way, there will come a point when the user wishes to solve one of their own problems, and so must use Java to write at least one ChromosomeEvaluator derived class.

10.4 GENETIC ALGORITHM OPTIMIZATION SOFTWARE SURVEY

This is a brief survey of some of the genetic algorithm optimization software products that can be found on the Internet. This survey includes all types of software – not only object-oriented class libraries, but also software that performs a similar role to typical class libraries.

10.4.1 EO: Evolutionary Computation Framework and Library

Written in C++ EO provides a toolbox of template-based classes for evolutionary computation (especially GAs) that define interfaces for many classes of algorithm and also provides examples that use those interfaces (there are over eighty classes in all!). It is available from

`http://geneura.ugr.es/~jmerelo/eo` and

`ftp://geneura.ugr.es/pub/eo`

The following are the major features of EO:

- Users can derive their own classes from the supplied base classes (such as EOBase for Chromosomes, EOop for genetic operators, EOBreed for breeding strategies, EOReplace for replacement procedures)
- **Chromosomes**: supplied variations include binary, floating point, bi-dimensional, string and value plus sigma; all algorithms will operate on any kind of chromosome
- **Genetic operators**: operators can be unary, binary or n-ary (orgy) operators; many floating point, bit-string, and generic operators are defined, such as mutation and crossover
- **Selection/elimination** procedure: steady-state and rank-based variations are provided
- **Reproduction** procedure: random, lottery, and rank-based are provided
- **Replacement** procedure: eliminate-all, eliminate-worst are provided
- **Termination** conditions: fitness and generation-based
- **Algorithms**: Easy GA, Goldberg's SimpleGA, Evolution Strategies and simulated annealing with user-definable cooling schedule
- Some **utilities**: command-line parsing, random-number generation
- **Examples**: examples that test all features and classes, plus genetic "Mastermind"

10.4.2 EVOLC: The Evolutionary Program of EEAAX at CMAP – Ecole Polytechnique

This is a general evolutionary product, written in C++. The features that it has are:

- Easy parameterization (through parameter file or command-line arguments)

- Large choice of selection/replacement procedures (including standard GAs, ESs, EP and SSGA popular schemes) through the parameters; you can even build your own without recompiling!
- Many standard operators on binary and real representations, including ES self-adaptive mutations
- On-line graphical monitoring of population statistics
- Restart facilities

10.4.3 Eiffel Genetic Algorithm Classes

Written in Eiffel (Meyer (1990)), this set of classes uses a chromosome hierarchy (termed individuals) and a population hierarchy (for generational and steady state GAs). The individuals include the evaluation function as a method of the class, so that for each problem there will be a new individual class. (A number of implementations for various problems are supplied.) Some specific derivations of the population classes are also provided (containing various selection and replacement strategies). It is available from

`http://cs.ru.ac.za/homes/g93i0527/Projects/GAEiffel`

10.4.4 GAJIT: Genetic Algorithm Java Implementation Toolkit

The author (Faupel (1998)) has based this class library on the classes in GAGS (not the application generator), translating the classes to Java and evolving them. The population and chromosome classes correspond approximately to those in GAGS, but genetic operators have also been put into classes. A *GenOp* abstract class is used for deriving genetic operator classes and several such classes are provided; furthermore, a set of *View* classes provides the means to display the contents of chromosomes.

10.4.5 GALOPPS

GALOPPS is an acronym: Genetic ALgorithm Optimized for Portability and Parallelism System. It is written in C and based on Goldberg's Simple Genetic Algorithm (SGA) (Goldberg (1989)), being designed to give users the capability to learn and extend it via source code. Parallelism can be used in one of three forms: a single PC simulating parallel processing, a multi-processor PC, and multiple PCs on a network. It is available from Michigan State University's Genetic Algorithms Research and Applications Group (GARAGe) on

`http://garage.cps.msu.edu/software/galopps/index.html`

GALOPPS extends the capabilities of the SGA in a large number of ways:

- (optional) A new Graphical User Interface, based on TCL/TK, for Unix users, allowing easy running of GALOPPS 3.2 (single or multiple subpopulations) on one or more processors; GUI writes/reads "standard" GALOPPS input and master files, and displays graphical output (during or after run) of user-selected variables

- Five selection methods are implemented: roulette wheel, stochastic remainder sampling, tournament selection, stochastic universal sampling (SUS), linear-ranking-then-SUS
- Random or superuniform initialization of "ordinary" (non-permutation) binary or non-binary chromosomes; random initialization of permutation-based chromosomes; or user-supplied initialization of arbitrary types of chromosomes
- Binary or non-binary alphabetic fields on value-based chromosomes, including different user-definable field sizes
- Three crossovers for value-based representations: 1-point, 2-point, and uniform, all of which operate at field boundaries if a non-binary alphabet is used
- Four crossovers for order-based representations: PMX, order-based, uniform order-based, and cycle
- Four types of mutation: fast bitwise, multiple-field, swap and random sublist scramble
- Fitness scaling: linear scaling, Boltzmann scaling, sigma truncation, window scaling, ranking
- (optional) Automatic control of selection pressure, using Boltzmann scaling with automatic control of the β parameter, coupled with SUS
- (optional) DeJong-style crowding replacement with (optional) incest-reduction for mating restriction (negative assortative mating bias)
- Other replacement strategy options: child-replaces-parent (from SGA) or child-replaces-random
- Elitism is optional
- Convergence: "lost", "converged", "percent converged" and other measures
- Performance measures: on-line and off-line (local and global across subpopulations), plus current local best, best ever, global best ever
- (optional) Inversion operation on entire subpopulations, with migration among them automatically handled
- Allow users to define multiple (different) representations in subpopulations, with migration among them
- Provide a "beta" implementation of Falkenauer's Grouping Genetic Algorithm (GGA) (Falkenauer (1998)) representation and operators, providing tools for efficient solution of a variety of "grouping-type" combinatorial problems
- Uses SGA philosophy of one template file for the user to modify, but enriched with many additional user callbacks, for added flexibility and extensibility

- All runs are restartable or seedable from automatic checkpoint files
- Output easily suppressed to one of three reduced levels, including user-callback-driven outputs; input is from file or keyboard, and files allow optional keywords to permit automatic diagnosis of missing parameters, etc.
- Communication topology for parallel runs is easily specified in a single "master" file

10.4.6 GAMeter/X-GAMeter

GAMeter is available from the University of East Anglia on

`http://www.sys.uea.ac.uk/Teaching/Staff/gds.html`

Its main list of features includes:

- Full GUI available
- Binary, integer and floating point representations handled
- Eight selection mechanisms
- Four crossover operators
- Up to four mutation operators (depending on representation)
- Five replacement mechanisms
- Special permutation-based operators including 18 crossover and three mutation operators
- Minimization/Maximization problems handled
- Any fitness type allowed, including arbitrarily complex types
- Dynamic parameter alteration
- Experimental mode available for batch processing
- Ability to view and manipulate any chromosome in the current population
- Graphing facilities updated automatically
- Various mechanisms available to seed the initial population
- Fully extendable
- Portable to any platform supporting a C compiler

X-GAMeter is basically the same application but with a graphical user interface written using X-windows on UNIX. This interface can display statistics, populations and graphs, all in real-time. The interface allows the choice of selection method (eight variations), genetic operators (up to 21 variations, depending on the representation) and merge strategies (five variations), in addition to allowing a variety of parameters to be modified (such as the crossover rate and population size). The choices of strategy and the associated parameters can be changed at any time – even during a search.

10.4.7 GAVin: Genetic Algorithms Visual Interface

GAVIn (Burke and Varley (1998)) was designed as a tutorial tool, focusing on numerical function optimization problems. It is written in C++ (running under Microsoft Windows 95 or compatible). The aims are to enable layman users to learn and understand the basic principles and more advanced operations of GAs, to act as a research tool for the analysis of operator performance, and to perform as an interface for further work. Multiple runs using different variations of a GA can be compared to show their effects.

In contrast to some of the products discussed above, its functionality has been deliberately limited to avoid confusion – only binary strings are supported, for example. Despite these limitations, the user can set various parameters, including chromosome length, population size, termination conditions, selection method, genetic operators and reproduction techniques. The genetic operators can be configured, including the ability to turn them off – allowing the user to appreciate their effects clearly. In addition to the variety of methods, a selection of ten common test functions has been supplied. The GUI provides access to settings, text and graphical output from a search run and an "explore" window which displays all generated data in hierarchical form in addition to providing more detailed information on any item selected.

10.4.8 Generator

This is a commercial product from New Light industries, obtainable from

`http://myweb.iea.com/~nli/main.htm`

It is very specific in that it interacts with Microsoft Excel, the user defining the problem in a standard worksheet. A few uses of Generator are:

- Solving complex routing and scheduling problems
- Designing electronic circuits
- Curve fitting
- Improving factory efficiency
- Solving coupled sets of nonlinear partial differential equations

What distinguishes Generator from other commercial genetic algorithm software is that users can fine-tune their parameters on the fly. In addition, Generator has special hill-climbing features, recombination operators and mate selection processes to help to find high-quality solutions rapidly. It uses roulette wheel selection, two-point crossover and has a few variations of mutation. Variations of the mutation operator include two types of hill-climbing: random and directional. It uses a survival of the fittest replacement strategy: members of the new population may be replaced by members of the old population if the old members are fitter. Representation types include integer, real and permutations.

The main difference between this product and non-commercial products is that it interacts with a very common third-party application with which many business users are already familiar. However, while it is easy to install and start using with little effort, in order to use the software effectively it will require some knowledge as to what effect the parameters and strategy choices might have on a particular problem.

10.4.9 GENESIS

GENESIS (GENEtic Search Implementation System) was one of the earliest general-purpose GA systems (Grefenstette (1990)), and was written to promote the study of genetic algorithms for function optimization. Since it treats GAs as task independent optimizers, the user must provide only an evaluation function that returns a value when given a particular point in the search space. GENESIS is written in C, with source code and makefiles supplied. Chromosomes can be represented as bit strings or as vectors of real numbers. It includes rank-based or proportional selection and the optional use of Gray code. Parameters are stored in files generated by a "Setup" program, which asks questions (such as representation type and number of genes), most of which have default answers for use by those with limited knowledge of the software and GAs.

10.4.10 GENOCOP

GENOCOP (Genetic Algorithm for Numerical Optimization for COnstrained Problems) is another early program, described by Michalewicz (1999). It specializes in optimizing a function subject to (any number of) linear constraints (equalities and inequalities) and is also written in C. It does not accept integer variables. GENOCOP is capable of solving linear programming problems using a GA, although of course it is the potential for optimizing non-linear objectives that is more interesting.

10.4.11 LIBGA

This is another program in C (Corcoran (1993)). Features include:

- Generational or steady state GA
- Selection can be uniform random, roulette or rank linear biased
- Bit string crossover operators include simple and uniform
- Order-based integer crossover operators include order1, order2, position, cycle, PMX, uniform, relative and asexual
- Mutation for bit strings are simple invert or random bit value and, for either representation, random element swap
- Replacement strategies include append, rank, replace first weaker member and replace weakest member

10.4.12 OOGA

OOGA (Object-Oriented GA) is a GA designed for industrial use by Lawrence "Dave" Davis, one of the early pioneers in applying GAs to real-world problems. Each of the techniques employed by the GA is an object that may be modified, displayed or replaced in object-oriented fashion. OOGA is especially well-suited for those who wish to modify the basic GA techniques or tailor them to new domains. OOGA is designed to abstract the algorithm itself away from the specifics of the individuals under evolution. The general architecture of the system is a GA "black box" into which modules defining particular evo-tasks can be plugged. The GA code does not know anything about the individuals, and the individuals need only support a basic set of genetic operations that the GA expects to be able to perform on them. In this way, the black box is portable across any evo-task. OOGA thus provides a general, reusable framework for solving evo-tasks that is intended to encourage greater use of GAs to solve problems.

10.4.13 PGAPack: Parallel Genetic Algorithm Library

PGAPack is a general-purpose, data-structure-neutral, parallel genetic algorithm library (Levine (1996)). It is intended to provide most capabilities desired in a GA library, in an integrated, seamless, and portable manner. Key features in PGAPack V 1.0 include:

- Callable from Fortran or C
- Runs on uniprocessors, parallel computers, and workstation networks
- Binary-, integer-, real-, and character-valued native data types
- Full extensibility to support custom operators and new data types
- Easy-to-use interface for novice and application users
- Multiple levels of access for expert users
- Parameterized population replacement
- Multiple crossover, mutation, and selection operators
- Easy integration of hill-climbing heuristics
- Extensive debugging facilities
- Large set of example problems
- Detailed user's guide

10.4.14 SUGAL

SUGAL is the SUnderland Genetic ALgorithm system (Hunter (1995)). Trajan Software Ltd. claim that "SUGAL is the most sophisticated genetic algorithms simulator available today, and has over 2,000 users worldwide". It is written in C with accessible source code with the aim of supporting research in and implementation of GAs on a common software platform. As such, SUGAL supports many variants of GAs, and has extensive features to support customization and extension. Some of its many features are as follows:

- Supports multiple datatypes seamlessly: bit strings, integers, real numbers, symbols (from arbitrarily sized alphabets), permutations; also supports mixed datatypes
- Uses a powerful and general GA
- Generates extensive statistics on population fitness and diversity
- Interface can be extended and/or customized using simple registration routines and screen layout files
- Conducts multiple experiments and calculates aggregate statistics
- Users can add their own configuration file parameters and/or add additional options to existing parameters

The detailed variations are also extensive:

- Initialization: uniform, Gaussian, loaded bit, or from file
- Selection: roulette, expected value model, tournament and uniform
- Fitness Normalization: biased, inversion, linear and mean-linear, rank linear and rank geometric (all for either minimization or maximization)
- Crossover: uniform, one-point, two-point and arbitrary n-point crossover, variable crossover rates, multiple operator application
- Mutation: bit inversion, gene reinitialization, step-delta, uniform-delta and Gaussian-delta Mutation; time-decay mutation sizes; variable mutation rates, multiple operator application
- Replacement: uniform, rank-based, tournament, crowding and parental replacement; replacement can be unconditional, simple conditional (i.e. if improved) or randomized (*à la* simulated annealing); variable replacement rate (generation gap); elitism
- Stopping conditions: number of generations, fitness target achieved, fitness or diversity convergence level

The SUGAL genetic algorithm subsumes most of the well-known GA variants as sub-sets of its functionality, and can be extended to model those it does not subsume. For example, the several Holland/Goldberg/DeJong versions of a GA, Whitley's Genitor and Fogel's real parameter GA are all covered (and can be extended to arbitrary datatypes), along with other more obscure versions. SUGAL breaks the features of the various algorithms into separate parts, so that an extremely wide range of hybrids is also available. However, aspects of Evolution Strategies are not covered (e.g. self-adaptive mutation rates).

The major features that give SUGAL its power and flexibility are its use of *SUGAL Parameters* and *SUGAL Actions*. The Parameters form a mechanism which allows control values (integers, doubles and strings) to be supplied by configuration, command line, and/or GUI – with a simple declaration and routine call, user-defined parameters can be fully integrated into the SUGAL system. Most of its major operations are performed using SUGAL Actions, which are calls to routines through pointers, rather than directly. Consequently, users can provide additional routines that overrule parts of the SUGAL algorithm. SUGAL Actions and Parameters are fully integrated, so that users can provide routines which implement additional options for existing SUGAL Actions – for example, to perform initialization, mutation, crossover, replacement, fitness normalization and selection in the GA.

10.5 CONCLUSIONS

10.5.1 The Class Libraries

GAGS allows a user with a fairly limited knowledge of object-oriented programming, development and design to use a class library, thus gaining all of the benefits that object-orientation provides. As long as the user can code an objective function, GAGS will generate the required object-oriented application. On the other hand, many variations of GAs are supplied and their parameters can be altered using configuration files. Thus, users with more advanced skills can modify and extend the supplied classes. Integration of an application built using GAGS into other applications should be fairly trivial, since source code is generated, and can be altered to suit any needs.

GAlib is more advanced in that it has fewer limits and more built-in functionality than GAGS. It provides a reasonably comprehensive class library for GA optimization tasks. A little more work is required for simple applications: users need to construct their own main function to create the class instances and call their methods, although this can be fairly straightforward. Again, more advanced users can take full advantage of the class library and extend it as they wish.

Templar goes beyond the class library definition and provides an application framework. It is more complex to use than the other systems, but it is also the most flexible. Users will require a good knowledge of how Templar works to use the product effectively, but the greater functionality on offer may outweigh the initial learning.

Each system is based around a class library and each has different objectives. This demonstrates the fact that there is no right way of providing a GA class library. Object-oriented programming can be very complex and so it is common to target the objectives at a specific aspect for specific requirements. This is demonstrated by the systems presented here, in that they are all very different, having various objectives and being aimed at different levels of user, yet designed for a very similar task. Nevertheless, all of these systems have achieved their goals, fitting into a particular niche in the genetic algorithm class library arena.

JDEAL is more-or-less the Java equivalent of a combined GAGS and GAlib class library. It provides a mechanism for non-programmers to use the class library like GAGS, it can operate in a distributed manner like GAlib and both of the C++ libraries, and it contains many traditional variations of GA methods. It does not contain all of the rich functionality of the Templar framework, but it does have its own unique functionality, such as a useful event handling mechanism, and, being written in Java, it is readily portable.

10.5.2 Future Directions

There is a well-known fundamental problem with class libraries (Sing (1997)). Without the source code they can be very hard to use and especially difficult to debug. If source code is supplied, the vendor has the problem of plagiarism and faces difficulties associated with tampering with the code and support of the software if rebuilt – possibly built on an unknown system with an unknown compiler/linker and unknown options! Thus, the commercial market for class libraries has never really taken off, although for academic applications the provision of source code is usually not a problem (since there is usually little support provided anyway).

The future appears to be in the use of components and distributed objects, in which ready-made and tested code can be assembled together to create applications (Szyperski (1998)). These building blocks can be used in other contexts to build alternative applications. This will only work effectively if there is a simple, effective supporting infrastructure. Currently this is provided by three main approaches, each from a different vendor: Microsoft supplies DCOM, OLE and ActiveX, while OMG supply CORBA and OMA, and Sun provides Java and JavaBeans.

In future, the preferred choice of programming language may shift from C/C++ to Java. For a start, Java forces the programmer to use object-oriented techniques, in that classes must be used, whereas C++ will let the programmer use the C subset in which there is no requirement to use classes at all. Java code has been designed to be portable when compiled, simply requiring a compatible Java Virtual Machine on the system that it is to be run on. C++ does not contain portable code once it has been compiled, and even when source is supplied, there can be many problems trying to build the class library on different systems. Java also has many classes available as standard (including a range of GUI classes). One of the few areas in which C++ outperforms Java, is that C++ code is generally quicker to execute than its Java counterpart. However, Java can call C or C++ functions for time critical requirements, and the Java JIT (Just In Time) compilers are becoming more efficient as they are further developed. Thus we would speculate that future developments of GA software for optimization will more and more tend to be based on Java.

Abbreviations

ABACUS	A Branch-And-CUt System
AI	Artificial Intelligence
BARON	Branch And Reduce Optimization Navigator
BT	British Telecom
CAP	Capacity Requirement
COM	Component Object Model
CORBA	Common Object Request Broker Architecture
CP	Constraint Programming
CPU	Central Processing Unit
CSP	Constraint Satisfaction Problem
DCOM	Microsoft Distributed Component Object Model
DERA	British Defence Evaluation and Research Agency
EA	Evolutionary Algorithm
EMOSL	Entity Modelling and Optimisation Subroutine Library
EO	Evolutionary Computation Framework and Library
EP	Evolutionary Program
ES	Evolutionary Strategy
EVOLC	Evolutionary Program of EEAAX at CMAP
GA	Genetic Algorithm
GAGS	Genetic Algorithm from Granada, Spain
GALOPPS	Genetic Algorithm Optimized for Portability and Parallelism System
GAJIT	Genetic Algorithm Java Implementation Toolkit
GC	Graph Coloring
GENOCOP	Genetic Algorithm for Numerical Optimization for COnstrained Problems
GGA	Grouping Genetic Algorithm
GLS	Guided Local Search

GRASP	Greedy Randomized Adaptive Search Procedure
GRG	Generalized Reduced Gradient
GUI	Graphical User Interface
HC	Hill Climbing
HS	Heuristic Search
HSF	Heuristic Search Framework
IDL	Interface Description Language
IL	Invariant Library
IT	Information Technology
JDEAL	Java Distributed Evolutionary Algorithms Library
LDS	Limited Discrepancy Search
LGO	Lipschitz Global Optimizer
LP	Linear Programming
MIMD	Multiple Instruction, Multiple Data
MIP	Mixed Integer Programming
MMC	Multiple Markov Chain
MP	Mathematical Programming
MPI	Message Passing Interface
OCL	OptQuest Callable Library (Chapter 7)
OCL	Object Constraint Language (Chapter 4)
OPL	Optimization Programming Language
OS	Operating System
PE	Processing Element
PMX	Partially Matched Crossover
POSIX	IEEE Portable Operating System Interface
PSA	Parallelized Simulated Annealing
PSU	Potential Speed Up
PVM	Parallel Virtual Machine
QAP	Quadratic Assignment Problem
RAM	Random Access Memory
RCS	Residual Cancellation Sequence
REM	Reverse Elimination Method
RNG	Random Number Generator
RTTI	Run-Time-Type Information
SA	Simulated Annealing

SAT	Satisfiability
SGA	Simple Genetic Algorithm
SSGA	Steady-State Genetic Algorithm
SUGAL	SUnderland Genetic ALgorithm
SUS	Stochastic Universal Sampling
TOMS	ACM Transactions on Mathematical Software
Tr	Prefix for identifier within Templar
TS	Tabu Search
TSP	Travelling Salesman Problem
TSPO	Open Travelling Salesman Problem
UIS	Union of Independent Sets (graph coloring specific crossover)
UML	Unified Modeling Language
VM	Virtual Machine
VNS	Variable Neighborhood Search
VRP	Vehicle Routing Problem
XML	Extensible Markup Language

References

Aarts, E. and Lenstra, J., editors (1997). *Local Search in Combinatorial Optimization.* Wiley, Chichester.

Aarts, E. and Verhoeven, M. (1997). Local search. In Dell'Amico, M., Maffioli, F., and Martello, S., editors, *Annotated Bibliographies in Combinatorial Optimization*, pages 163–180. Wiley, Chichester.

ABACUS (2001). A Branch-And-CUt System. `http://www.informatik.uni-koeln.de/ls_juenger/projects/abacus.html`.

Adams, J., Balas, E., and Zawack, D. (1988). The shifting bottleneck procedure for job shop scheduling. *Management Science*, 34:391–401.

Aksit, M., Wakita, K., Bosch, J., Bergmans, L., and Yonezawa, A. (1994). Abstracting object interactions using composition filters. In *Proceedings of the European Conference on Object-Based Distributed Programming 1993*.

Allard, M. (1998). Object technology: Overcoming cultural barriers to the adoption of object technology. *Information Systems Management*, 15 (3):82–85.

Alpern, B., Hoover, R., Rosen, B., Sweeney, P., and Zadeck, F. (1990). Incremental evaluation of computational circuits. In *Proceedings of ACM SIGACT-SIAM'89*, pages 32–42.

AMPL (2001). AMPL: A modeling language for mathematical programming. `http://www.ampl.com`.

Anderson, E., Bai, Z., Bischof, C., Demmel, J., and Dongarra, J. (1995). *LAPACK Users Guide, 2nd Ed.* SIAM (Society for Industrial & Applied Mathematics), Philadelphia.

Andreatta, A. (1998). *A Framework for the Development of Local Search Heuristics with an Application to the Phylogeny Problem (in Portuguese)*. PhD thesis, Catholic University of Rio de Janeiro, Computer Science Department.

Andreatta, A., Carvalho, S., and Ribeiro, C. (1998). An object-oriented framework for local search heuristics. In *Proceedings of the 26th Conference on Technology of Object-Oriented Languages and Systems (TOOLS USA'98)*, pages 33–45. IEEE, Piscataway.

Andreatta, A. and Ribeiro, C. (2001). Heuristics for the phylogeny problem. To appear in: Journal of Heuristics.

Applegate, D. and Cook, W. (1991). A computational study of the job-shop scheduling problem. *ORSA Journal on Computing*, 3:149–156.

Ayala, F. (1995). The myth of eve: Molecular biology and human origins. *Science*, 270:1930–1936.

Azencott, R. and Graffigne, C. (1992). Parallel annealing by periodically interacting multiple searches: Acceleration rates. In Azencott, R., editor, *Simulated Annealing – Parallelization Techniques*, pages 81–90. Wiley, New York.

Bäck, T., Fogel, D., and Michalewicz, Z., editors (1997). *Handbook of Evolutionary Computation*. Institute of Physics Publishing, Bristol.

Bacon, J. (1997). *Concurrent Systems: Operating Systems, Database and Distributed Systems: An Integrated Approach*. Addison-Wesley, Reading, 2nd edition.

Ball, M. and Datta, A. (1997). Managing operations research models for decision support systems applications in a database environment. *Annals of Operations Research*, 72:151–182.

BARON (2001). The Branch-And-Reduce Optimization Navigator, global optimization software.
`http://archimedes.scs.uiuc.edu/baron.html`.

Barr, R., Golden, B., Kelly, J., Resende, M., and Stewart, W. (1995). Designing and reporting on computational experiments with heuristic methods. *Journal of Heuristics*, 1:9–32.

Battiti, R. (1996). Reactive search: Toward self-tuning heuristics. In Rayward-Smith, V., Osman, I., Reeves, C., and Smith, G., editors, *Modern Heuristic Search Methods*, pages 61–83. Wiley, Chichester.

Battiti, R. and Tecchiolli, G. (1994). The reactive tabu search. *ORSA Journal on Computing*, 6:126–140.

Beasley, J. E. (1988). An algorithm for solving large capacitated warehouse location problems. *European Journal of Operational Research*, 33:314–325.

Beasley, J. E. (1990). OR-Library: Distributing test problems by electronic mail. *Journal of the Operational Research Society*, 41:1069–1072.

Beldiceanu, N. and Contjean, E. (1994). Introducing global constraints in CHIP. *Mathematical and Computer Modelling*, 12:97–123.

Bertsekas, D., Tsitsiklis, J., and Wu, C. (1997). Rollout algorithms for combinatorial optimization. *Journal of Heuristics*, 3:245–262.

Bessiere, C. and Regin, J.-C. (1997). Arc consistency for general constraint networks: Preliminary results. In *Proceedings of the 15th IJCAI*, pages 398–404.

Biskup, U. (2000). Entwicklung einer Konfigurationsanwendung basierend auf den Anforderungen von HotFrame: Grafische Benutzeroberfläche, Software-Generator, Konfigurationssprache. Diploma thesis, Braunschweig University of Technology.

Bisschop, J. and Meeraus, A. (1982). On the development of a general algebraic modeling system in a strategic planning environment. *Mathematical Programming Study*, 20:1–29.

Bodlaender, H., Fellows, M., and Warnow, T. (1992). Two strikes against the perfect phylogeny problem. *Lecture Notes in Computer Science*, 577:273–283.

Bosch, J., Molin, P., Mattsson, M., Bengtsson, P., and Fayad, M. (1999). Framework problems and experiences. In Fayad, M., Schmidt, D., and Johnson, R., editors, *Building Application Frameworks: Object-Oriented Foundations of Framework Design*, pages 55–82. Wiley, Chichester.

Böse, J., Reiners, T., Steenken, D., and Voß, S. (2000). Vehicle dispatching at seaport container terminals using evolutionary algorithms. In Sprague, R., editor, *Proceedings of the 33rd Annual Hawaii International Conference on System Sciences*, volume DTM-IT, pages 1–10. IEEE, Piscataway.

Briot, J., Guerraoui, R., and Lohr, K.-P. (1998). Concurrency and distribution in object-oriented programming. *ACM Computing Surveys*, 30(3):291–329.

Burkard, R. (1984). Quadratic assignment problems. *European Journal of Operational Research*, 15:283–289.

Burke, K. and Varley, D. (1998). A genetic algorithms tutorial tool for numerical function optimisation. Technical report, Automated Scheduling and Planning Group, Department of Computer Science, University of Nottingham, UK.

Buschmann, F., Meunier, R., Rohnert, H., and Sommerlad, P. (1996a). *Pattern-Oriented Software Development*. Wiley, Chichester.

Buschmann, F., Meunier, R., Rohnert, H., Sommerlad, P., and Stal, M. (1996b). *Pattern-Oriented Software Architecture*. Wiley, Chichester.

Campos, V., Glover, F., Laguna, M., and Martí, R. (1999a). Experimental evaluation of a scatter search for the linear ordering problem. Technical report, Boulder, CO.

Campos, V., Laguna, M., and Martí, R. (1999b). Scatter search for the linear ordering problem. In Corne, D., Dorigo, M., and Glover, F., editors, *New Ideas in Optimization*, pages 331–339. McGrawHill, London.

Campos, V., Laguna, M., and Martí, R. (2001). Context-independent scatter and tabu search for permutation problems. Technical report, University of Valencia.

Caseau, Y. and Laburthe, F. (1995). The CLAIRE documentation. Technical Report LIENS Report 96-15, École Normale Supérieure.

Caseau, Y. and Laburthe, F. (1996). CLAIRE: Combining objects and rules for problem solving. In Chakravarty, M., Guo, Y., and Ida, T., editors, *Multi-Paradigm Logic Programming: Proceedings of the JICSLP'96 Post-Conference Workshop*, pages 105–114, Technische Universität Berlin, Fachbereich Informatik, Report No. 96-28.

Caseau, Y. and Laburthe, F. (1998). SALSA: A language for search algorithms. In Maher, M. and Puget, J.-F., editors, *Proceedings of the Fourth International Conference on Principles and Practice of Constraint Programming (CP '98)*, volume 1520 of *Lecture Notes in Computer Science*, pages 310–324. Springer, Berlin.

Caseau, Y., Laburthe, F., and Silverstein, G. (1999). A meta-heuristic factory for vehicle routing problems. In Jaffar, J., editor, *Principles and Practice of Constraint Programming – CP '99*, volume 1713 of *Lecture Notes in Computer Science*, pages 144–158. Springer, Berlin.

Charon, I. and Hudry, O. (1993). The noising method: A new method for combinatorial optimization. *Operations Research Letters*, 14:133–137.

CLAIRE (1999). The CLAIRE programming language. (1999 date last checked) `http://www.ens.fr/~laburthe/claire.html`.

Clarke, G. and Wright, J. (1964). Scheduling of vehicles from a central depot to a number of delivery points. *Operations Research*, 12:568–581.

Codenie, W., Hondt, K. D., Steyaert, P., and Vercammen, A. (1997). From custom applications to domain-specific frameworks. *Communications of the Association of Computing Machinery*, 40(10):71–77.

Colmerauer, A. (1990). An Introduction to Prolog III. *Communications of the Association of Computing Machinery*, 28(4):412–418.

Condognet, P. and Diaz, D. (1996). Compiling constraints in `clp(fd)`. *Journal of Logic Programming*, 27(3):185–226.

Coplien, J. (1999). *Multi-Paradigm Design for C++*. Addison-Wesley, Reading.

Corberán, A., Fernández, E., Laguna, M., and Martí, R. (2000). Heuristic solutions to the problem of routing school buses with multiple objectives. Technical report, University of Valencia.

Corcoran, A. (1993). Libga. Department of Computer Science, Clemson University, South Carolina.

Costa, J., Silva, P., and Lopes, N. (1999). JDEAL Java Distributed Evolutionary Algorithms Library. Technical report, Institute of Computer Technology, University of Lisbon, Portugal.

CPLEX (2001). ILOG CPLEX, Mathematical programming software. `http://www.cplex.com`.

Crainic, T., Toulouse, M., and Gendreau, M. (1997). Toward a taxonomy of parallel tabu search heuristics. *INFORMS Journal on Computing*, 9:61–72.

Culberson, J. (1998). On the futility of blind search: An algorithmic view of "no free lunch". *Evolutionary Computation*, 6:109–127.

Czarnecki, K. and Eisenecker, U. (2000). *Generative Programming: Methods, Tools, and Applications*. Addison-Wesley, Reading.

Dammeyer, F. and Voß, S. (1993). Dynamic tabu list management using the reverse elimination method. *Annals of Operations Research*, 41:31–46.

Davis, L. (1991). *Handbook of Genetic Algorithms*. Van Nostrand Reinhold, New York.

Day, W., Johnson, D., and Sankoff, D. (1986). The computational complexity of inferring rooted phylogenies by parsimony. *Mathematical Biosciences*, 81:33–42.

de Backer, B. and Furnon, V. (1999). Local search in constraint programming: Experiments with tabu search on the vehicle routing problem. In Voss, S., Martello, S., Osman, I., and Roucairol, C., editors, *Meta-heuristics: Advances and Trends in Local Search Paradigms for Optimization*, pages 63–76. Kluwer, Boston.

de Backer, B., Furnon, V., and Shaw, P. (1999). An object model for meta-heuristic search in constraint programming. In *CP-AI-OR'99: Workshop on Integration of AI and OR Techniques in Constraint Programming for Combinatorial Optimization Problems*.

de Backer, B., Furnon, V., Shaw, P., Kilby, P., and Prosser, P. (2000). Solving vehicle routing problems with constraint programming and metaheuristics. *Journal of Heuristics*, 6:501–523.

Decker, K. (1987). Distributed problem-solving techniques: A survey. *IEEE Transactions on Systems, Man, and Cybernetics*, 17(5):729–740.

DeJong, K. (1975). *An Analysis of the Behavior of a Class of Genetic Adaptive Systems*. PhD thesis, University of Michigan, Ann Arbor.

Di Gaspero, L. and Schaerf, A. (2000). EasyLocal++: An object-oriented framework for flexible design of local search algorithms. Technical Report UDMI/13/2000/RR, Universita degli Studi di Udine.
`http://www.diegm.uniud.it/schaerf/projects/local++.`

Di Gaspero, L. and Schaerf, A. (2001). Easylocal++: An object-oriented framework for the flexible design of local search algorithms and metaheuristics. In *Proceedings of the 4th Metaheuristics International Conference (MIC 2001)*.

Dincbas, M., van Hentenryck, P., Simonis, H., Aggoun, A., Graf, T., and Berthier, F. (1988). The constraint programming language CHIP. In *Proceedings on the International Conference on Fifth Generation Computer Systems FGCS-88*, pages 693–702. Ohmsha Publishers.

Dodd, N. (1990). Slow annealing versus multiple fast annealing runs – An empirical investigation. *Parallel Computing*, 16:269–272.

Dorigo, M., Maniezzo, V., and Colorni, A. (1996). Ant system: Optimization by a colony of cooperating agents. *IEEE Transactions on Systems, Man, and Cybernetics*, B - 26:29–41.

Dorne, R. and Hao, J.-K. (1998). A new genetic local search algorithm for graph coloring. In Eiben, A. E., Bäck, T., Schoenauer, M., and Schwefel, H.-P., editors, *Parallel Problem Solving from Nature – PPSN V*, pages 745–754. Springer, Berlin.

Dowsland, K. (1993). Simulated annealing. In Reeves, C., editor, *Modern Heuristic Techniques for Combinatorial Problems*, pages 20–69. Halsted, Blackwell.

Dueck, G. and Scheuer, T. (1990). Threshold accepting: A general purpose optimization algorithm appearing superior to simulated annealing. *Journal of Computational Physics*, 90:161–175.

Duin, C. and Voß, S. (1994). Steiner tree heuristics - A survey. In Dyckhoff, H., Derigs, U., Salomon, M., and Tijms, H., editors, *Operations Research Proceedings 1993*, pages 485–496, Berlin. Springer.

Duin, C. and Voß, S. (1999). The pilot method: A strategy for heuristic repetition with application to the Steiner problem in graphs. *Networks*, 34:181–191.

Durfee, E., Lesser, V., and Corkill, D. (1989). Cooperative distributed problem solving. In Cohen, P. and Feigenbaum, E., editors, *The Handbook of Artificial Intelligence*, volume 4, pages 83–137. Addison-Wesley, Reading.

Eclipse (2001). Constraint logic programming system.
`http://www.icparc.ic.ac.uk/eclipse/.`

Falkenauer, E. (1998). *Genetic Algorithms and Grouping Problems*. Wiley, Chichester.

Faupel, M. (1998). GAJIT: Genetic Algorithm Java Implementation Toolkit.
`http://www.angelfire.com/ca/Amnesiac/gajit.html.`

Fayad, M. and Schmidt, D. (1997a). Object-oriented application frameworks. *Communications of the Association of Computing Machinery*, 40(10):32–38.

Fayad, M. and Schmidt, D., editors (1997b). *Special Issue: Object-Oriented Application Frameworks*. Communications of the Association of Computing Machinery 40 (10):32–87.

Feo, T. and Resende, M. (1995). Greedy randomized adaptive search procedures. *Journal of Global Optimization*, 6:109–133.

Feo, T., Resende, M., and Smith, S. (1994). A greedy randomized adaptive search procedure for maximum independent set. *Operations Research*, 42:860–878.

Ferland, J., Hertz, A., and Lavoie, A. (1996). An object-oriented methodology for solving assignment-type problems with neighborhood search techniques. *Operations Research*, 44:347–359.

Fernandez, A. and Hill, P. (2000). A comparative study of eight constraint programming languages over the boolean and finite domains. *Constraints*, 5(3):275–301.

Fink, A. (2000). *Software-Wiederverwendung bei der Lösung von Planungsproblemen mittels Meta-Heuristiken*. Shaker, Aachen.

Fink, A., Schneidereit, G., and Voß, S. (2000). Solving general ring network design problems by meta-heuristics. In Laguna, M. and González Velarde, J., editors, *Computing Tools for Modeling, Optimization and Simulation*, pages 91–113. Kluwer, Boston.

Fink, A. and Voß, S. (1998). Building reusable software components for heuristic search. In *Modern Heuristics for Decision Support*, London. UNICOM Seminars Ltd.

Fink, A. and Voß, S. (1999a). Applications of modern heuristic search methods to pattern sequencing problems. *Computers & Operations Research*, 26:17–34.

Fink, A. and Voß, S. (1999b). Generic metaheuristics application to industrial engineering problems. *Computers & Industrial Engineering*, 37:281–284.

Fink, A. and Voß, S. (2001). Solving the continuous flow-shop scheduling problem by metaheuristics. Technical report, Braunschweig University of Technology.

Fink, A., Voß, S., and Woodruff, D. L. (1999a). An adoption path for intelligent heuristic search componentware. In Rolland, E. and Umanath, N., editors, *Proceedings of the 4th INFORMS Conference on Information Systems and Technology*, pages 153–168. INFORMS, Linthicum.

Fink, A., Voß, S., and Woodruff, D. L. (1999b). Building reusable software components for heuristic search. In Kall, P. and Lüthi, H.-J., editors, *Operations Research Proceedings 1998*, pages 210–219. Springer, Berlin.

Fink, A., Voß, S., and Woodruff, D. L. (2001). Optimization software libraries. In Pardalos, P. and Resende, M., editors, *Handbook of Applied Optimization*. Oxford University Press, New York. In print.

Flanagan, D. (1999). *Java in a Nutshell*. O'Reilly & Associates, 3rd edition.

Fleurent, C. and Ferland, J. (1996). Object-oriented implementation of heuristic search methods for graph coloring, maximum clique, and satisfiability. In Johnson, D. and Trick, M., editors, *Cliques, Coloring, and Satisfiability: Second DIMACS Implementation Challenge*, volume 26 of *DIMACS Series in Discrete Mathematics and Theoretical Computer Science*, pages 619–652. AMS, Princeton.

Fogel, D. (1993). On the philosophical differences between evolutionary algorithms and genetic algorithms. In Fogel, D. and Atmar, W., editors, *Proceedings of the Second Annual Conference on Evolutionary Programming*, pages 23–29. Evolutionary Programming Society, La Jolla.

Fogel, D. (1995). *Evolutionary Computation: Toward a New Philosophy of Machine Intelligence*. IEEE Press, New York.

Foulds, L. and Graham, R. (1982a). The Steiner problem in phylogeny is NP-complete. *Advances in Applied Mathematics*, 3:43–49.

Foulds, L. and Graham, R. (1982b). Unlikelihood that minimal phylogenies for a realistic biological study can be constructed in reasonable computational time. *Mathematical Biosciences*, 60:133–142.

Fourer, R. (1998). Extending a general-purpose algebraic modeling language to combinatorial optimization: A logic programming approach. In Woodruff, D., editor, *Advances in Computational and Stochastic Optimization, Logic Programming, and Heuristic Search*, pages 31–74. Kluwer, Boston.

Fourer, R. (2001). Software survey: Linear programming. *OR/MS Today*, 28(4):58–68.

Fourer, R., Gay, D., and Kernighan, B. (1993). *AMPL: A Modeling Language for Mathematical Programming*. The Scientific Press, San Francisco.

French, A., Robinson, A., and Wilson, J. (1997). A hybrid genetic / branch and bound algorithm for integer programming. In *Proceedings of the International Conference on Artificial Neural Nets and Genetic Algorithms*, Norwich, UK.

Frühwirth, T. (1998). Theory and practice of constraint handling rules. *Journal of Logic Programming, Special Issue on Constraint Logic Programming*, 37:95–138.

GAGS (2001). GAGS: Genetic Algorithm from Granada, Spain.
`http://kal-el.ugr.es/gags.man.html.`

GAlib (2001). A C++ Library of Genetic Algorithm Components.
`http://lancet.mit.edu/ga/.`

Gallego, R., Alves, A., Monticelli, A., and R.Romero (1997). Parallel simulated annealing applied to long term transmission network expansion planning. *IEEE Transactions on Power Systems*, 12(1):181–186.

Gamma, E., Helm, R., Johnson, R., and Vlissides, J. (1995). *Design Patterns – Elements of Reusable Object-Oriented Software*. Addison-Wesley, Reading.

GAMS (2001). Modeling language.
`http://www.gams.com.`

Garey, M. and Johnson, D. (1979). *Computers and Intractability: A Guide to the Theory of NP-Completeness*. Freeman, San Francisco.

Geist, A., Bequelin, A., and Dongarra, J. (1994). *PVM: Parallel Virtual Machine: A Users' Guide and Tutorial for Networked Parallel Computing Scientific and Engineering Computation Series*. MIT Press, Cambridge.

Genitor (2001). A genetic algorithm library.
`http://www.cs.colostate.edu/~genitor/.`

Gillam, R. (1998). The assignment operator revisited. *C++ Report*, 10 (5):28–34, 41.

Glover, F. (1977). Heuristics for integer programming using surrogate constraints. *Decision Sciences*, 8:156–166.

Glover, F. (1986). Future paths for integer programming and links to artificial intelligence. *Computers & Operations Research*, 13:533–549.

Glover, F. (1990). Tabu search – Part II. *ORSA Journal on Computing*, 2:4–32.

Glover, F. (1994). Optimization by ghost image processing in neural networks. *Computers & Operations Research*, 8:801–822.

Glover, F. (1995). Scatter search and star-paths: Beyond the genetic metaphor. *OR Spektrum*, 17:125–137.

Glover, F. (1997). Tabu search and adaptive memory programming – Advances, applications and challenges. In Barr, R., Helgason, R., and Kennington, J., editors, *Interfaces in Computer Science and Operations Research: Advances in Metaheuristics, Optimization, and Stochastic Modeling Technologies*, pages 1–75. Kluwer, Boston.

Glover, F., editor (1998a). *Tabu Search Methods for Optimization*. European Journal of Operational Research 106:221–692. Elsevier, Amsterdam.

Glover, F. (1998b). A template for scatter search and path relinking. In Hao, J.-K., Lutton, E., Ronald, E., Schoenauer, M., and Snyers, D., editors, *Artificial Evolution*, volume 1363 of *Lecture Notes in Computer Science*, pages 13–54. Springer, Berlin.

Glover, F. and Laguna, M. (1997). *Tabu Search*. Kluwer, Boston.

Glover, F., Laguna, M., and Martí, R. (1999). Scatter search. In Ghosh, A. and Tsutsui, S., editors, *Theory and Applications of Evolutionary Computation: Recent Trends*. Springer, Berlin. To appear.

Glover, F., Laguna, M., and Martí, R. (2000a). Fundamentals of scatter search and path relinking. *Control and Cybernetics*, 39:653–684.

Glover, F., Løkketangen, A., and Woodruff, D. L. (2000b). Scatter search to generate diverse MIP solutions. In Laguna, M. and Velarde, J. L. G., editors, *Computing Tools for Modeling, Optimization and Simulation*, pages 299–320. Kluwer, Boston.

Gnuplot (2001). Gnuplot [graphical output]. `ftp://ftp.dartmouth.edu-/pub/gnuplot/gnuplotxxx.tar.Z.`

Goldberg, D. (1989). *Genetic Algorithms in Search, Optimization, and Machine Learning*. Addison-Wesley, Reading.

Graffigne, C. (1992). Parallel annealing by periodically interacting multiple searches: An experimental study. In Azencott, R., editor, *Simulated Annealing – Parallelization Techniques*, pages 47–79. Wiley, New York.

Greening, D. R. (1990). Parallel simulated annealing techniques. *Physica D*, 42:293–306.

Grefenstette, J. (1990). *A User's guide to GENESIS Version 5.0.* Department of Computer Science, Vanderbilt University, Nashville.

Gu, J. (1999). The Multi-SAT algorithm. *Discrete Applied Mathematics*, 96–97:111–126.

Gutenschwager, K., Niklaus, C., and Voß, S. (2001). Dispatching of an electronic monorail system: Applying meta-heuristics to an online pickup and delivery problem. Technical report, University of Technology Braunschweig.

Hancock, P. (1994). An empirical comparison of selection methods in evolutionary algorithms. In Fogarty, T., editor, *Evolutionary Computing: AISB Workshop, Leeds, UK, April 1994: Selected Papers*, volume 865 of *Lecture Notes in Computer Science*. Springer, Berlin.

Hansen, P. and Mladenović, N. (1999). An introduction to variable neighborhood search. In Voß, S., Martello, S., Osman, I., and Roucairol, C., editors, *Meta-Heuristics: Advances and Trends in Local Search Paradigms for Optimization*, pages 433–458. Kluwer, Boston.

Hart, J. and Shogan, A. (1987). Semi-greedy heuristics: An empirical study. *Operations Research Letters*, 6:107–114.

Harvey, W. and Ginsberg, M. (1995). Limited discrepancy search. In *Proceedings of the 14th IJCAI*, pages 607–615, San Mateo. Morgan Kaufmann.

Heipcke, S. (1999). Comparing constraint programming and mathematical programming approaches to discrete optimisation – the change problem. *Journal of the Operational Research Society*, 50:581–595.

Heitkötter, J. and Beasley, D. (2001). The hitch-hiker's guide to evolutionary computation (FAQ for comp.ai.genetic). Issue 9.1, 12 April 2001 `http://surf.de.uu.net/encore/www/`.

Heller, D. and Ferguson, P. (1991). *Motif Programming Manual*, volume 6A. O'Reilly & Associates, Inc.

Henz, M., Smolka, G., and Wurtz, J. (1993). Oz – a programming language for multiagent systems. In *Proceedings of the 13th IJCAI*, pages 404–409, San Mateo. Morgan Kaufmann.

Henz, M., Smolka, G., and Wurtz, J. (1995). Object-oriented concurrent constraint programming in Oz. In Saraswat, V. and van Hentenryck, P., editors, *Principles and Practice of Constraint Programming*, chapter 2, pages 29–48. MIT Press, Cambridge, MA.

Hertz, A. and Kobler, D. (2000). A framework for the description of evolutionary algorithms. *European Journal of Operational Research*, 126:1–12.

Hoffmeister, F. and Bäck, T. (1991). Genetic algorithms and evolution strategies: Similarities and differences. In Schwefel, H.-P. and Männer, R., editors, *Parallel Problem Solving from Nature – PPSN I*, number 496 in Lecture Notes in Computer Science, pages 455–469. Springer, Berlin.

Holland, J. (1975). *Adaptation in Natural and Artificial Systems*. The University of Michigan Press, Ann Arbor.

Hooker, J. (1994). Needed: An empirical science of algorithms. *Operations Research*, 42:201–212.

Hooker, J. (1998). Constraint satisfaction methods for generating valid cuts. In Woodruff, D., editor, *Advances in Computational and Stochastic Optimization, Logic Programming, and Heuristic Search*, pages 1–30. Kluwer, Boston.

Hoos, H. and Stützle, T. (2000). Local search algorithms for SAT. *Journal of Automated Reasoning*, 24:421–481.

Hoover, R. (1987). *Incremental Graph Evaluation*. PhD thesis, Department of Computer Science, Cornell University, Montpellier, New York.

Horstmann, C. and Cornell, G. (1998). *Core Java 1.2 Volume 1: Fundamentals*. Prentice Hall, Englewood Cliffs.

HotFrame (2001). Heuristic OpTimization FRAMEwork. `http://www.winforms.phil.tu-bs.de/winforms/research/hotframe.html`.

Howard, C. and Rayward-Smith, V. (1998). Knowledge discovery from low quality databases. In *IEE Digest 98/483, Colloquium on Knowledge Discovery and Data Mining*.

Hudson, S. (1991). Incremental attribute evaluation: A flexible algorithm for lazy update. *ACM Transactions on Programming Languages and Systems*, 13:315–341.

Hunter, A. (1995). *SUGAL User Manual V2.1*. Sunderland University, UK.

IBM Open Source Software (2001). OpenTS. `http://oss.software.ibm.com/developerworks/opensource/coin/OpenTS/`.

IC-PARC (2001). *The ECLiPSe Library Manual Release 5.2*. `http://www.icparc.ic.ac.uk/eclipse`.

ILOG (2000a). *ILOG Concert Technology User's Manual, Version 1.1*. ILOG S.A., 9, Rue de Verdun, Gentilly, France.

ILOG (2000b). *ILOG Solver User's Manual, Version 5.1*. ILOG S.A., 9, Rue de Verdun, Gentilly, France.

ILOG (2001). Constraint logic programming libraries. `http://www.ilog.com`.

IMSL (2001). Mathematics and Statistics Libraries. `http://www.vni.com/products/imsl/index.html`.

Jaffar, J., editor (1999). *Principles and Practice of Constraint Programming – CP '99*, volume 1713 of *Lecture Notes in Computer Science*. Springer, Berlin.

JDEAL (2001). `http://laseeb.ist.utl.pt/sw/jdeal/`.

Jiang, Y., Kautz, H., and Selman, B. (1995). Solving problems with hard and soft constraints using a stochastic algorithm for MAX-SAT. Technical report, AT&T Bell Laboratories.

Johnson, D. S., Aragon, C. R., McGeoch, L. A., and Schevon, C. (1989). Optimization by simulated annealing: An experimental evaluation; part i, graph partitioning. *Operations Research*, 37:865–892.

Johnson, R. (1997). Frameworks = components + patterns. *Communications of the Association of Computing Machinery*, 40(10):39–42.

Johnson, R. and Foote, B. (1988). Designing reusable classes. *Journal of Object-Oriented Programming*, 1 (2):22–35.

Jones, C. (1994). Visualization and optimization. *ORSA Journal on Computing*, 6:221–257.

Jones, C. (1996). *Visualization and Optimization*. Kluwer, Boston.

Jones, M. (2000). *An Object-Oriented Framework for the Implementation of Search Techniques*. PhD thesis, University of East Anglia.

Jones, M., McKeown, G., and Rayward-Smith, V. (1998). Templar: An object-oriented framework for distributed combinatorial optimzation. In *Proceedings of the UNICOM Seminar on Modern Heuristics for Decision Support*. UNICOM Ltd, Brunel University, UK.

Jünger, M. and Thienel, S. (1997). The design of the branch-and-cut system ABACUS3. Technical Report TR97.263, University of Cologne, Dept. of Computer Science.

Jünger, M. and Thienel, S. (2000). The ABACUS system for branch-and-cut-and-price algorithms in integer programming and combinatorial optimization. *Software – Practice and Experience*, 30:1325–1352.

Karisch, S., Burkard, R., and Rendl, F. (1997). QAPLIB a quadratic assignment problem library. *Journal of Global Optimization*, 10:391–403.

Kautz, H. and Selman, B. (1996). Pushing the envelope: Planning, propositional logic, and stochastic search. In *Proceedings of the Thirteenth National Conference on Artificial Intelligence (AAAI-96)*, pages 1194–1201. MIT Press, Cambridge, MA.

Kelley, A. and Pohl, I. (1990). *A Book on C*. Benjamin/Cummings, Redwood City, CA, 2nd edition.

Kilby, P., Prosser, P., and Shaw, P. (2000). A comparison of traditional and constraint-based heuristic methods on vehicle routing problems with side constraints. *Constraints*, 5:389–414.

Kirkpatrick, S., Gelatt Jr., C., and Vecchi, M. (1983). Optimization by simulated annealing. *Science*, 220:671–680.

Laburthe, F. and Caseau, Y. (1998). SALSA: A language for search algorithms. In Maher, M. and Puget, J.-F., editors, *Principles and Practice of Constraint Programming – CP98*, number 1520 in Lecture Notes in Computer Science, pages 310–324. Springer, Berlin.

Laguna, M. (2001). Scatter search. In Pardalos, P. and Resende, M., editors, *Handbook of Applied Optimization*. Oxford University Press, New York. To appear.

Laguna, M. and Armentano, V. (2001). Lessons from applying and experimenting with scatter search. In Rego, C. and Alidaee, B., editors, *Adaptive Memory and Evolution: Tabu Search and Scatter Search*. Kluwer, Boston. To appear.

Laguna, M., Lino, P., Pérez, A., Quintanilla, S., and Valls, V. (2000). Minimizing weighted tardiness of jobs with stochastic interruptions in parallel machines. *European Journal of Operational Research*, 127:444–457.

Laguna, M. and Martí, R. (2000). Experimental testing of advanced scatter search designs for global optimization of multimodal functions. Technical report, University of Valencia.

Laporte, G. and Osman, I., editors (1996). *Metaheuristics in Combinatorial Optimization*. Annals of Operations Research 63. Baltzer, Amsterdam.

Lauria, M. and Chien, A. (1997). MPI-FM: High performance MPI on workstation clusters. *Journal of Parallel and Distributed Computing*, 40:4–18.

Lawler, E., Lenstra, J., Rinnooy Kan, A., and Shmoys, D., editors (1985). *The Traveling Salesman Problem: A Guided Tour of Combinatorial Optimization*. Wiley, Chichester.

Lee, E. (2000). What's ahead for embedded software? *IEEE Computer*, 33:18–26.

Lee, F.-H. (1995). *Parallel Simulated Annealing on a Message-Passing Multi-Computer*. PhD thesis, Electrical Engineering, Logan, Utah.

Lee, S.-Y. and Lee, K. (1996). Synchronous and asynchronous parallel simulated annealing with multiple markov chains. *IEEE Transactions on Parallel and Distributed Systems*, 7(10):993–1007.

Levine, D. (1996). *Users Guide to the PGAPack Parallel Genetic Algorithm Library*. Mathematics and Computer Science Division, Argonne National Laboratory.

Liddle, S. and Hansen, J. (1997). An object-oriented method and language for implementing cooperative distributed problem solving. *Annals of Operations Research*, 75:147–169.

Lin, S. (1965). Computer solutions of the traveling salesman problem. *Bell Systems Tech. J.*, 44:2245–2269.

Lin, S. and Kernighan, B. (1973). An effective heuristic algorithm for the traveling-salesman problem. *Operations Research*, 21:498–516.

LINDO (2001). Mathematical programming software. `http://www.lindo.com.`

Liston, G. (1993). 'Channel 4 Television are using advanced optimization technology based on genetic algorithms'. *Computer World*, 20th December.

Lobo, F. and Goldberg, D. (1996). Decision making in a hybrid genetic algorithm. Technical Report IlliGAL 96009, University of Illinois at Urbana-Champaign, Urbana, IL.

Mackworth, A. K. (1977). Consistency in networks of relations. *Artificial Intelligence*, 8:99–118.

Mamrak, S. and Sinha, S. (1999). A case study: Productivity and quality gains using an object-oriented framework. *Software – Practice and Experience*, 29(6).

Mann, J. (1995a). X-GAmeter devloper's manual. Technical report, University of East Anglia.

Mann, J. (1995b). X-SAmson developer's manual. Technical report, University of East Anglia.

Marriott, K. and Stuckey, P. J. (1998). *Programming with Constraints: An Introduction*. MIT Press, Cambridge, MA.

Marsaglia, G. and Zaman, A. (1994). Some portable very-long-period random number generators. *Computers in Physics*, 8(1):117–121.

Martin, R. (1997). Design patterns for dealing with dual inheritance hierarchies in C++. *C++ Report*, 9 (4):42–48.

Marzetta, A. (1998). *A Library of Parallel Search Algorithms and its Use in Enumeration and Combinatorial Optimization*. PhD thesis, Zurich.

MATLAB (2001). An integrated technical computing environment. `http://www.matlab.com.`

Matteis, A. D. and Pagnutti, S. (1995). Controlling correlations in parallel monte carlo. *Parallel Computing*, 21:73–84.

Mautor, T. and Michelon, P. (1997). MIMAUSA: A new hybrid method combining exact solution and local search. In *Second Meta-Heuristics International Conference–MIC'97*. INRIA, Sophia Antipolis.

McAllester, D., Selman, B., and Kautz, H. (1997). Evidence for invariants in local search. In *Proceedings of the Fourteenth National Conference on Artificial Intelligence (AAAI-97)*, 321–326. AAAI Press/MIT Press, Cambridge.

McAloon, K., Tretkoff, C., and Wetzel, G. (1997). Sport League Scheduling. In *Proceedings of the 3th Ilog International Users Meeting*, Paris, France.

McBryan, O. (1994). An overview of message passing environments. *Parallel Computing*, 20:417–444.

Mens, T., Lucas, C., and Steyaert, P. (1999). Supporting disciplined reuse and evolution of UML models. In Bézivin, J. and Muller, P.-A., editors, *The Unified Modeling Language – UML'98: Beyond the Notation*, number 1618 in Lecture Notes in Computer Science, pages 378–392. Springer, Berlin.

Merelo, J. (1994). *GAGS 0.94: User's Manual*. Department of Electronics and Computer Technology, Granada University, Spain.

Merelo, J. (1995). *libGAGS 0.94 Programmer's Manual*. Department of Electronics and Computer Technology, Granada University, Spain.

Meyer, B. (1990). Eiffel – The Language. In *Series in Computer Science*. Prentice-Hall, Englewood Cliffs.

Meyer, B. (1997). *Object-Oriented Software Construction*. Prentice Hall, Englewood-Cliffs, 2nd edition.

Michalewicz, Z. (1999). *Genetic Algorithms + Data Structures = Evolution Programs*. Springer, Berlin, 3rd (corrected reprinting) edition.

Michel, L. and van Hentenryck, P. (1997). Localizer: A modeling language for local search. In Smolka, G., editor, *Proceedings of the Third International Conference on Principles and Practice of Constraint Programming (CP '97)*, pages 237–251. Springer, Berlin.

Michel, L. and van Hentenryck, P. (1998). Localizer. Technical Report CS-98-02, Brown University, Providence, Rhode Island 02912.

Michel, L. and van Hentenryck, P. (1999). LOCALIZER: A modeling language for local search. *INFORMS Journal on Computing*, 11:1–14.

Michel, L. and van Hentenryck, P. (2000). Localizer. *Constraints*, 5:43–84.

Michel, L. and van Hentenryck, P. (2001a). Localizer++: An open library for local search. Technical Report CS-01-02, Computer Science Department, Brown University.

Michel, L. and van Hentenryck, P. (2001b). Modeler++: A Modeling Layer for Constraint Programming Libraries. In *CP-AI-OR'2001*, Wye College (Imperial College), Ashford, Kent UK.

MINTO (1999). A Mixed INTeger Optimizer.
`http://akula.isye.gatech.edu/~mwps/projects/minto.html`.

Minton, S., Johnston, M. D., Philips, A. B., and Laird, P. (1992). Minimizing conflicts: A heuristic repair method for constraint satisfaction and scheduling problems. *Artificial Intelligence*, 58:161–205.

Mladenović, N. and Hansen, P. (1996). Variable neighbourhood search. *Computers & Operations Research*, 24:1097–1100.

MOPS (2001). Mathematical programming software.
`http://www.mops.tu-berlin.de`.

More, J. and Wright, S. (1993). *Optimization Software Guide (Frontiers in Applied Mathematics, Vol. 14)*. Society for Industrial & Applied Mathematics (SIAM), Philadelphia.

Moscato, P. (1993). An introduction to population approaches for optimization and hierarchical objective functions: A discussion on the role of tabu search. *Annals of Operations Research*, 41:85–121.

MPI (1995). *MPI: A Message-Passing Interface Standard*. ARPA, NSF and Esprit.

MPL (2001). Maximal software, inc. `http://www.maximal-usa.com`.

Mühlenbein, H. (1997). Genetic algorithms. In Aarts, E. and Lenstra, J., editors, *Local Search in Combinatorial Optimization*, pages 137–171. Wiley, Chichester.

Musser, D. and Saini, A. (1996). *STL Tutorial and Reference Guide: C++ Programming with the Standard Template Library*. Addison-Wesley, Reading.

Musser, D. and Stepanov, A. (1994). Algorithm-oriented generic libraries. *Software – Practice and Experience*, 24:623–642.

Muth, J. and Thompson, G., editors (1963). *Industrial scheduling*. Prentice-Hall, Englewood Cliffs.

Myers, B., Giuse, D., Dannenberg, R., Zanden, B., Kosbie, D., Pervin, E., Mickish, A., and Marchal, P. (1990). Comprehensive support for graphical, highly-interactive user interfaces. *IEEE Computer*, 23:71–85.

Myers, B., McDaniel, R., Miller, R., Ferrency, A., Faulring, A., Kyle, B., Mickish, A., Klimovitski, A., and Doane, P. (1997). The amulet environment: New models for effective user interface software development. *IEEE Transactions on Software Engineering*, 23:347–365.

NAG (2001). Numerical Libraries, Numerical Algorithms Group. `http://www.nag.com`.

Nareyek, A. (2001). *Constraint-Based Agents*. Number 2062 in Lecture Notes in Artificial Intelligence. Springer, Berlin.

NEOS (2001). NEOS guide to optimization software. `http://www.mcs.anl.gov/otc/Guide/SoftwareGuide/`.

NETFLOW (2001). Network flow packages. `ftp://dimacs.rutgers.edu/pub/netflow`.

Nichols, B., Buttlar, D., and Farrell, J. (1996). *Pthreads Programming*. O'Reilly and Associates.

Nievergelt, J. (1994). Complexity, algorithms, programs, systems: The shifting focus. *Journal of Symbolic Computation*, 17:297–310.

Nye, A. (1995). *Xlib Programming Manual*, volume 1. O'Reilly & Associates, Inc.

Nye, A. and O'Reilly, T. (1995). *X Toolkit Intrinsics Programming Manual*, volume 4. O'Reilly & Associates, Inc.

OptQuest (2000). *OptQuest Callable Library User's Manual*. Boulder, CO. `http://www.opttek.com`.

OSL (2001). The IBM Optimization Subroutine Library. `http://www6.software.ibm.com/sos/osl/optimization.htm`.

Osman, I. H. and Kelly, J. P., editors (1996). *Meta-Heuristics: Theory and Applications*. Kluwer, Boston.

Panda, D. and Ni, L. (1997a). Special issue on workstation clusters and network-based computing. *Journal of Parallel and Distributed Computing*, 40:1–3.

Panda, D. and Ni, L. (1997b). Special issue on workstation clusters and network-based computing. *Journal of Parallel and Distributed Computing*, 43:63–64.

Pearl, J. (1984). *Heuristics: Intelligent Search Strategies for Computer Problem Solving*. Addison-Wesley, Reading.

PERL (2001). PERL [scripting language]. `ftp://ftp.cis/ufl.edu-/pub/perl/src/xxx`.

Pesant, G. and Gendreau, M. (1996). A view of local search in constraint programming. In Freuder, E. C., editor, *Proceedings of the Second International Conference on Principles and Practice of Constraint Programming (CP '96)*, pages 353–366. Springer, Berlin.

Pesant, G. and Gendreau, M. (1999). A constraint programming framework for local search methods. *Journal of Heuristics*, 5:255–279.

Pesch, E. and Voß, S., editors (1995). *Applied Local Search*. OR Spektrum 17:55–225. Springer, Berlin.

Pinter, J. D. (1996). *Global Optimization in Action: Continuous and Lipschitz Optimization – Algorithms, Implementations and Applications*. Kluwer, Boston.

Polya, G. (1945). *How to Solve it*. Princeton University Press, Princeton.

Pree, W. (1994). *Design Patterns for Object Oriented Software Development*. Addison-Wesley, Reading.

Press, W. H., Teukolsky, S. A., Vetterling, W. T., and Flannery, B. P. (1993). *Numerical Recipes in C: The Art of Scientific Computing, 2nd Ed.* Cambridge University Press, Cambridge.

Prestwich, S. (2000). A hybrid search architecture applied to hard random 3-SAT and low-autocorrelation binary sequences. In *Proceedings of the Sixth International Conference on Principles and Practice of Constraint Programming (CP '00)*, pages 337–352. Springer, Berlin.

Puget, J.-F. (1995). Applications of constraint programming. In Montanari, U. and Rossi, F., editors, *Principles and Practice of Constraint Programming*, pages 647–650. Springer. Berlin.

Puget, J.-F. and Leconte, M. (1995). Beyond the glass box: constraints as objects. In Lloyd, J., editor, *Proceedings of the International Symposium on Logic Programming*, pages 513–527. MIT Press, Cambridge, MA.

PVM (2001). Parallel Virtual Machine.
`http://www.netlib.org/pvm3/index.html`
`http://www.epm.ornl.gov/pvm/pvm\_home.html`.

Radcliffe, N. and Surry, P. (1994). Formal memetic algorithms. In Fogarty, T., editor, *Evolutionary Computing: AISB Workshop*. Springer, Berlin.

Raines, P. and Tranter, J. (1999). *TCL/TK In A Nutshell*. O'Reilly & Associates, CA.

Ram, D., Sreenivas, T., and Subramaniam, K. (1996). Parallel simulated annealing algorithms. *Journal of Parallel and Distributed Computing*, 37:207–212.

Rayward-Smith, V., editor (1995). *Applications of Modern Heuristic Methods*. Waller, Henley-on-Thames.

Rayward-Smith, V., Osman, I., Reeves, C., and Smith, G., editors (1996). *Modern Heuristic Search Methods*. Wiley, Chichester.

Redmond, F. (1997). *DCOM: Microsoft Distributed Component Object Model*. IDG Books Worldwide.

Reeves, C., editor (1993). *Modern Heuristic Techniques for Combinatorial Problems*. Blackwell, Oxford. Re-issued by McGraw-Hill, London (1995).

Reeves, C. (1997). Genetic algorithms for the operations researcher. *INFORMS Journal on Computing*, 9:231–250.

Reeves, C. and Rowe, J. (2002). *Genetic Algorithms: Principles and Perspectives*. Kluwer, Boston. To appear.

Régin, J.-C. (1994). A filtering algorithm for constraints of difference in CSPs. In *Proceedings of the Twelfth National Conference on Artificial Intelligence (AAAI-94)*, pages 362–367. MIT Press, Cambridge, MA.

Régin, J.-C. (1996). Generalized arc consistency for global cardinality constraint. In *Proceedings of the Thirteenth National Conference on Artificial Intelligence (AAAI-96)*, pages 209–215. MIT Press, Cambridge, MA.

Régin, J.-C. (1998). Sport league scheduling. In *INFORMS*, Montreal, Canada.

Regin, J.-C. and Puget, J.-F. (1997). A filtering algorithm for global sequencing constraints. In Smolka, G., editor, *Proceedings of the Third International Conference on Principles and Practice of Constraint Programming (CP '97)*, pages 32–46. Springer, Berlin.

Reinelt, G. (1991). TSPLIB - A travelling salesman problem library. *ORSA Journal on Computing*, 3:376–384.

Reinelt, G. (1994). *The Traveling Salesman: Computational Solutions for TSP Applications*. Springer, Berlin.

Resende, M., Pardalos, P., and Li, Y. (1996). Algorithm 754: Fortran subroutines for approximate solution of dense quadratic assignment problems using GRASP. *ACM Transactions on Mathematical Software*, 22(1):104–118.

Resende, M. and Ribeiro, C. (1997). A GRASP for graph planarization. *Networks*, 29:173–189.

Ribeiro, C. and Hansen, P., editors (2002). *Essays and Surveys in Metaheuristics*. Kluwer, Boston.

Ribeiro, C., Uchoa, E., and Werneck, R. (2000). A hybrid GRASP with perturbations for the Steiner problem in graphs. Technical report, Department of Computer Science, Catholic University of Rio de Janeiro.

Rich, E. and Knight, K. (1991). *Artificial Intelligence*. McGraw-Hill, 2nd edition.

Rochat, Y. and Taillard, E. D. (1995). Probabilistic diversification and intensification in local search for vehicle routing. *Journal of Heuristics*, 1:147–167.

Rogers, E. (1995). *Diffusion of Innovations*. The Free Press, New York, 4th edition.

Rossi, F. (2000). Constraint logic programming. In *Proceedings of the ERCIM/Compulog Net Workshop on Constraints, Lecture Notes on Artificial Intelligence 1865*. Springer, Berlin.

Rousseau, L.-M., Gendreau, M., and Pesant, G. (2000). Using constraint-based operators to solve the vehicle routing problem with time windows. Technical report, CRT, University of Montreal, Canada.

Rumbaugh, J. (1995a). OMT: The dynamic model. *Journal of Object Oriented Programming*, 7:6–12.

Rumbaugh, J. (1995b). OMT: The object model. *Journal of Object Oriented Programming*, 7:21–27.

Rumbaugh, J., Blaha, M., Premerlani, W., Eddy, F., and Lorensen, W. (1991). *Object-Oriented Modelling and Design*. Prentice Hall, Englewood-Cliffs.

Ryan, M., Debuse, J., Smith, G., and Whittley, I. (1999). A hybrid genetic algorithm for the fixed channel assignment problem. In *Proceedings of the 1999 Genetic and Evolutionary Computation Conference (GECCO '99)*.

Sakawa, M. (2001). *Genetic Algorithms and Fuzzy Multiobjective Optimization*. Kluwer, Boston.

Sankoff, D. and Rousseau, P. (1975). Locating the vertices of a Steiner tree in an arbitrary metric space. *Mathematical Programming*, 9:240–246.

Savage, S. (1997). Weighing the pros and cons of decision technology in spreadsheets. *OR/MS Today*, 24 (1):42–45.

Schaerf, A. (1999). A survey of automated timetabling. *Artificial Intelligence Review*, 13(2):87–127.

Schaerf, A., Lenzerini, M., and Cadoli, M. (1999). LOCAL++: A C++ framework for local search algorithms. In *TOOLS Europe '99: Technology of Object Oriented Languages and Systems*, pages 152–161.

Schaffer, J. and Eshelman, L. (1996). Combinatorial optimization by genetic algorithms: The value of the genotype/phenotype distinction. In Rayward-Smith, V., Osman, I., Reeves, C., and Smith, G., editors, *Modern Heuristic Search Methods*, pages 85–97. Wiley, Chichester.

Schmidt, D. (1995). An OO encapsulation of lightweight OS concurrency mechanisms in the ACE toolkit. Technical Report WUCS-95-31, Washington University, St. Louis.

Schmidt, D. and Coplien, J., editors (1995). *Pattern Languages of Program Design (Vol. 1)*. Addison-Wesley, Reading.

Schmidt, D. and Fayad, M. (1997). Lessons learned building reusable OO frameworks for distributed software. *Communications of the Association of Computing Machinery*, 40(10):85–87.

Schmidt, D., Fayad, M., and Johnson, R. (1996). Software patterns. *Communications of the Association of Computing Machinery*, 39:37–39.

Schuurmans, D. and Southey, F. (2000). Local search characteristics of incomplete SAT procedures. In *Proceedings of the 17th National Conference on Artificial Intelligence (AAAI-2000)*, pages 297–302.

Schwefel, H.-P. and Bäck, T. (1998). Artificial evolution: How and why? In Quagliarella, D., Périaux, J., Poloni, C., and Winter, G., editors, *Genetic Algorithms and Evolution Strategy in Engineering and Computer Science: Recent Advances and Industrial Applications*, pages 1–19. Wiley, Chichester.

Seetharaman, K. (1998). The CORBA connection. *Communications of the Association of Computing Machinery*, 41(10):34–36.

Selman, B., Kautz, H., and Cohen, B. (1994). Noise strategies for improving local search. In *Proceedings of the Twelfth National Conference on Artificial Intelligence (AAAI-94)*, pages 337–343. American Association for Artificial Intelligence, AAAI Press / MIT Press, Cambridge, MA.

Selman, B., Levesque, H., and Mitchell, D. (1992). A new method for solving hard satisfiability problems. In *Proceedings of the 9th National Conference on Artificial Intelligence (AAAI-92)*, pages 440–446. MIT Press, Cambridge, MA.

Shaw, P. (1998). Using constraint programming and local search methods to solve vehicle routing problems. In Maher, M. and Puget, J.-F., editors, *Fourth International Conference on Principles and Practice of Constraint Programming (CP '98)*, pages 417–431. Springer, Berlin.

Shaw, P., Furnon, V., and de Backer, B. (2000). A lightweight addition to CP frameworks for improved local search. In Junker, U., Karisch, S. E., and Fahle, T., editors, *Proceedings of CP-AI-OR 2000*. Published as Technical report TR-001-2000,

Paderborn Center for Parallel Computing. Extended version submitted to *Annals of Operations Research*.

Simonis, H., Beldiceanu, N., and Kay, P. (1995). *The CHIP System*. `http://www.cosytec.fr`.

Simos, M. and Anthony, J. (1998). Weaving the model web: A multi-modeling approach to concepts and features in domain engineering. In Devanbu, P. and Poulin, J., editors, *Proceedings of the Fifth International Conference on Software Reuse*, pages 94–102. IEEE Computer Society Press, Los Alamitos.

Sing, L. (1997). *Professional Visual C++ 5 ActiveX/COM Control Programming*. Wrox Press Ltd.

Smith, K. (1999). Neural networks for combinatorial optimisation: A review of more than a decade of research. *INFORMS Journal on Computing*, 11:15–34.

Smolka, G. (1995). The Oz programming model. In van Leeuwen, J., editor, *Computer Science Today*, volume 1000 of *Lecture Notes in Computer Science*. Springer, Berlin.

Sober, E. (1987). Parsimony, likelihood and the principle of the common cause. *Philosophy of Science*, 54:465–469.

SoPlex (2001). Mathematical programming software. `http://www.zib.de/Optimization/Software/Soplex/`.

Spinellis, D. and Papadopoulos, C. (2001). Modular production line optimization: The exPLOre architecture. *Mathematical Problems in Engineering*, 6:527–541.

Storer, R., Wu, S., and Vaccari, R. (1995). Problem and heuristic space search strategies for job shop scheduling. *ORSA Journal on Computing*, 7:453–467.

Stroustrup, B. (1997). *The C++ Programming Language*. Addison-Wesley, Reading, 3rd edition.

Stützle, T. (1998). Applying iterated local search to the permutation flow shop problem. Technical Report AIDA-98-04, FG Intellektik, TU Darmstadt.

Swofford, D. and Olsen, G. (1990). Phylogeny reconstruction. In Hillis, D. and Moritz, C., editors, *Molecular Systematics*, pages 411–501, Sinauer.

Szyperski, C. (1998). *Component Software: Beyond Object-Oriented Programming*. Addison-Wesley, Reading.

Taillard, E. (2000). An introduction to ant systems. In Laguna, M. and Gonzalez-Velarde, J., editors, *Computing Tools for Modeling, Optimization and Simulation*, pages 131–144. Kluwer, Boston.

Taillard, E., Gambardella, L., Gendreau, M., and Potvin, J.-Y. (2001). Adaptive memory programming: A unified view of meta-heuristics. *European Journal of Operational Research*, 135:1–16.

Taillard, E. and Voß, S. (2002). POPMUSIC - Partial optimization metaheuristic under special intensification conditions. In Ribeiro, C. and Hansen, P., editors, *Essays and Surveys in Metaheuristics*, pages 613–629. Kluwer, Boston.

TCL/TK (2001). TCL/TK [Tool command language - pronounced 'tickle']. `ftp://ftp.cs.berkeley.edu/ucb/tcl`.

Templar (2001). The Templar Framework. `http://www.sys.uea.ac.uk/~msj/templar/`.

Thienel, S. (1997). ABACUS - A Branch-And-CUt System, version 2.0, user's guide and reference manual. Technical report, Universität Köln. No. 97.298.

Tian, P., Ma, J., and Zhang, D.-M. (1999). Application of the simulated annealing algorithm to the combinatorial optimisation problem with permutation property: An investigation of generation mechanism. *European Journal of Operational Research*, 118:81–94.

Tödter, K., Hammer, C., and Struckman, W. (1995). Parc++: A parallel C++. *Software – Practice and Experience*, 25(6):623–636.

Tsang, E. (1993). *Foundations of Constraint Satisfaction*. Academic Press, New York.

Ugray, Z., Lasdon, L., Plummer, J., Glover, F., Kelly, J., and Martí, R. (2001). A multistart scatter search heuristic for smooth NLP and MINLP problems. In Rego, C. and Alidaee, B., editors, *Adaptive Memory and Evolution: Tabu Search and Scatter Search*. Kluwer, Boston. To appear.

Vaessens, R., Aarts, E., and Lenstra, J. (1998). A local search template. *Computers & Operations Research*, 25:969–979.

van Hentenryck, P. (1989). *Constraint Satisfaction in Logic Programming*. MIT Press, Cambridge, MA.

van Hentenryck, P. (1995). Constraint solving for combinatorial search problems: A tutorial. In Montanari, U. and Rossi, F., editors, *Principles and Practice of Constraint Programming – CP '95*, volume 976 of *Lecture Notes in Computer Science*, pages 564–587. Springer, Berlin.

van Hentenryck, P. (1999). *The OPL Optimization Programming Language*. MIT Press, Cambridge, MA. With contributions by I. Lustig, L. Michel, and J.-P. Puget.

van Hentenryck, P. and Michel, L. (2000). *OPL Script: Composing and Controlling Models*, volume 1865 of *Lecture Notes in Artificial Intelligence*. Springer, Berlin.

van Hentenryck, P., Michel, L., Laborie, P., Nuijten, W., and Rogerie, J. (1999a). Combinatorial Optimization in OPL Studio. In *Proceedings of the 9th Portuguese Conference on Artificial Intelligence International Conference (EPIA'99)*, Evora, Portugal. (Invited Paper).

van Hentenryck, P., Michel, L., Perron, L., and Regin, J. (1999b). Constraint Programming in OPL. In *Proceedings of the International Conference on the Principles and Practice of Declarative Programming (PPDP'99)*, Paris, France. (Invited Paper).

Verhoeven, M. and Aarts, E. (1995). Parallel local search techniques. *Journal of Heuristics*, 1:43–65.

Vidal, R., editor (1993). *Applied Simulated Annealing*. Number 396 in Lecture Notes in Economics and Mathematical Systems. Springer, Berlin.

Vlissides, J., Coplien, J., and Kerth, N., editors (1996). *Pattern Languages of Program Design (Vol. 2)*. Addison-Wesley, Reading.

Voß, S. (1993a). Intelligent Search. Manuscript, Technische Hochschule Darmstadt.

Voß, S. (1993b). Tabu search: Applications and prospects. In Du, D.-Z. and Pardalos, P., editors, *Network Optimization Problems*, pages 333–353. World Scientific, Singapore.

Voß, S. (1995). Solving quadratic assignment problems using the reverse elimination method. In Nash, S. and Sofer, A., editors, *The Impact of Emerging Technologies on Computer Science and Operations Research*, pages 281–296. Kluwer, Boston.

Voß, S. (1996). A reverse elimination approach for the p-median problem. *Studies in Locational Analysis*, 8:49–58.

Voß, S. (2001). Meta-heuristics: The state of the art. In Nareyek, A., editor, *Local Search for Planning and Scheduling*, volume 2148 of *Lecture Notes in Artificial Intelligence*, pages 1–23. Springer, Berlin.

Voß, S., Martello, S., Osman, I. H., and Roucairol, C., editors (1999). *Meta-Heuristics: Advances and Trends in Local Search Paradigms for Optimization*. Kluwer, Boston.

Voudouris, C., Dorne, R., Lesaint, D., and Liret, A. (2001). iOpt: A software toolkit for heuristic search methods. *7th International Conference on Principles and Practice of Constraint Programming (CP '2001)*.

Voudouris, C. and Tsang, E. (1995a). Function optimization using guided local search. Technical Report CSM-249, Department of Computer Science, University of Essex, Colchester.

Voudouris, C. and Tsang, E. (1995b). Guided local search. Technical Report CSM-247, Department of Computer Science, University of Essex, Colchester.

Wall, L., Christiansen, T., and Schwartz, R. (1991). *Programming Perl*. O'Reilly & Associates, CA.

Wall, M. (1996). GAlib: A C++ Library of Genetic Algorithm Components. Technical report, Mechanical Engineering Department, Massachusetts Institute of Technology.

Wallace, M. (1996). Practical applications of constraint programming. *Constraints*, 1:139–168.

Walser, J. (1999). *Integer Optimization by Local Search*, volume 1637 of *Lecture Notes in Artificial Intelligence*. Springer, Berlin.

Weihe, K. (1997). Reuse of algorithms: Still a challenge to object-oriented programming. *ACM SIGPLAN Notices*, 32 (10):34–48. Proceedings of the 1997 ACM SIGPLAN Conference on Object-Oriented Programming, Systems, Languages & Applications (OOPSLA'97).

Welch, B. (1997). *Practical Programming in Tcl and Tk*. Prentice-Hall, Englewood Cliffs.

Wiley, E., Siegel-Causey, D., Brooks, D., and Funk, V. (1991). The compleat cladist - A primer of phylogenetic procedures. Special publication no. 19.

Witte, E., Chamberlain, R., and Franklin, M. (1991). Parallel simulated annealing using speculative computation. *IEEE Transactions on Parallel and Distributed Systems*, 2(4):483–494.

Wolpert, D. and Macready, W. (1997). No free lunch theorems for optimization. *IEEE Transactions on Evolutionary Computation*, 1:67–82.

Woodruff, D. L. (1995). Ghost image processing for minimum covariance determinant estimators. *ORSA Journal on Computing*, 7:468–473.

Woodruff, D. L. (1997). A class library for heuristic search optimization. *INFORMS Computer Science Technical Section Newsletter*, 18 (2):1–5.

Woodruff, D. L. (1998). Proposals for chunking and tabu search. *European Journal of Operational Research*, 106:585–598.

Woodruff, D. L. (1999). A chunking based selection strategy for integrating metaheuristics with branch and bound. In Voß, S., Martello, S., Osman, I., and Roucairol, C., editors, *Meta-Heuristics: Advances and Trends in Local Search Paradigms for Optimization*, pages 499–511. Kluwer, Boston.

Woodruff, D. L. (2001). Metrics for solution variety. In *Proceedings of 4th Metaheuristics International Conference (MIC 2001, 16-20 July 2001, Porto, Portugal)*, pages 511–514.

Woodruff, D. L. and Zemel, E. (1993). Hashing vectors for tabu search. In Glover, F., Taillard, E., Laguna, M., and de Werra, D., editors, *Tabu Search*, Annals of Operations Research 41, pages 123–137. Baltzer, Amsterdam.

Wright, S. (2001). Optimization software packages. In Pardalos, P. and Resende, M., editors, *Handbook of Applied Optimization*. Oxford University Press, New York. In print.

Wu, P.-C. and Wang, F.-J. (1996). On efficieny and optimization of C++ programs. *Software – Practice and Experience*, 26(4):453–465.

XPRESS-MP (2001). Mathematical programming software. `http://www.dash.co.uk`.

Yang, F.-C. and Ho, Y.-K. (1997). An object-oriented cooperative distributed problem solving shell with groupware management ability. *Software – Practice and Experience*, 27(11):1307–1334.

Zanden, B., Halterman, R., Myers, B., McDaniel, R., Miller, R., Szekely, P., Giuse, D., and Kosbie, D. (1999). Lessons learned about one-way, dataflow constraints in the Garnet and Amulet graphical toolkits. *Manuscript in preparation*.

Zanden, B., Myers, B., and Szekely, D. G. P. (1994). Integrating pointer variables into one-way constraint models. *ACM Transactions on Computer-Human Interaction*, 1:161–213.

Index